Practical Astronomy

Springer
London
Berlin
Heidelberg
New York
Barcelona
Budapest
Hong Kong
Milan
Paris
Santa Clara
Singapore
Tokyo

Other titles in this series

The Modern Amateur Astronomer
Patrick Moore (Ed.)

The Observational Amateur Astronomer
Patrick Moore (Ed.)

Telescopes and Techniques: An Introduction to Practical Astronomy
C.R. Kitchin

Small Astronomical Observatories
Patrick Moore (Ed.)

The Art and Science of CCD Astronomy
David Ratledge (Ed.)

Seeing Stars
C.R. Kitchin and R.W. Forrest

Photo-guide to the Constellations
C.R. Kitchin

The Sun in Eclipse
Michael Maunder and Patrick Moore

The Observer's Year

366 Nights of the Universe

Patrick Moore

Springer

Dr Patrick Moore, CBE, FRAS

Cover illustrations: M13; The Pleiades; M34; Mars; Auriga; star trails; The Moon; Comet Hale-Bopp; Scorpio; M1; Jupiter; Saturn.

ISBN 3-540-76147-0 Springer-Verlag Berlin Heidelberg New York

British Library Cataloguing in Publication Data
Moore, Patrick, 1923–
The observer's year : 366 Nights of the Universe
– (Practical astronomy)
1.Astronomy – Observers' manuals
I.Title
522
ISBN 3540761470

Library of Congress Cataloging-in-Publication Data
A catalog record for this book is available from the Library of Congress

Typeset by EXPO Holdings, Malaysia
Printed and bound by Interprint Limited, Malta
58/3830-5432 Printed on acid-free paper

Contents

Introduction

It was once said that 'the night sky always looks much the same'. In fact, nothing could be further from the truth. There are 365 days in each year (366 in a Leap Year!), and from an astronomical point of view no two are alike.

What I aim to do, in this book, is to go through a complete year and point out some special items of interest for each night. It may be a double star, a variable star or a nebula; it may be a planet, or even the Moon in some particular aspect – there is plenty of variety. (Anyone unfamiliar with astronomical terms should consult the Glossary at the end of this book.)

Let it be said at once that you do not need a large and expensive telescope. A surprising amount can be seen with the naked eye, and binoculars give increased range; indeed, it is probably fair to say that good binoculars are ideal for the beginner, and are far better than very small telescopes.

Telescopes are of two types: refractors, and reflectors. A refractor collects its light by means of a glass lens known as an object-glass (OG); the light passes down the telescope tube and is brought to focus, where an image is formed and is then magnified by a second lens, termed an eyepiece. Note that it is the eyepiece which is responsible for the actual magnification, and each telescope should be equipped with several different eyepieces to give different powers. The function of the OG is to collect the light in the first place – and the larger the OG, the brighter the image, and the higher the magnification which can be used. In general, it is probably unwise to spend much money on a refractor with an OG less than three inches in diameter. (Beware the Japanese $2\frac{1}{2}$-inch refractor, on a spidery mounting!)

With a reflector, the light passes down an open tube and falls upon a curved mirror or 'speculum'. The light is sent back up the tube on to a second, small mirror or 'flat'; the rays are then diverted into the side of the tube, where the image is formed and magnified as before. (This is known as the Newtonian form, because it was invented by Sir Isaac Newton; there are other optical systems, but these need not concern us for the moment.) Inch for inch, a lens is more effective than a mirror, and with a Newtonian reflector the minimum really useful aperture for the main mirror is six inches. Of course, smaller telescopes are far better than nothing at all, but they are bound to be limited.

The cost of a telescope may seem high – for a good instrument you need an outlay of several hundred pounds. Yet, if properly cared for, a telescope will last for a lifetime, and the price seems less when compared with, say, the cost of a couple of British Rail tickets between London and Edinburgh!

A pair of binoculars consists of two small refractors joined together, so that both eyes can be used. They do not cost a great deal, and astronomically they are remarkably useful. Their main disadvantage is the lack of sheer magnification, but a pair of, say, 7 × 50 binoculars (magnification 7 times, each OG 50 millimetres in diameter) will give endless enjoyment.

In the descriptions which follow, I have limited myself to objects within the range of the naked eye or either binoculars or small telescopes. However, there are problems. Light pollution is one; city-dwellers of today can never enjoy the beauty of the Milky Way. There is also the Moon, which when near full will drown all but the brighter stars. It may therefore be useful to give the phases of the Moon for the period covered in this book: 1998 to 2003; it is then easy to work out whether or not moonlight is going to be a serious hindrance.

I have also listed specific events for the same period, and in a few cases I have added historical notes as well. So I hope that you will enjoy our trip through the year – beginning fittingly enough, with New Year's Day, January 1.

Phases of the Moon 1998–2003

1998

New: 28 Jan, 26 Feb, 28 Mar, 26 Apr, 25 May, 24 Jne, 23 Jly, 22 Aug, 20 Sept, 20 Oct, 19 Nov, 18 Dec.
1st Qr: 5 Jan, 3 Feb, 5 Mar, 3 Apr, 3 May, 2 Jne, 1 Jly, 31 Jly, 30 Aug, 28 Sept, 28 Oct, 27 Nov, 26 Dec
Full: 12 Jan, 11 Feb, 13 Mar, 11 Apr, 11 May, 10 Jne, 9 Jly, 8 Aug, 6 Sept, 5 Oct, 4 Nov, 3 Dec.
Last Qr: 20 Jan, 19 Feb, 21 Mar, 19 Apr, 19 May, 17 Jne, 16 Jly, 14 Aug, 13 Sept, 12 Oct, 11 Nov, 10 Dec.

1999

New: 17 Jan, 16 Feb, 17 Mar, 16 Apr, 15 May, 13 Jne, 13 Jly, 11 Aug, 9 Sept, 9 Oct, 8 Nov, 7 Dec.
1st Qr: 24 Jan, 23 Feb, 24 Mar, 22 Apr, 22 May, 20 Jne, 20 Jly, 19 Aug, 17 Sept, 17 Oct, 16 Nov, 16 Dec.
Full: 2 Jan, 31 Jan, 2 Mar, 31 Mar, 30 Apr, 30 May, 28 Jne, 28 Jly, 26 Aug, 25 Sept, 24 Oct, 23 Nov, 22 Dec.
Last Qr: 9 Jan, 8 Feb, 10 Mar, 9 Apr, 8 May, 7 Jne, 6 Jly, 4 Aug, 2 Sept, 2 Oct, 31 Oct, 29 Nov, 29 Dec.

2000

New: 6 Jan, 5 Feb, 6 Mar, 4 Apr, 4 May, 2 Jne, 1 Jly, 31 Jly, 29 Aug, 27 Sept, 27 Oct, 25 Nov, 25 Dec.
1st Qr: 14 Jan, 12 Feb, 13 Mar, 11 Apr, 10 May, 9 Jne, 8 Jly, 7 Aug, 5 Sept, 5 Oct, 4 Nov, 4 Dec.
Full: 21 Jan, 19 Feb, 20 Mar, 18 Apr, 18 May, 16 Jne, 16 Jly, 15 Aug, 13 Sept, 13 Oct, 11 Nov, 11 Dec.
Last Qr: 28 Jan, 27 Feb, 28 Mar, 26 Apr, 26 May, 25 Jne, 24 Jly, 22 Aug, 21 Sept, 20 Oct, 18 Nov, 18 Dec.

2001

New: 24 Jan, 23 Feb, 25 Mar, 23 Apr, 23 May, 21 Jne, 20 Jly, 19 Aug, 17 Sept, 16 Oct, 15 Nov, 14 Dec.
1st Qr: 2 Jan, 1 Feb, 3 Mar, 1 Apr, 30 Apr, 29 May, 28 Jne, 27 Jly, 25 Aug, 24 Sept, 24 Oct, 22 Nov, 22 Dec.
Full: 9 Jan, 8 Feb, 9 Mar, 8 Apr, 7 May, 6 Jne, 5 Jly, 4 Aug, 2 Sept, 2 Oct, 1 Nov, 30 Nov, 30 Dec.
Last Qr: 16 Jan, 15 Feb, 16 Mar, 15 Apr, 15 May, 14 Jne, 13 Jly, 12 Aug, 10 Sept, 10 Oct, 8 Nov, 7 Dec.

2002

New: 13 Jan, 12 Feb, 14 Mar, 12 Apr, 12 May, 10 Jne, 10 Jly, 8 Aug, 7 Sept, 6 Oct, 4 Nov, 4 Dec.
1st Qr: 21 Jan, 20 Feb, 22 Mar, 20 Apr, 19 May, 18 Jne, 17 Jly, 15 Aug, 13 Sept, 13 Oct, 11 Nov, 11 Dec.
Full: 28 Jan, 27 Feb, 28 Mar, 27 Apr, 26 May, 24 Jne, 24 Jly, 22 Aug, 21 Sept, 21 Oct, 20 Nov, 19 Dec.
Last Qr: 6 Jan, 4 Feb, 6 Mar, 4 Apr, 4 May, 3 Jne, 2 Jly, 1 Aug, 31 Aug, 29 Sept, 29 Oct, 27 Nov, 27 Dec.

2003

New: 2 Jan, 1 Feb, 3 Mar, 1 Apr, 1 May, 31 May, 29 Jne, 29 Jly, 27 Aug, 26 Sept, 25 Oct, 23 Nov, 23 Dec.
1st Qr: 10 Jan, 9 Feb, 11 Mar, 9 Apr, 9 May, 7 Jne, 7 Jly, 5 Aug, 3 Sept, 2 Oct, 1 Nov, 30 Nov, 30 Dec.
Full: 18 Jan, 16 Feb, 18 Mar, 16 Apr, 16 May, 14 Jne, 13 Jly, 12 Aug, 10 Sept, 10 Oct, 9 Nov, 8 Dec.
Last Qr: 25 Jan, 23 Feb, 15 Mar, 23 Apr, 23 May, 21 Jne, 21 Jly, 20 Aug, 18 Sept, 18 Oct, 17 Nov, 16 Dec.

SOUTH
NORTH
WEST
Leibnitz Mountains
Mare Australe
MARE NECTARIS
MARE FŒCUNDITATIS
MARE TRANQUILLITATIS
MARE CRISIUM
MARE SERENITATIS
MARE VAPORUM
LACUS SOMNIORUM
MARE SMYTHII
MARE SPUMANS
MARE UNDARUM
MARE MARGINIS
MARE ANGUIS
MARE NOVUM
MARE HUMBOLDTIANUM
Palus Somnii
Palus Nebularum
Haemus Mts
Caucasus Mts
Alpine Valley
Petavius
Wrottesley
Phillips
Borda
Biot
Santbech
Fracastorius
Beaumont
Rosse
Monge
Holden
Cook
Colombo
Macluro
Bonneberger
Behaim
Ansgarius
Crozier
Bellot
Magelhaens
Goclenius
Gutenberg
Capella
Isidorus
Gaudibert
Mädler
Theophilus
Langrenus
La Peyrouse
Kastner
Lubbock
Pickering
Messier
Maclaurin
Webb
Secchi
Taruntius
Censorinus
Maskelyne
Apollonius
Firmicus
Schubert
Azout
Condorcet
Neper
Hansen
Alhazen
Barker's Quadrangle
Lick
Glaisher
Yerkes
Picard
Cape Agarum
Cape Lavinium
Cape Olivium
Proclus
Peirce
Tisserand
Macrobius
Cauchy
Da Vinci
Lyell
Franz
Vitruvius
Maraldi
Cape Argaeus
Littrow
Le Monnier
Römer
Cleomedes
Tralles
Newcomb
Kirchhoff
Debes
Delmotte
Eimmart
Oriani
Plutarch
Seneca
Hahn
Burckhardt
Geminus
Berosus
Bernoulli
Timoleon
Gauss
Messala
Berzelius
Franklin
Hooke
Cepheus
Oersted
Maury
Schumacher
Schuckburg
Carrington
Chevallier
Struve
Zeno
Mercurius
Atlas
Hercules
Williams
Endymion
De la Rue
Strabo
Thales
Schwabe
Cusanus
Petermann
Baillaud
Meton
Gärtner
Democritus
Arnold
Moigno
Peters
Kane
Mayer
W.C. Bond
Barrow
Scoresby
Challis
Main
Euctemon
Gioja
Galle
Sheepshanks
Archytas
Protagoras
Egede
Trouvelot
Aristoteles
Mitchell
Bailly
Burg
Lacus Mortis
Mason
Grove
Plana
Eudoxus
Lamech
Alexander
Calippus
Cassini
Cape Agassiz
Theaetetus
Aristillus
Autolycus
Cape Fresnel
Linné
Hadley
Chacornac
Posidonius
Luther
G.P. Bond
Hall
Daniell
Deseilligny
Bessel
Sulpicius Gallus
Aratus
Conon
Bradley
Menelaus
Taquet
Auwers
Plinius
Dawes
Jansen
Ross
Maclear
Julius Caesar
Manilius
Sinas
Sosigenes
Arago
Lamont
Manners
Ariadaeus
Cayley
De Morgan
Dionysius
Ritter
Sabine
Schmidt
D'Arrest
Whewell
Tempel
Silberschlag
Hyginus
Schneckenberg
Boscovich
Agrippa
Godin
Dembowski
Chladni
Triesnecker
Ukert
Moltke
Delambre
Theon Junior
Theon Senior
Lade
Rhaeticus
Torricelli
Hypatia
Alfraganus
Zöllner
Taylor
Saunder
Pickering
Horrocks
Hipparchus
Seeliger
Halley
Hind
Müller
Kant
Descartes
Dollond
Andel
Ritchey
Albategnius
Burnham
Abulfeda
Almanon
Tacitus
Geber
Abenezra
Airy
Argelander
Vogel
Parrot
Faye
Donati
Delaunay
Lacaille
Blanchinus
Playfair
Krusenstern
Apianus
Azophi
Pontanus
Poisson
Goodacre
Aliacensis
Werner
Walter
Nonius
Fernelius
Miller
Kaiser
Stöfler
Faraday
Maurolycus
Pons
Fermat
Sacrobosco
Polybius
Altai Scarp
Catharina
Cyrillus
Pavlov
Weinek
Neander
Rheita
Stiborius
Piccolomini
Rothmann
Lindenau
Zagut
Wilkins
Rabbi Levi
Riccius
Buch
Büsching
Gemma Frisius
Nicolai
Lockyer
Janssen
Fabricius
Mallet
Young
Reimarus
Vega
Furnerius
Marinus
Adams
Hase
Snellius
Stevinus
Biela
Steinheil
Watt
Pitiscus
Spallanzani
Dove
Hommel
Baco
Barocius
Asclepi
Idelor
Brillak
Vlacq
Rosenberger
Nearch
Pontécoulant
Hagecius
Mutus
Manzinus
Tannerus
Nasmyth
Jacobi
Kinau
Tillius
Heraclitus
Lilius
Cuvier
Licetus
Clairaut
Moretus
Cysatus
Boussingault

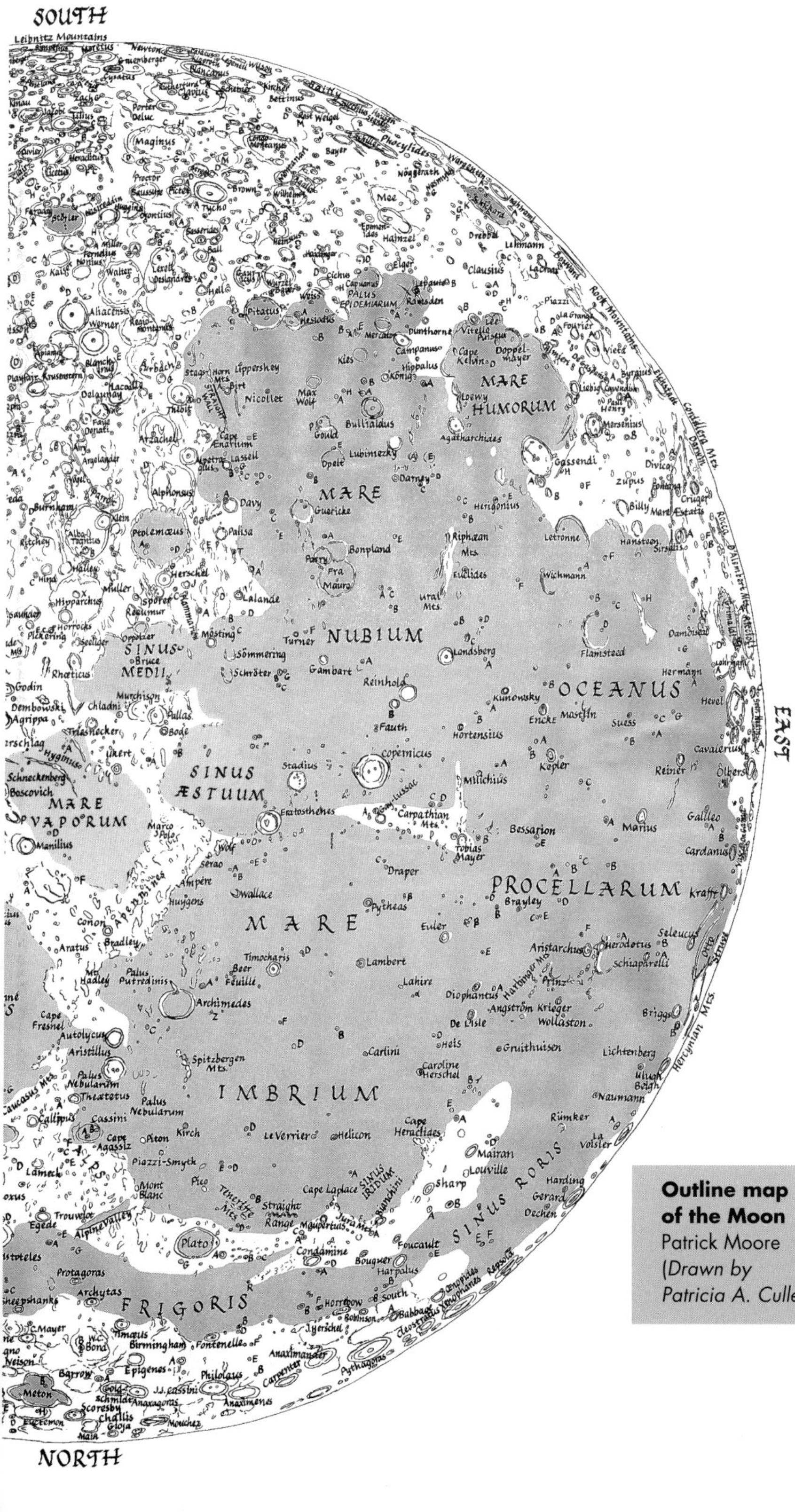

Outline map of the Moon
Patrick Moore
(*Drawn by Patricia A. Cullen*)

January

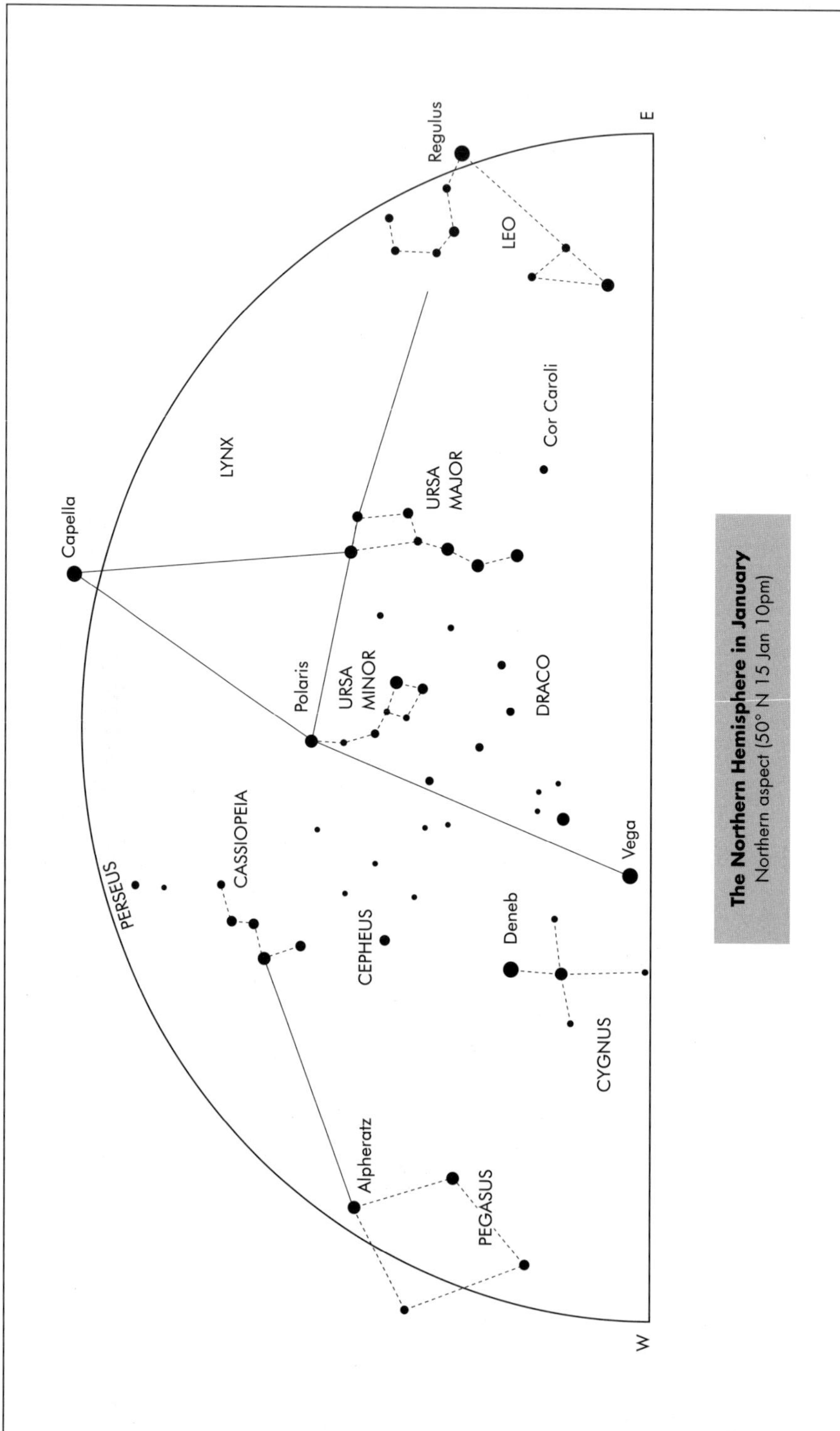

The Northern Hemisphere in January
Northern aspect (50° N 15 Jan 10pm)

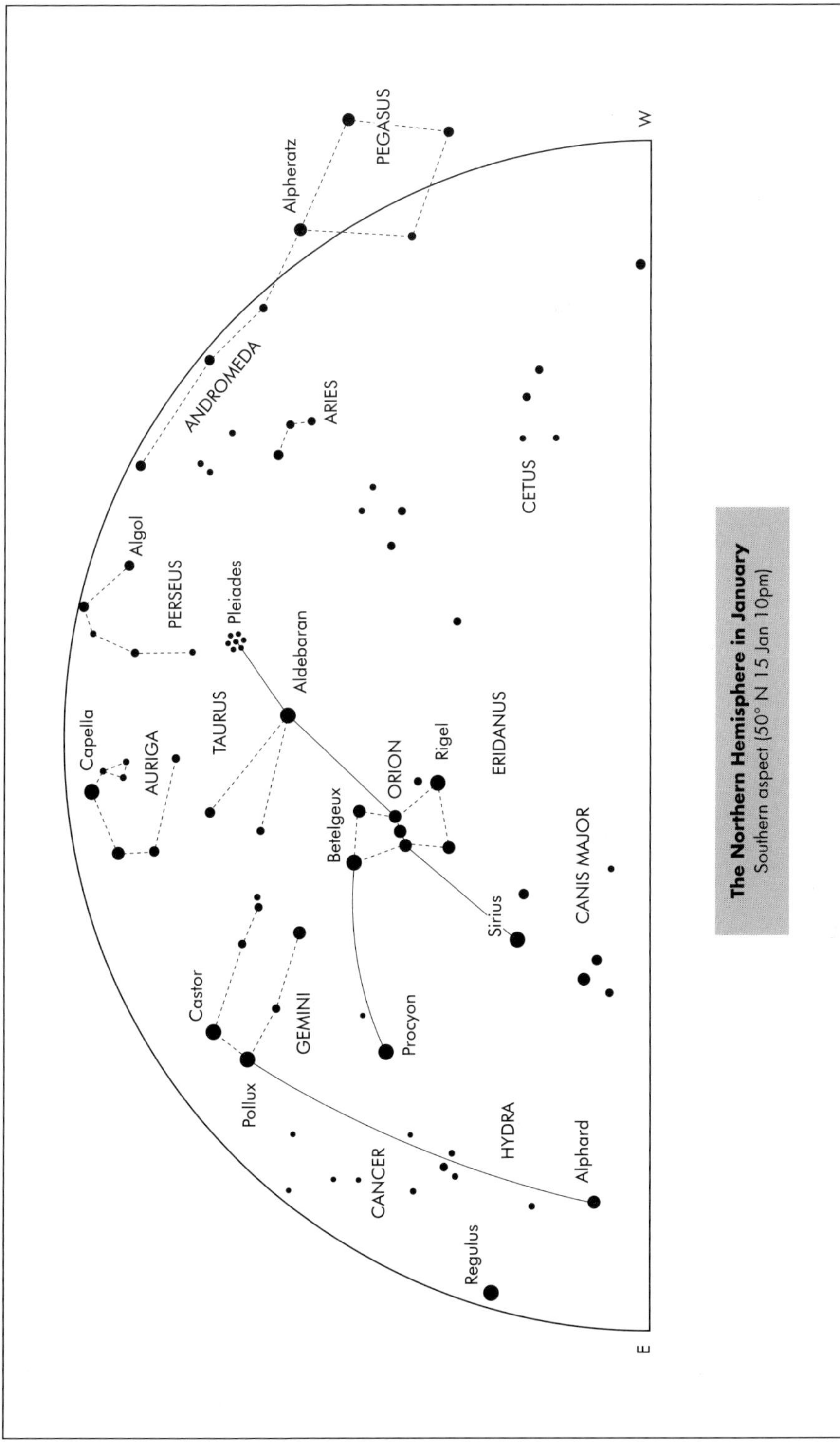

The Northern Hemisphere in January
Southern aspect (50° N 15 Jan 10pm)

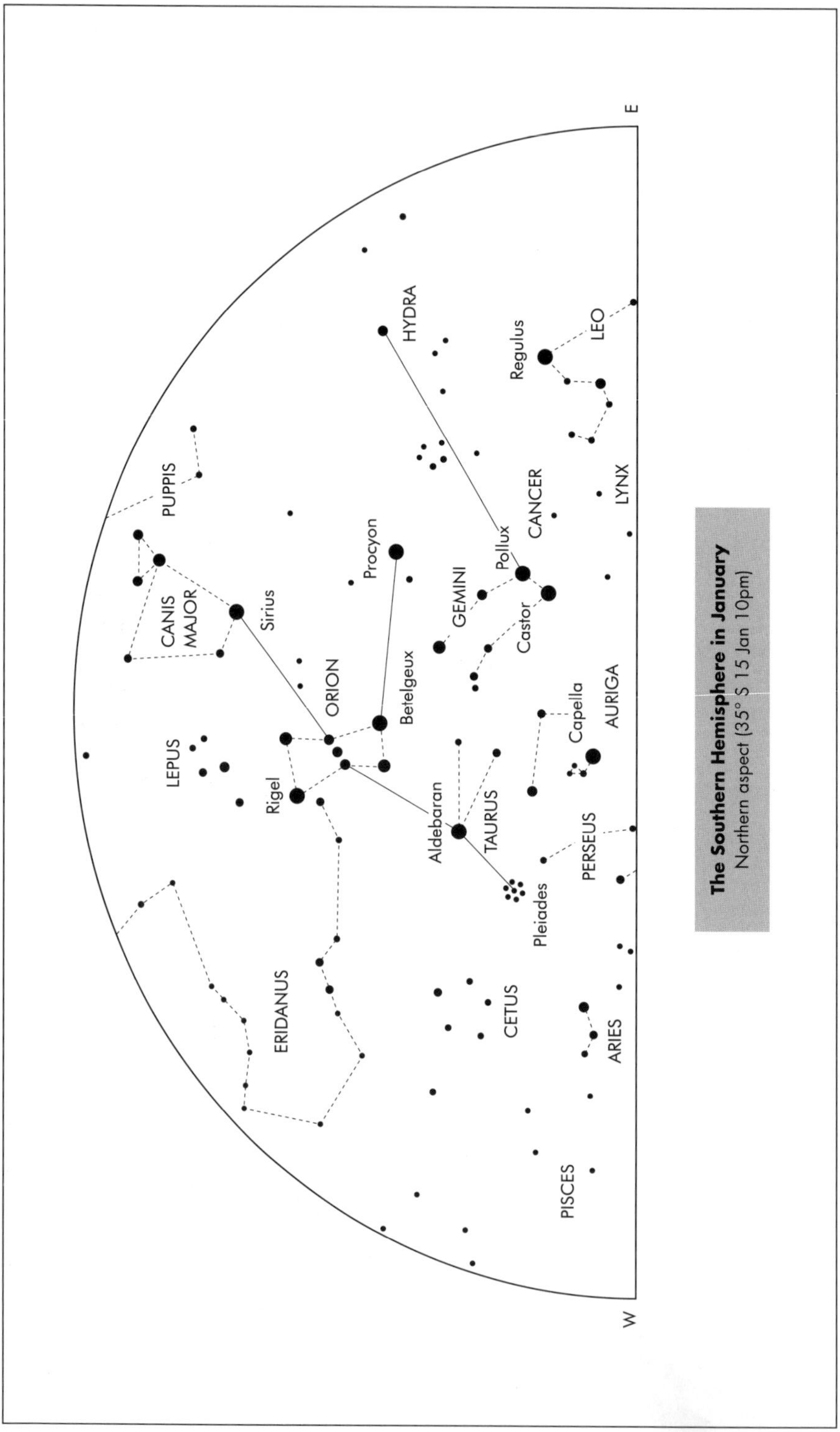

The Southern Hemisphere in January
Northern aspect (35° S 15 Jan 10pm)

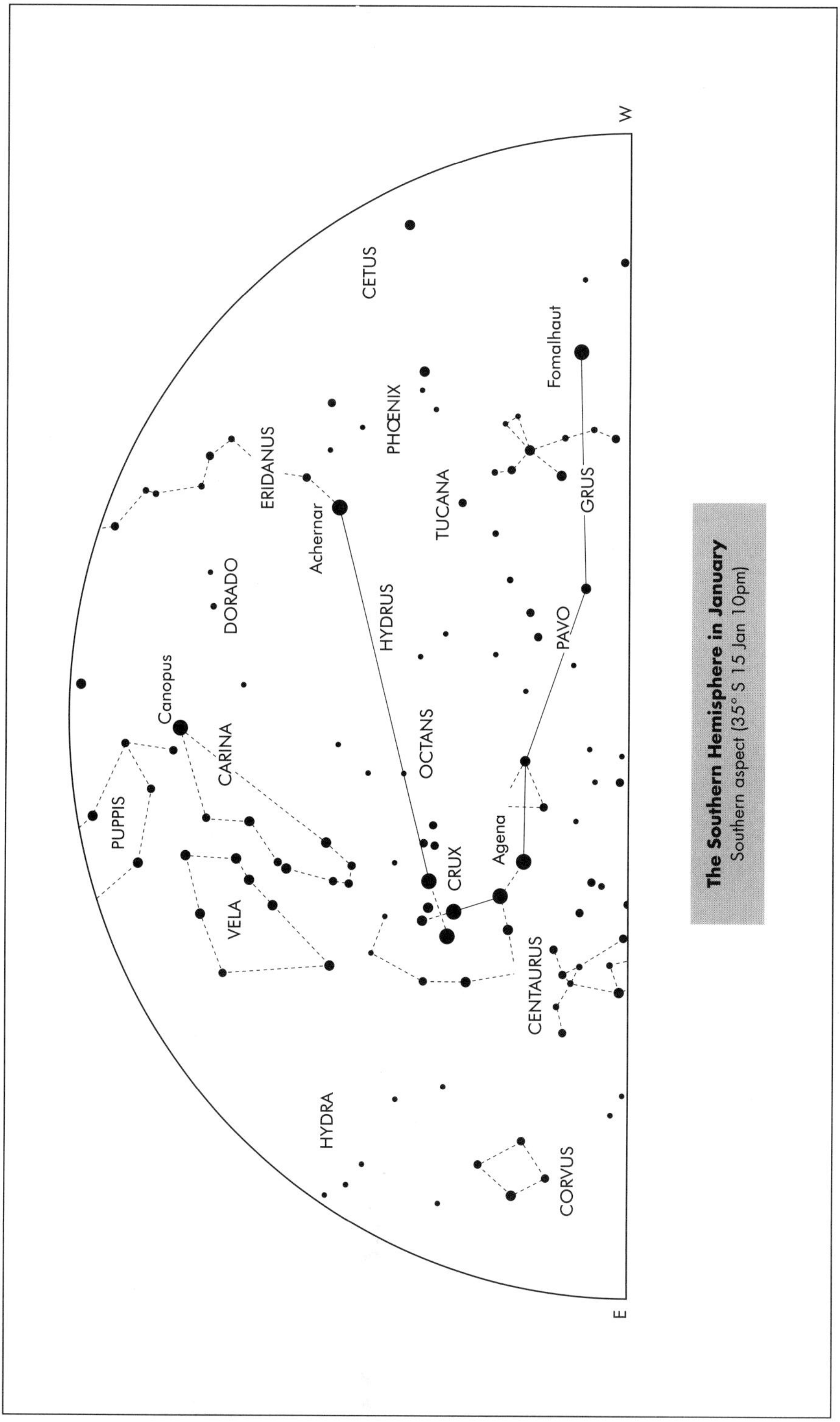

The Southern Hemisphere in January
Southern aspect (35° S 15 Jan 10pm)

January 1

The January Sky

When you start to learn your way around the night sky, the first thing to remember is that the stars always stay in the same relative positions – or virtually so; the constellations which we see today are to all intents and purposes the same as those which must have been seen by Julius Cæsar or the warriors of the Trojan War. It is only our near neighbours, the members of the Solar System, which wander around from one group to another. This makes star recognition very much easier, particularly as there are only a few thousands of stars visible with the naked eye.

All in all, 88 constellations are recognized by the International Astronomical Union (a full list of them is given in the Appendix). They are very diverse in size and brightness, but they are purely arbitrary; the patterns have no real significance, because the stars lie at very different distances from us, and we are dealing merely with line of sight effects. Still, the patterns are distinctive enough, and during January evenings we can see one of the most striking of them all: Orion, the celestial Hunter, with his two brilliant stars (Betelgeux and Rigel) and his Belt and Sword. Orion is crossed by the celestial equator, and can therefore be seen from every inhabited country. As a 'signpost in the sky' Orion is unrivalled; from it we can locate many other stars and groups – such as the brilliant Sirius, which lies in line with the three stars of the Belt, and far outshines any other star in the sky. In the other direction the Belt stars show the way to the orange-red Aldebaran, in the Bull; and so on.

Almost equally useful is Ursa Major, the Great Bear, whose seven main stars make up the pattern known as the Plough or the Big Dipper. From it we can identify Polaris, the northern Pole Star. Observers in Australia or South Africa will not see Ursa Major during January evenings, and will never see our Pole Star; but they do have the glorious Southern Cross, together with the Pointers, Alpha and Beta Centauri.

The charts given here show the evening sky in January as seen from latitude 52° N, approximately that of London, and 35° S, corresponding to Sydney or Cape Town. The best way to start learning is to find a few distinctive groups, and use them as guides to the rest. If you have the patience to follow me through the year, I assure you that you will end up with a very sound working knowledge of the sky.

Anniversary

On 1 January 1801 Piazzi, at Palermo, discovered Ceres, the largest of the minor planets or asteroids. Unfortunately it is never bright enough to be visible with the naked eye.

Future Points of Interest

2002: Opposition of Jupiter.

Anniversary

On 2 January 1959 the Russians launched their first unmanned rocket toward the Moon: Luna 1. It passed the Moon at 3700 miles on 4 January, and sent back information showing, for example, that the Moon has no overall magnetic field, so that your magnetic compass will not work there.

January 2

The Two Most Famous Constellations

Despite the splendour of Orion, there is little doubt that the two best-known constellations in the sky are Ursa Major (the Great Bear) in the northern hemisphere, and Crux Australis (the Southern Cross) in the southern. Both are on view tonight – but which you will see depends upon where you are!

From latitudes such as those of Britain, Ursa Major will be in the north-east after dark. Its seven main stars make up the Plough or Dipper pattern; though they are not outstandingly bright, they cannot be mistaken, particularly as from Britain or the northern United States they never set. They are:

Ursa Major

Greek letter	Name	Magnitude	Luminosity (Sun=1)	Distance (light-years)
α Alpha	Dubhe	1.8	60	75
β Beta	Merak	2.4	28	62
γ Gamma	Phad	2.4	50	75
δ Delta	Megrez	3.3	17	65
ϵ Epsilon	Alioth	1.8	60	62
ζ Zeta	Mizar	2.1	56+11	59
η Eta	Alkaid	1.9	450	108

A little clarification is needed here. A star's apparent magnitude is a measure of how bright it looks; the lower the magnitude, the brighter the star – thus Alkaid (1.9) is brighter than Phad (2.4). Distances are measured in light-years. One light-year is the distance travelled by a ray of light in one year; roughly 5.8 million million miles. The table shows that Alkaid is the most distant of the seven stars; it and Dubhe are travelling through space in a direction opposite to that of the other five.

The stars in any constellation are allotted Greek letters – a system devised as long ago as 1603, and which has stood the test of time. The Greek alphabet is given in Appendix C.

From Australia or South Africa you will not see the Plough, but the Southern Cross will be high – looking more like a kite than an X. Its leaders are:

Crux Australis

Greek letter	Name	Magnitude	Luminosity (Sun=1)	Distance (light-years)
α Alpha	Acrux	0.8	3200+2000	360
β Beta	–	1.2	8200	425
γ Gamma	–	1.6	120	88
δ Delta	–	2.8	1320	260

The stars of the Cross are much brighter than those of the Plough, and are closer together; adjoining them are two more brilliant stars, Alpha and Beta Centauri. Of the stars in the Cross, three are bluish white, but the fourth (Gamma Crucis) is orange-red; this is easily seen with the naked eye, and binoculars bring out the contrast beautifully. I will have more to say about star colours later.

January 3

The Poles of the Sky

In learning your way around the sky, it is obviously useful to become familiar at an early stage with the stars which can always be seen from your observing site. Stars which never set are said to be 'circumpolar'. Thus the Great Bear is circumpolar from Britain, while the Southern Cross is circumpolar from New Zealand.

Just as the Earth's equator divides the globe into two hemispheres, so the celestial equator divides the sky into two hemispheres – north and south. Northward, the Earth's axis of rotation points to the celestial pole, marked within one degree by the fairly bright star Polaris in Ursa Minor (the Little Bear). Polaris seems to remain almost motionless in the sky, with everything else – including the Sun and Moon – revolving round it once in 24 hours. It is easy to locate, but in case of any problem it may be found by using two of the stars in the Plough, Merak and Dubhe, as pointers.

To an observer at the Earth's north pole, Polaris would be directly overhead; from the Earth's equator Polaris lies on the horizon, and south of the equator it can never be seen at all. There is a corresponding south celestial pole, but unfortunately it is not well marked; the nearest naked-eye star, Sigma in the constellation of the Octant, is below the fifth magnitude, so that it is none too easy to identify. Any mist or artificial lighting will drown it. The whole of the south-polar region is in fact remarkably barren; the pole itself lies around midway between the Southern Cross and the brilliant star Achernar, in Eridanus (the River).

It is a pity that Sigma Octantis is so insignificant. Southern navigators are always envious of the northern hemisphere's convenient Pole Star.

From Sydney or Cape Town, the Southern Cross is circumpolar, while Achernar just grazes the horizon when at its lowest. During January evenings the Cross is in the south-east, while Achernar is high in the south-west.

January 4

The Quadrantid Meteors

The night of January 3–4 is always an interesting one, because it marks the peak of a major meteor shower – that of the Quadrantids.

A meteor is a tiny piece of material, usually smaller than a grain of sand, moving round the Sun; it is part of the dusty trail left by a comet. If one of these particles dashes into the Earth's upper air, moving at a rate which may reach 45 miles per second, it rubs against the air particles, and becomes so heated by friction that it burns away in the streak of radiance which we call a shooting-star. It burns away before it has penetrated to around 40 miles above sea-level, and finishes its groundward journey in the form of ultra-fine dust. What we see as a shooting-star is not the tiny particle itself, but the effects which it produces during its headlong plunge through the atmosphere.

Meteors tend to move round the Sun in shoals. Each time the Earth passes through a shoal we collect vast numbers of particles, and the result is a shower of shooting-stars. The meteors of any particular shoal seem to radiate from one definite point in the sky, known as the radiant of the shower. This is purely an effect of perspective: the shower meteors are travelling through space in parallel paths.

The meteors of tonight's shower radiate from a point in the constellation of Boötes, the Herdsman, not far from the Plough; they are known as the Quadrantids because the stars in this region were once grouped into a separate constellation, Quadrans (the Quadrant) which has been deleted from modern maps. The shower has a very short maximum, though a few Quadrantids can sometimes be seen as early as January 1 or as late as January 6.

The richness of a shower is measured by what is termed its ZHR, or zenithal hourly rate. This is defined as being the number of naked eye meteors which could be seen by an observer under ideal conditions, with the radiant at the zenith. In practice these conditions are never met, so that the actual observed rate is always less than the theoretical ZHR; all the same, it is a useful guide. The average ZHR of the Quadrantids is about 60, and there may be blue meteors with splendid trains; but because of the shortness of the maximum, the observer has to be very much on the alert. Peak activity lasts for only a few hours at most.

January 5

The Horse and his Rider

Look carefully at Mizar or Zeta Ursæ Majoris, the second star in the tail of the Great Bear (or, if you like, the handle of the Plough). Close beside it is a much fainter star, Alcor, of the fourth magnitude. Its angular separation from Mizar is over 700 seconds of arc, and it is very easy to see with the naked eye whenever the sky is reasonably dark and clear. Telescopically, Mizar itself is found to be double, with one component decidedly brighter than the other. The separation between the two is just over 14 seconds of arc; this is rather beyond the range of binoculars, but even a small telescope will give an excellent view. The Mizar–Alcor pair has often been nicknamed 'the Horse and his Rider'.

Double stars are of two types. With optical pairs there is no genuine association; one star merely happens to lie in almost the same direction as seen from Earth, so that its companion appears well in the background. However, in most cases we are dealing with physically associated or binary systems, with the components moving round their common centre of gravity much as the two bells of a dumbbell will do when twisted by the arm joining them. Mizar is of this type. The components are a long way apart – well over 30 000 million miles – and the revolution period amounts to thousands of years. Alcor is also a member of the group, but must be at least a quarter of a light-year away from the main pair.

It is rather surprising to find that the Arabs of a thousand years ago regarded Alcor as a difficult naked-eye object. There is nothing difficult about it today, so that we have a minor mystery on our hands. One suggested solution is that the star referred to by the Arabs was not Alcor at all, but a much fainter star (of the 8th magnitude) which lies between Alcor and the main pair; it has no association with the group, and is much further away. Apparently it was first seen in 1671 by an astronomer named Georg Eimmart. It was reported again in 1723 by a German observer whose name has not been preserved, and who regarded it as new; he named it Sidus Ludovicianum in honour of Ludwig V, Landgrave of Hesse.

If Ludwig's star were brighter, it would indeed be a naked-eye test, but the evidence is against its being variable, because the group has been so consistently observed during the past 200 years that any fluctuations would have been noticed. So the mystery remains, and we still do not know why the keen-eyed Arabs regarded Alcor as so elusive. If the sky is clear tonight, go out and look at it for yourself!

Future Points of Interest

Around 3–4 January each year the Earth is at its closest to the Sun: 91.5 million miles, even though it is winter in the northern hemisphere. We are at our greatest distance from the Sun (94.5 million miles) in June.

January 6

The Faintest Star in the Plough

Even a casual glance at the Plough shows that one of the seven stars, Megrez or Delta Ursæ Majoris, is much fainter than the rest. Its magnitude is well below 3; all the others are above 2.5.

Magnitude, as we have seen, is a measure of a star's apparent brightness (nothing directly to do with its real luminosity,

Future Points of Interest

1998: Mercury at western elongation.

because the stars are at different distances from us – a star may be brilliant either because it is relatively close, or because it is intrinsically very luminous). The brightest star in the sky, Sirius, is of magnitude –1.5; the faintest stars normally visible with the naked eye are of magnitude +6, while electronic equipment used with giant telescopes can delve down to +30. Conventionally, the 21 brightest stars are classed as being of the first magnitude. On the same scale, the magnitude of the Sun is about –27, and the full moon about –12.7.

The naked eye can easily detect a difference of one-tenth of a magnitude, and the relative faintness of Megrez compared with the other Plough stars is very striking. Yet in some of the old catalogues, Megrez was given as equal to the rest. Can there have been any real change?

Almost certainly the answer is 'no'. Megrez is a normal star, and appears to be absolutely stable, so that it is not likely to show any permanent or secular change over an immense period of time. Of course, many variable stars are known, but Megrez is not one of them. It is never wise to place too much reliance on old naked-eye estimates, though in the case of Megrez it seems more probable that we are dealing with an error in interpretation rather than sheer observation.

January 7

The Little Bear – and the Guardians of the Pole

Once you have found the Great Bear and the Pole Star, it is easy to trace the outline of the Little Bear, Ursa Minor. In form it is not unlike a very faint and distorted version of the Plough; the pattern curves from Polaris in the general direction of Mizar and Alkaid.

The seven 'Little Bear' stars are:

The Little Bear

Greek letter	Name	Magnitude	Luminosity (Sun=1)	Distance (light-years)
α Alpha	Polaris	2.0	6000	680
β Beta	Kocab	2.1	110	95
γ Gamma	Pherkad Major	3.0	230	225
δ Delta	Yildun	4.4	28	143
ϵ Epsilon	–	4.2	65	200
ζ Zeta	Alifa	4.3	16	108
η Eta	Alasco	4.9	8	90

I have given the old proper names, but apart from Polaris and, occasionally, Kocab they are almost never used. No stars

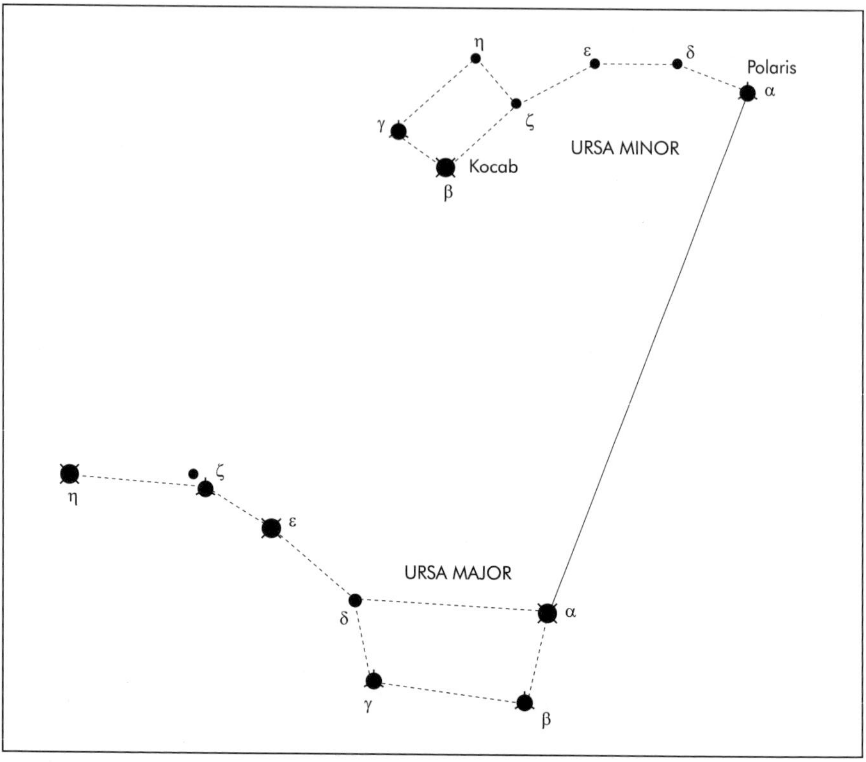

are named today. There are various unscrupulous agencies which claim to be able to name stars (on payment of a sum of money, of course!), but these agencies are completely bogus, and the names they give are not recognized anywhere. Unfortunately, many people have been taken in, and it is not too strong to say that these schemes are nothing more than confidence tricks.

Polaris is very slightly variable, but its fluctuations amount to less than a tenth of a magnitude, and are too small to be detected with the naked eye. Telescopically it is seen to have a 9th-magnitude companion, at an angular separation of over 18 seconds of arc; a three-inch refractor shows it easily, but it is well beyond the range of binoculars.

Beta and Gamma are known popularly as the Guardians of the Pole. Beta is notable because of its colour; with the naked eye it is seen to be somewhat orange in hue, and binoculars bring out the colour very clearly. This means that its surface is decidedly cooler than that of our yellow Sun, though it is much larger and more luminous.

Apart from Polaris and the Guardians, the stars of the 'Little Plough' are so dim that mist or strong moonlight will hide them completely. There is nothing else of immediate interest in Ursa Minor.

As we have seen, there is no bright south pole star, and neither are there any 'Guardians'. The nearest reasonably

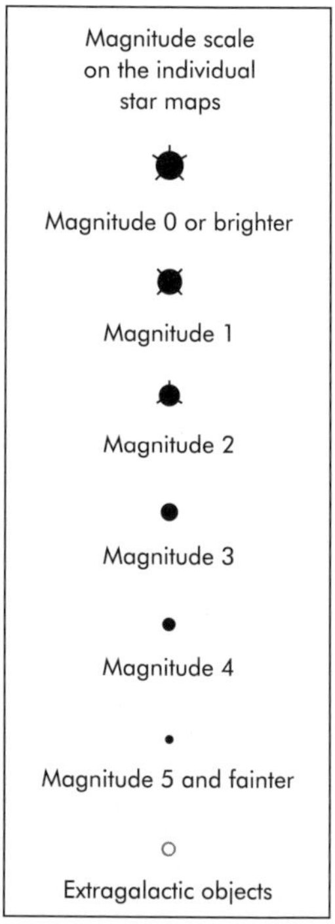

bright star to the south polar point is Beta Hydri, in the constellation of the Little Snake; it is of magnitude 2.8, but is more than a dozen degrees from the actual pole.

January 8

The Legend of the Bears

It has been said that the night sky is a vast picture-book. The constellation patterns which we use are basically those of the Greeks; if we had followed, say, the Chinese or the Egyptian systems our maps would look very different, though of course the stars themselves would be exactly the same. The names are quite arbitrary, and few of the constellations have the slightest resemblance to the objects they are meant to represent – which is one of many reasons why the pseudoscience of astrology is such rubbish.

Ptolemy, last of the great astronomers of Classical times, listed 48 constellations, and all of these are still in use, though in most cases with altered boundaries.

Many of Ptolemy's groups are mythological, and have legends attached to them. One of these concerns the two Bears. It was said that Callisto, daughter of King Lycaon of Arcadia, was very beautiful – indeed, even lovelier than Juno, the queen of Olympus. Juno was so jealous that she ill-naturedly changed Callisto into a bear. Years later Arcas, Callisto's son, was out hunting and came across the bear. Naturally he had no idea that it was his long-lost mother, and was about to shoot when Jupiter, the king of the gods, intervened. He changed Arcas into a bear also, snatched up both animals by their tails and swung them up, to the safety of the sky, where they still live. That is why both the Bears have tails of decidedly un-ursine length!

As we have noted, the stars in any particular constellation are not genuinely associated with each other, and we are dealing with nothing more significant than line-of-sight effects. Still, the old stories are attractive, and are always worth retelling.

If we had followed the Egyptian system, there would be no Bears in the sky. However, we would at least have a Crocodile and a Hippopotamus!

January 9

Future Points of Interest

2001: Total eclipse of the Moon.

Eclipses of the Moon

The Moon is known officially as the Earth's satellite, though there are grounds for claiming that it should really be regarded as a companion planet. It is 2160 miles in diameter, and moves round the Earth at a mean distance of 239 000 miles in a period of 27.3 days.

The Moon shines by reflected sunlight; the phases, from new to full, depend upon how much of the sunlit hemisphere is turned in our direction. When the Moon is full, the Sun, Earth and Moon are lined up, with the Earth in the mid-position, so that the whole of the daylit hemisphere faces us. If the alignment is exact, the Moon passes into the cone of shadow cast by the Earth, and its supply of direct sunlight is cut off. However, the Moon does not (generally) vanish completely, because some of the Sun's rays are bent or refracted on to its surface by way of the shell of atmosphere surrounding the Earth. All that happens is that the Moon turns a dim, often coppery colour until it passes out of the shadow again. Some eclipses are no more than partial, but the 2001 eclipse is total, with the Moon wholly immersed in the Earth's shadow for half an hour.

Obviously, a lunar eclipse can happen only at full moon, and can be seen from any point from which the Moon is above the horizon at the time. Eclipses do not happen at every full moon, because the Moon's orbit is tilted to ours at an angle of just over 5° – and on most occasions the full moon passes either above or below the cone of shadow, and escapes eclipse. There are six total eclipses during the period from 1997 to 2003, though not all of them are visible from Britain.

No two eclipses are alike. All the light reaching the shadowed Moon has to pass through the Earth's atmosphere, and everything depends on the state of the upper air; if it contains a great deal of volcanic ash or dust, for instance, the eclipse will be 'dark' – as happened during the 1990s following the violent eruption of Mount Pinatubo, in the Philippine Islands.

It cannot honestly be said that lunar eclipses are of much scientific importance, but they are fascinating to watch. Mid-eclipse on 9 January 2001 falls at 20 h 22 m GMT; the partial phase extends over a total of 1 hour 38 minutes.

January 10

Colours of the Stars

It is very easy to see the difference in colour between the two Pointers to the Pole Star, Merak and Dubhe; Merak is pure white, Dubhe decidedly orange. As we have seen, differences in colour indicate differences in surface temperature, so that Dubhe is cooler than our yellow Sun, while the Sun is cooler than Merak. This brings us on to that vital tool of astronomical research: the spectroscope. Just as a telescope collects light, so a spectroscope splits it up.

Light is not so simple as it might appear; for example a beam of sunlight is a mixture of all the colours of the

Spectral Classification			
Type	Colour	Surface temperature (°C)	Examples
W	Bluish-white	Up to 80 000	Zeta Puppis (rare)
O	Bluish-white	40 000–35 000	Regor (rare)
B	Bluish	25 000–12 000	Rigel, Alkaid
A	White	10 000–8000	Merak, Vega
F	Yellowish	7500–6000	Polaris, Procyon
G	Yellow	6000–5000	The Sun, Capella
K	Orange	5000–3400	Arcturus, Aldebaran
M	Orange-red	3400–3000	Betelgeux, Antares
R	Reddish	2600	V Arietis
N	Reddish	2500	R Leporis
S	Reddish	2600	Chi Cygni

rainbow. Moreover, light is a wave motion, and the colour of the light depends upon its wavelength – that is to say, the distance between two successive wave-crests. For visible light, red has the longest wavelength and violet the shortest, with orange, yellow, green and blue coming in between.

Pass a beam of sunlight through a glass prism, and it will be split up; there will be a rainbow band, from red at the long-wave end through to violet at the shortwave end. This band will be crossed by dark lines, each one of which is the trademark of one particular substance. For instance, there are two dark lines in the yellow part of the band which are due to sodium (one of the main elements in common salt) and cannot be reproduced by anything else; therefore, we can tell that there is sodium in the Sun.

The Sun is an ordinary star, and all the stars produce spectra, but they are not all alike – and again everything depends upon their surface temperatures. Conventionally, the stars are divided into certain definite types, each denoted by a letter of the alphabet. These spectral types are shown in the table above.

Most of the stars are contained in the sequence from B to M. Each type is divided into subdivisions, from 0 to 9; thus Polaris is of type F8 – three-quarters of the way from F to G.

Future Points of Interest

2002: Eastern elongation of Mercury.

2003: Western elongation of Venus.

January 11

Phases of the Inferior Planets

By mid-January there has been a definite change in the sky since the start of the year. At the same time in the evening, northern observers will see that the Great Bear is a little higher and the Square of Pegasus a little lower, while to southern observers the Cross will have gained in altitude. Of course, the positions of any visible planets will have shifted against the starry background.

Mercury and Venus have their own way of behaving. They are known as the Inferior Planets, not because of any reduced status but because they are the only planets in the Solar System which are closer to the Sun than we are.

The diagram shows the orbits of Venus and the Earth – not to scale, but for our present purpose this does not matter. In position 1, Venus is more or less between the Earth and the Sun; its dark side faces us, so that it is 'new', and we cannot see it. As it moves along it will appear in the morning sky, in the east before dawn, and will be a crescent, increasing until it reaches position 2, when it will be the shape of a half-moon; this is termed western elongation, and the actual time of half-phase is termed dichotomy. It then becomes three-quarters in phase (gibbous); by the time it reaches position 3 it is full, though it is then on the far side of the Sun and is to all intents and purposes out of view. It then reappears as a gibbous object in the western sky after sunset, reaching half-phase (eastern elongation) at position 3, and then a narrowing crescent until it returns to new at position 1.

The phases of Venus are detectable with binoculars, or with any telescope; very keen-sighted people can see the crescent form with the naked eye.

Mercury behaves in the same way, but is much less obtrusive, and the phases are well beyond the range of binoculars. A telescope of fair size is needed to show them really well.

Venus is almost as large as the Earth, and is surrounded by a dense atmosphere which reflects sunlight well, so that it is very brilliant; at its best it can cast a strong shadow. Mercury, on the other hand, is never prominent, and remains inconveniently close to the Sun, so that there are many people who have never seen it at all.

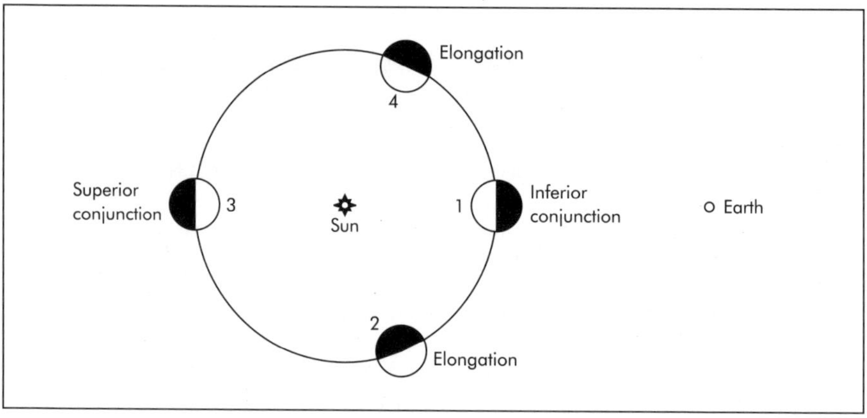

January 12

Orion, the Hunter

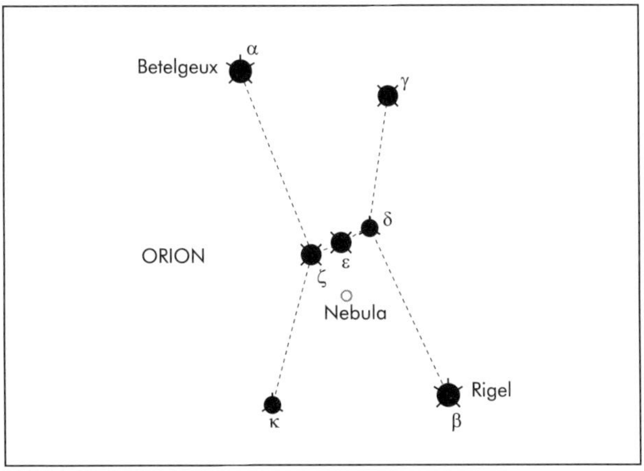

Up to now we have concentrated upon stars which are in the far north of the sky, and are therefore inaccessible from the far south. Let us now turn to Orion, which is arguably the most magnificent constellation in the entire sky. It is at its best in January, and it is visible from anywhere on Earth, because the celestial equator passes through it. To British observers, Orion is high in the south after dark; from Australia, New Zealand or South Africa it is high in the north.

There are seven bright stars, of which two, Rigel and Betelgeux, are among the most brilliant stars in the sky. The leaders are shown in the table below.

The two leaders are very different. Rigel is (usually) considerably the brighter of the two, though it is lettered Beta rather than Alpha; it is bluish-white, and is a true cosmic searchlight. Betelgeux is orange-red, and though i is cooler than Rigel it makes up for this by its huge size – its diameter is of the order of 250 000 000 miles, so that it could swallow

Orion

Greek letter	Name	Magnitude	Luminosity (Sun=1)	Distance (light-years)	Spectrum
α Alpha	Betelgeux	0.4, variable	15 000	310	M2
β Beta	Rigel	0.1	60 000	910	B8
γ Gamma	Bellatrix	1.6	2200	360	B2
δ Delta	Mintaka	2.2	22 000	2350	O9.5
ϵ Epsilon	Alnilam	1.7	23 000	1200	B0
ζ Zeta	Alnitak	1.8	19 000	1100	O9.5
κ Kappa	Saiph	2.1	49 000	2100	B0.5

up the entire path of the Earth round the Sun. It is decidedly variable, and though it can sometimes almost match Rigel it is usually several tenths of a magnitude fainter.

Alnilam, Alnitak and Mintaka make up the Hunter's Belt. All are bluish-white; it is interesting to compare Alnilam with Alnitak – Alnilam is perceptibly the brighter of the two, though the difference is only one-tenth of a magnitude. Below the Belt (as seen from northern latitudes) is the misty Sword, marked by the Great Nebula, about which I will have much more to say later.

The celestial equator passes close to Mintaka, thereby more or less cutting the Hunter in half. As seen from the Earth's equator (for instance, from Singapore) Orion can at times pass directly overhead.

January 13

Orion as a Guide

Because Orion is so conspicuous, and because it is so convenient inasmuch as it can be seen from all parts of the Earth, it is invaluable as a 'sky guide'. This may therefore be the time to do

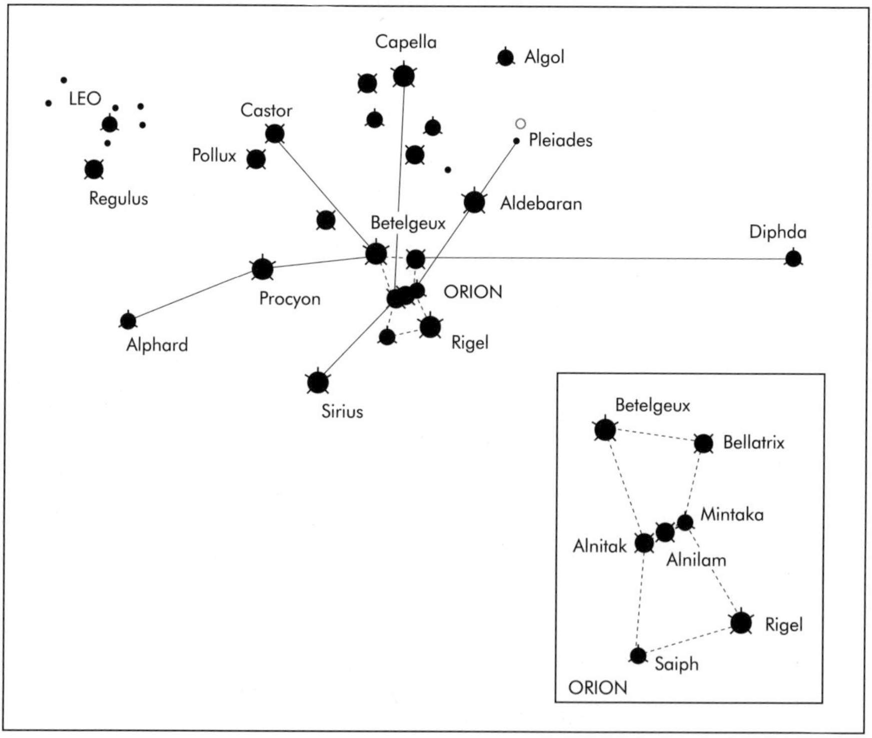

some serious star recognition, because Orion is so well placed – and the stars become much more interesting when you know which is which.

Let us begin with the sky as seen from northern-hemisphere observers, with Orion high in the south. Remember that because its stars are so bright, they can be seen even in strong moonlight; they are far more brilliant than those of the Plough.

Downward, the three stars of the Belt show the way to Sirius in Canis Major (the Great Dog), which is much the brightest of all the stars even though it cannot compete with Venus, Jupiter, or Mars at its best. It stands out at once; its magnitude is −1.5, which means that it is more than 1.5 magnitudes brighter than Rigel (though as so often in astronomy, appearances are deceptive; Sirius owes its eminence to the fact that at a mere 8.6 light-years, it is one of the very nearest of all our stellar neighbours). In the opposite direction – upward – the Belt shows the way to Aldebaran in Taurus (the Bull), of magnitude 0.8; Aldebaran, like Betelgeux, is orange-red, though it is not nearly so large or so powerful. Continue the line from the Belt through Aldebaran, and you will come to the lovely star-cluster of the Pleiades or Seven Sisters, which looks at first sight like a misty glow, though more careful inspection shows that it is made up of stars.

Next, take a line from Bellatrix through Betelgeux and curve it slightly; it will lead to Procyon in Canis Minor (the Little Dog), which is of magnitude 0.4. Procyon is white, and like Sirius is a relatively near neighbour; its distance from us is 11.4 light-years.

A line from Rigel through Betelgeux, and extended, will reach Castor and Pollux, the Heavenly Twins, which are close together; Pollux (magnitude 1.1) is appreciably brighter than Castor (1.6), and is orange, while Castor is white.

Almost overhead during winter evenings is the very brilliant Capella, in Auriga (the Charioteer); its magnitude is 0.1, and it is only marginally brighter than Rigel but as seen from Britain it is far more prominent simply because it is higher up. It is yellow, like the Sun; close beside it is a small triangle of stars, known collectively as the Hædi or Kids – two members of which are very remarkable objects, as we will see later. Capella is one of only two first-magnitude stars which can reach the zenith as seen from Britain; Vega is the other.

From the southern hemisphere the patterns of the stars are the same, but now the Belt points upward to Sirius and downward to Aldebaran, while Capella is always very low; from the southernmost part of New Zealand it does not rise at all. However, the brilliant Canopus, inferior in brilliance only to Sirius, is not far from the overhead point. Finally, take note of the Milky Way, which can be seen from either hemisphere and which runs across the sky from one horizon to the other. City-dwellers, alas, will not see it, because it is so overpowered by artificial lights, but when the sky is dark the Milky Way is truly magnificent.

January 14

The Legend of Orion

Since Orion is one of Ptolemy's original 48 constellations, it is only natural that there should be a legend about it. In fact there are several, and in each version Orion is represented as a great hunter.

Orion's skill could not be doubted, but he was also a boaster, and it was his claim that no creature on the face of the Earth was a match for him. This offended the Olympian gods, who decided to make an example of him. Certainly he could deal with animals – but what about insects? So a scorpion crawled out of the ground, stung Orion in the heel and killed him.

However, the gods could at times be merciful, and it was so on this occasion. Orion was brought back to life and swung up into the sky, where he still shines down in all his glory. It was perhaps fitting that the scorpion too should be elevated to celestial rank – and this was done, but the scorpion was placed so far from Orion that the two can never meet again; from Britain, Orion and the scorpion can never be above the horizon at the same time.

Of course we must always remember that the main stars of Orion have no genuine connection with each other. According to the Cambridge catalogue, Betelgeux is just over 300 light-years from us, Rigel just over 900. It is true that other catalogues give rather different values, and place Betelgeux at over 500 light-years, but in any case it is clear that Rigel is just as far away from Betelgeux as we are. If we ware observing from a different vantage point in the Galaxy, Rigel and Betelgeux could well be on opposite sides of the sky.

The Orion stars are so remote that their individual or 'proper' motions are very slight see 21 August, but the pattern will not persist indefinitely, and neither would it have been recognizable a million years ago. At that time the brightest star in the entire sky was Saiph, shining with a brilliancy greater than that of Venus today – and it would have been in the far north, whereas now it lies well south of the celestial equator.

January 15

Betelgeux

Betelgeux may not be the most brilliant star in Orion, but it is certainly the most beautiful. Its orange-red colour is striking even with the naked eye, and is well brought out with binoculars. The nineteenth-century astronomer William Lassell called it 'a most beautiful and brilliant gem. Singularly beautiful in colour, a rich topaz, in hue and brilliancy, different from any other star I have seen'.

It ranks as a supergiant star, but in spite of its immense size it is not so massive as might be thought, and is no more than around 15 times as massive as the Sun. This means that it is relatively rarefied, at least in its outer layers, and in fact large stars are always less dense than smaller ones. If you could put a stellar giant and a stellar dwarf on opposite sides of a vast pair of scales, it would be rather like balancing a lead pellet against a meringue!

It used to be thought that Betelgeux must be a very young star, but this is now known to be wrong. It is well advanced in its evolution, and qualifies as a stellar OAP. It has used up its original store of 'fuel', and is drawing on its reserves. It is also unstable; it swells and shrinks, changing its output of radiation as it does so. The variations are slow, but they are noticeable enough, and there is even a very rough period of the order of $5\frac{3}{4}$ years. At its best, Betelgeux can match Rigel or Capella; at its best, it is comparable with Aldebaran. Usually it is more or less equal to Procyon.

To estimate the magnitude of a variable star, the method is to compare it with other stars which do not fluctuate. Unfortunately, Betelgeux is an awkward star to estimate with naked-eye methods, because it has to be compared with another star which is not too unequal to it, and the only useful comparisons are Capella (magnitude 0.1), Rigel (also 0.1), Procyon (0.4) and Aldebaran (0.8). We have to allow for the effects of what is termed extinction. The light from a star which is low over the horizon will have to pass through a thick layer of the Earth's air, and will be weakened; the effect is very marked, as can be seen from the following table:

Extinction

Altitude (degrees)	Extinction (magnitudes)
1	3
2	2.5
4	2
10	1
17	0.5
20	0.3
43	0.1

Above 43°, extinction can safely be neglected.

For Betelgeux, it is not easy to find a comparison star at equal altitude; from Britain, for example, Betelgeux is always higher-up than Rigel, while from Australia it is always lower down. However, with a little practice it is possible to make reliable estimates, and it is always interesting to keep track of Betelgeux as it slowly brightens and fades.

Many red supergiants are similarly variable; I will have more to say about them later. Betelgeux is the brightest of them, because it is the closest star of its type.

January 16

Rigel

We have looked at the Giant's Shoulder. Now let us turn to his Foot, marked by the brilliant Rigel. This name too is Arabic, and comes from 'Rijl Jauzah al Yusră, which has been translated as 'the Left Leg of the Giant'.

Rigel is only the seventh apparently brightest star in the sky, but in sheer power it far outstrips most of its rivals, and is matched only by Canopus. However, it is very remote – just over 900 light-years away according to the Cambridge catalogue. Of course this value is bound to be uncertain to some extent, but it is certainly of the right order, in which case Rigel must be some 60 000 times more luminous than the Sun.

Rigel is pure white, with a surface temperature of about 10 000 °C. The temperature near the centre of its globe is very high indeed, and may soar to around 100 million degrees, so that Rigel is a very energetic star. Its life expectancy is much less than that of the Sun, because it is pouring out radiation at such a furious pace that in its present condition it cannot survive for more than a few million years at most, whereas it will be several thousands of millions of years before anything dramatic happens to our relatively mild Sun.

Rigel is not alone in space. It has a dim companion, of just above the 7th magnitude which would be an easy binocular object were it not so overpowered by the glare of its primary. The angular separation between the two is 9.5 seconds of arc. The two are genuinely associated, so that we are not dealing with a mere line-of-sight effect, but the real separation is thousands of times greater than the distance between the Sun and the Earth. A modest telescope (say a three-inch refractor) will show the companion easily; insignificant though it may look, it is well over 100 times as luminous as the Sun. It is actually a very close double.

Rigel is a true celestial searchlight – so powerful that it is able to illuminate clouds of gas and dust in space which are many hundreds of light-years away from it. It is not the sort of star which would be expected to be the centre of a planetary system, but if such a planet did exist its inhabitants would indeed have a truly glorious sun.

Anniversary

1786: Discovery of Encke's Comet.

Future Points of Interest

2001: Eastern elongation of Venus.

January 17

Comets

This is an 'anniversary day' – or, rather, night; in 1786 the French astronomer Pierre Méchain used a small telescope to discover one of the most famous of all comets. Let us, therefore, content ourselves for the moment by saying something about comets in general.

A comet is a member of the Solar System, but it is not a solid body similar to a planet. It consists mainly of a piece of ice, mixed with 'rubble', only a few miles across. When it is a long way from the Sun it is inert, but when it nears the Sun, and is warmed and the ices begin to evaporate; the comet produces a 'head' or coma of gas and dust, and very often a tail or tails stretching away from the Sun.

The planets move round the Sun in orbits which are not very different from circles (Pluto is an exception, but, as we will see, it is quite different from all the other planets of the Solar System, and seems to be in a class of its own.) Not so with comets. In general their orbits are very elliptical, and they have revolution periods ranging from a few years up to millions of years. Of course, comets of very long period cannot be predicted, and are always apt to take us by surprise; they are seen only once over many lifetimes – thus Comet Hyakutake, which made such a brave showing for some weeks during 1996, will not be back again for around 15 000 years. The even brighter Comet Hale-Bopp of 1997 was last in the inner Solar System about 4000 years ago. However, there are also many comets of short period which return regularly.

January 18

Hare in the Sky

Orion, so the legend tells us, was particularly fond of hunting hares, and so it seems appropriate that a hare – Lepus – should be placed next to him in the sky. The little animal lies below the Hunter's feet as seen from the northern hemisphere; to southerners, of course, Lepus is higher than Orion.

It is not bright, but it is easy enough to identify. Its brightest stars are shown in the table overleaf.

Binoculars will show that Epsilon is decidedly orange in hue. If you have a telescope, look at Gamma; you will find that it is double. The secondary star is of the sixth magnitude,

Lepus					
Greek letter	Name	Magnitude	Luminosity (Sun=1)	Distance (light-years)	Spectrum
α Alpha	Arneb	2.6	7500	945	F0
β Beta	Nihal	2.8	600	316	G2
γ Gamma	–	3.6	2	26	F6
δ Delta	–	3.8	58	145	G8
ε Epsilon	–	3.2	110	160	K5
ζ Zeta	–	3.5	17	78	A3
η Eta	–	3.7	17	65	F0
μ Mu	–	3.3	180	215	B9

and the separation is over 96 seconds of arc, so that this is a very easy pair.

Still with a telescope, follow a line from Alpha through Beta, and extend it until you come to the next reasonably bright star, which has no Greek letter and is known simply as 41 Leporis; its magnitude is 5.5. In the same telescopic field with it is a misty patch which proves to be what is termed a globular cluster – a spherical system of stars, perhaps a million in all; it was discovered in 1780 by Pierre Méchain. A small telescope will show it, and it is just within the range of strong binoculars. It is very remote; we see it as it used to be about 43 000 years ago.

In 1781 Charles Messier, Méchain's friendly rival, drew up a catalogue of over 100 star-clusters and nebulæ. The globular cluster in Lepus was No. 79 in his list, and we still refer to it as M79.

Perhaps the most interesting object in the Hare is the variable star R Leporis. Generally it is well below naked-eye visibility, but when at its best it can we well seen with binoculars, and it cannot be misidentified; its intensely red colour has led to its being called the Crimson Star. The period (that is to say, the interval between one maximum and the next) is 432 days, and the magnitude range is between 5.9 and 10.5, so

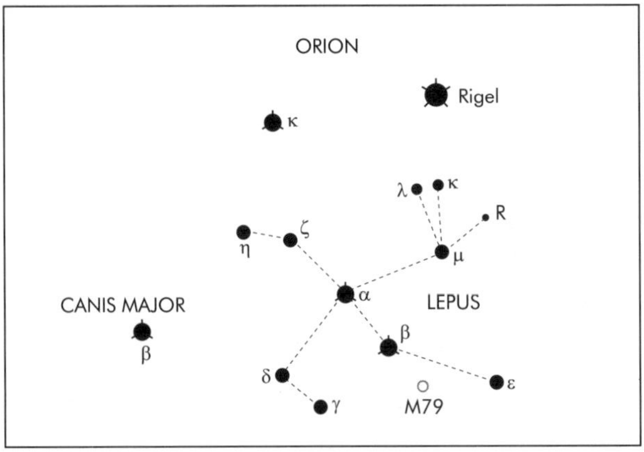

that the Crimson Star is always within the range of a three-inch telescope. It is over 100 light-years away, and at least 500 times as luminous as the Sun.

January 19

The Dove, the Graving Tool and Declination

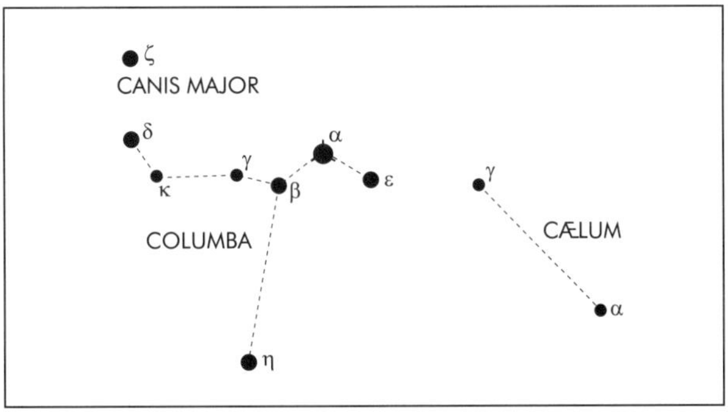

South of Lepus – below it as seen from the northern hemisphere, above it from the southern – is another small constellation, Columba, the Dove. Adjoining it is Cælum, the Graving Tool, which has no bright star, contains nothing of interest and is clearly unnecessary.

Columba is marked by an irregular line of stars, of which the brightest are Alpha (magnitude 2.6) and Beta (3.1). Well south of this line is Eta Columbæ, of magnitude 4.0, which never rises at all as seen from the British Isles. Its declination is 43° S, or –43°.

On Earth, the position of any particular point is given by its latitude and longitude. Latitude is the angular distance from the equator, reckoned from the centre of the globe; thus London is at approximately latitude +52° N, Athens +38°, Alexandria in Egypt +31°, Johannesburg in South Africa –28°, and so on; Singapore, at +1°, is practically on the equator. The latitude of the north pole is +90°, and of the south pole (where an observatory is now being built) –90°.

In the sky, the equivalent of latitude is declination, defined as the angular distance of the star (or other object) from the celestial equator. Thus Betelgeux is at declination +7° (in round numbers; actually it is +7° 24′), Rigel –8°, Mizar in the Plough +55°, and so on. If Polaris were exactly at the north pole its declination would be +90°; actually it is +89° 16′. The altitude of the celestial pole is always the same as the latitude of the observer on the surface of the Earth, so that from

London the pole will be 52° above the horizon; from Singapore it is virtually on the horizon, and from southern latitudes it is invisible, though of course the south celestial pole can be measured in the same way.

Next let us consider what is termed co-latitude, which is defined as the observer's latitude subtracted from 90°; thus for London the co-latitude will be 38° (since 90–52 = 38). Any star whose declination is north of +38° will be circumpolar, and will never set; any star south of declination –38° will never rise. Thus Mizar is circumpolar from London, but Betelgeux is not.

The brightest star in Columba, Alpha, is at declination –34°. This means that it can be seen from London, though it will always be very low. But Eta Columbæ, –43°, is below the 38° limit, and is permanently out of view.

Well south of Columba is the brilliant Canopus which, apart from Sirius, is the brightest of all the stars. During January evenings it is very high up as seen from Australia or South Africa, and from most of New Zealand it is circumpolar, but as its declination is –53° it can never be seen from London; it does not rise from anywhere north of latitude +37° (since 90–53 = 37). It can be seen from Alexandria, but never from Athens – and this was one of the earliest proofs that the Earth is a globe, rather than being as flat as a pancake.

January 20

The Belt and the Sword

One of the most striking features of Orion is the Belt, made up of three hot, bright stars – Delta (Mintaka), Episilon (Alnilam) and Zeta (Alnitak). All are many thousands of times more powerful than the Sun, and all have surface temperatures of the order of 20 000 °C. Alnilam is the brightest of them (though it is less than one-tenth of a magnitude superior to Alnitak). Mintaka has a sixth-magnitude companion at an angular separation of over 52 seconds of arc; the secondary star is beyond the range of binoculars, but is very easy to see with a small telescope.

Extending away from the Belt is the Sword, which is visible with the naked eye as a misty patch. This is the site of the Great Nebula, M42 (No. 42 in Charles Messier's catalogue), the finest example of its type.

'Nebula' is Latin for 'cloud', and the Sword is indeed a cloud in space, but it is not made up of water vapour; it consists of very tenuous gas – mainly hydrogen – together with 'dust'. It is over 1000 light-years away, and is well over 30 light-years in diameter. It is shining because of the hot stars at its Earth-turned edge. There are four main stars in the group, known officially as Theta Orionis but always nicknamed the Trapezium because of the way in which the four are arranged. They can be well seen with a small telescope, and there is nothing else quite like them in the sky.

The stars of the Trapezium are extremely hot, and their radiation not only lights up the Nebula but makes the nebular material send out a certain amount of light on its own account. A nebula of this sort is termed an 'emission nebula' or H II region. With other nebulæ, illuminated by stars which are less hot, the light is purely due to reflection; such is the nebulosity in the star-cluster of the Pleiades.

We cannot see through the Nebula, but we know that deep inside it there are very powerful stars which are permanently hidden from us. We also know that the Nebula is a stellar nursery, inside which fresh stars are being created. More than 4500 million years ago, our own Sun was born inside a nebula in just this way.

The gas making up the Nebula is incredibly thin, and is many millions of times more rarefied than the air which you and I are breathing. It has been calculated that if you could take a one-inch 'core sample' right through the cloud, from one siae to the other, the total weight of the material collected would be no more than enough to balance the weight of a fifty-pence coin. Yet the Nebula is so large that the total mass is very great indeed.

Many other gaseous nebulæ can be seen, but M42 is the supreme example. In fact it is merely the brightest part of a huge 'molecular cloud' which covers most of the constellation. It is fascinating to observe – and, predictably, it is a favourite target for astronomical photographers.

Future Points of Interest

2000: Total eclipse of the Moon. Mid-eclipse occurs at 4 h 45 m GMT, and totality lasts for an hour and a quarter.

January 21

Features of the Moon

During a lunar eclipse, the Moon must of course be full – but at times of non-eclipse, full moon is the very worst time to start observing, because the sunlight is coming 'straight down' on to the lunar surface, and there are virtually no shadows. This means that the only details easily visible are the dark plains or maria, still mis-called 'seas' even though there has never been any water in them, together with features which are either particularly brilliant (such as the crater Aristarchus) or particularly dark (such as the floor of the 60-mile crater Plato).

A crater is best seen when it is not far from the terminator, or boundary between the sunlit and night hemispheres. At such times the crater walls will cast shadows on to the sunken floor; a crater which is truly magnificent at such a time may be very hard to identify when the Sun is high above it. There are also the bright streaks or rays which extend from some of the craters, notably Tycho in the southern uplands and Copernicus in the Mare Nubium or Sea of Clouds. The rays are surface deposits, with no vertical relief, so that they are at their best near full phase, and are not seen under low illumination.

It takes time to learn one's way around the Moon. The best course is to make a list of suitable features, and then observe them under different angles of solar illumination. You may be surprised to see how quickly their appearances alter.

January 22

Right Ascension

We have seen that the declination of Betelgeux is (in round numbers) +7°, and that of Rigel –8°. Declination is the sky equivalent of terrestrial latitude. But we also need an equivalent of terrestrial longitude, and this is termed Right Ascension.

On Earth, the zero for longitude is the great circle on the terrestrial globe which passes through both the poles and also the Greenwich Observatory, in Outer London. Longitude is the angular distance of the site either east or west of the Greenwich meridian – thus the longitude of Athens is 24° E, Tokyo 139° E, New York 74° W, and so on. The actual zero is reckoned from one particular instrument at Greenwich, known as the Airy Transit Instrument. Straddle the line, and you will have one foot in the eastern hemisphere and the other in the western.

We need a zero point for the celestial equivalent of longitude, and fortunately there is one to hand. The Sun makes one full circuit of the sky in a year, but it does not travel along the celestial equator, because the Earth's axis of rotation is inclined to the perpendicular by an angle of $23\frac{1}{2}°$. Therefore the apparent yearly path of the Sun against the stars, termed the ecliptic, is inclined to the equator at the same angle: $23\frac{1}{2}°$. Around March 21 each year (the date is not quite constant, owing to the vagaries of our calendar) the Sun crosses the equator, travelling from south to north; this point is known as the Vernal Equinox, or First Point of Aries. (It is not marked by any bright star, but used to lie in the constellation of Aries, the Ram, hence the name.) Six months later, around September 22, the Sun again crosses the equator, this time moving from north to south; this is the autumn equinox, or First Point of Libra.

It is the Vernal Equinox which is taken as the zero for right ascension (RA), and the RA of any star or other object is defined as the angular distance from this point. However, it is generally given not in degrees, but in units of time. The

Vernal Equinox is bound to culminate – that is to say, reach its highest point in the sky – once every 24 hours. The right ascension of an object is given by the time which elapses between the culmination of the Vernal Equinox, and the culmination of the object. Betelgeux culminates 5 hours 55 minutes after the Vernal Equinox has done so; therefore the RA of Betelgeux is 5h 55m.

The Vernal Equinox is no longer in Aries. The position of the celestial equator shifts slowly, because of an effect known as precession, and the Equinox has now moved into the adjacent constellation of Pisces (the Fishes), though we still use the old name. The right ascensions and declinations of the stars alter only very slowly from year to year, but those of our nearer neighbours, the members of the Solar System, are changing all the time.

Anniversary

1962: Launch of Ranger 3 to the Moon. This was an early American attempt, but on January 28 it missed the Moon by 34 000 miles, and no pictures were received.

January 23

Clusters in Cassiopeia

Now let us return to the far-northern sky. The Great Bear is still in the north-east, and can be used to find Cassiopeia, which is also circumpolar as seen from Britain or the northern United States. To locate it, simply prolong the line from the Pointers through Polaris for about an equal distance on the far side. Cassiopeia cannot be mistaken, because even though its main stars are not particularly bright they make up a well-marked W or M pattern. They are listed in the table below.

The Milky Way runs through Cassiopeia, and the whole area is very rich in faint stars, so that it is well worth sweeping with binoculars. There are also several open or loose star-clusters, and although they are below naked-eye range they are easy to see with binoculars or a small telescope.

One of these is M52 (No. 52 in Charles Messier's catalogue) which lies in line with Alpha and Beta. Its leading stars are hot and white, so that by cosmical standards it is fairly young. Also in the Messier list is M103, in the same binocular field with Delta, but it is not particularly distinctive, and it is not obvious why Messier considered it to be worthy of inclusion.

Cassiopeia

Greek letter	Name	Magnitude	Luminosity (Sun=1)	Distance (light-years)	Spectrum
α Alpha	Shedir	2.2 (variable?)	200	120	K0
β Beta	Chaph	2.2	14	42	F2
γ Gamma	–	2 (variable)	6000	780	B0 pec
δ Delta	Ruchbah	2.7	11	62	A5
ε Epsilon	Segin	3.4	1200	520	B3

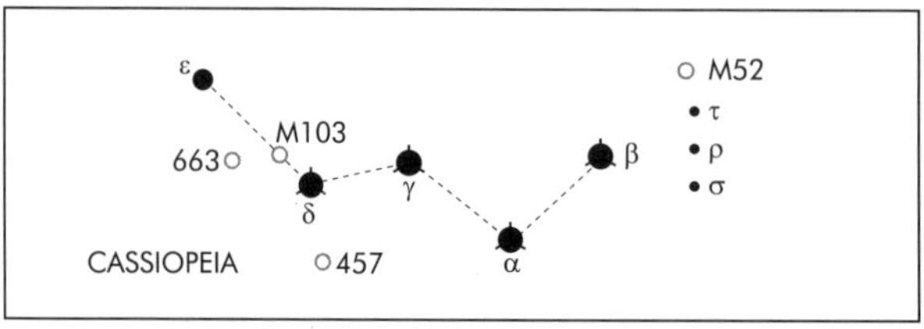

More than a century after Messier, the Danish astronomer J.L.E. Dreyer produced a much more extensive catalogue, the NGC or New General Catalogue, including all Messier's objects and many more. In 1995 I published the Caldwell Catalogue, which includes many of the bright objects omitted by Messier, and there seems no reason why the C numbers should not be used here.

The cluster NGC 663 (C10) is easy to find, in the same low-power binocular field with Delta and Epsilon. Of more interest is NGC 457 (C13), which lies between Delta and the fourth-magnitude Theta. Its distance seems to be of the order of 9000 light-years, and it is at least 30 light-years in diameter; it contains several thousands of stars. On its south-east edge is the star Phi Cassiopeiæ. If Phi is a genuine member of the cluster (and there is a definite doubt about this), it must be at least as powerful as Canopus. The cluster itself is an easy binocular object, in the same low-power field with Delta.

January 24

Variable Stars in Cassiopeia

Of the five stars in the W of Cassiopeia, one (Gamma) is definitely variable in light, and another (Alpha) probably is. Gamma has never had an accepted proper name, though it has sometimes been called Cih; but it is of special interest, and has given its name to a whole class of variables.

Normally it is just below second magnitude, but at times it can suffer outbursts which make it brighten appreciably. This happened in 1936, and by April 1937 the magnitude had risen to 1.6, much brighter than Polaris. It then declined, and by 1940 was below third magnitude. It then slowly recovered, and for the past few decades the magnitude has hovered around 2.2.

The spectrum is of type B, but is officially listed as 'peculiar' because the rainbow band is crossed by some bright lines as well as dark ones. What apparently happens is that during an outburst the star throws off shells of material, fading back to normal when the shell material dissipates. Other similar stars are known, but Gamma Cassiopeiæ is the brightest of them.

Anniversary

1986: The Voyager 2 probe passes the planet Uranus.

1990: Launch of Hagomoro, the first Japanese probe to the Moon.

1994: Launch of Clementine, the very successful US lunar mission.

Alpha, often still known by its proper name of Shedir, is of type K, and binoculars show it to be decidedly orange. It has long been suspected of slight variability. In official catalogues, its magnitude is given as constant at 2.2, but my own estimates indicate that it fluctuates slightly by a few tenths of a magnitude. Beta is a normal star which does not vary, and it is always worth comparing Alpha, Beta and Gamma; usually Gamma is the brightest of the three, but not always. As they are close together in the sky, there is no need to worry about extinction.

Also of note is a much fainter star, Rho Cassiopeiæ, which lies close to Beta, and is flanked to either side by Tau and Sigma, each of magnitude 4.9. Usually Rho is of about the fifth magnitude, but on rare, unpredictable occasions it can fall below naked-eye visibility, remaining dim for several months. This last happened as long ago as 1948, but a new minimum may start at any time.

The interesting point about Rho is that nobody knows quite what sort of variable it is; it does not seem to fit into any ordinary class. It is very remote – at least 1500 light-years away – and also highly luminous; it must be more than 100 000 times as powerful as the Sun. Compare it with Tau and Sigma, and estimate its brightness. This is the sort of observation which can usefully be made by amateurs; if Rho starts to decline, this will probably be first noticed by an amateur, and once minimum has started it will be carefully studied by professional astronomers with their sophisticated modern equipment.

January 25

Tycho's Star

We have looked at several variable stars in Cassiopeia. Now let us pause to say something about an extraordinary star which you will certainly not see this evening – though for a while in the year 1572 it was so brilliant that it could be seen with the naked eye even in broad daylight. It flared up near the star Kappa Cassiopeiæ, not far from the W. (Kappa is itself very remote and at least 50 000 times more luminous than the Sun, but there is nothing particularly notable about it.)

On 11 November 1572 a young Danish astronomer, Tycho Brahe, was taking a stroll when he looked up at Cassiopeia and saw something which caught his eye at once. In his own words:

> In the evening, after sunset, when, according to my habit, I was contemplating the stars in a clear sky, I noticed that a new and unusual star, surpassing all others in brilliancy, was shining almost directly above my head; and since I had,

> almost from boyhood, known all the stars of the heavens perfectly (there is no great difficulty in attaining that knowledge), it was quite evident to me that there had never before been any star in that place in the sky, to say nothing of a star so conspicuously bright as this. I was not ashamed to doubt the trustworthiness of my own eyes. But when I observed that others, too, having this place pointed out to them, could see that there really was a star there, I had no further doubts. A miracle indeed, either the greatest of all that have occurred in the whole range of nature since the beginning of the world, or one certainly that is to be classed with those attested by the Holy Oracles.

Naturally, Tycho had no real idea of the nature of the star. It remained very prominent for months, but finally it dropped below naked-eye visibility, and he could no longer track it. We now know that it was a supernova – a tremendous stellar outburst which resulted in the total destruction of a dwarf star. Supernovæ are of two distinct types, but we are sure that this was a Type I, because its remnants can still be identified in the form of wisps of gas and nebular material; the remnant sends out radio emissions, and in fact this is how it was first identified in modern times.

Supernovæ in our Galaxy are rare, and during the last thousand years only four have been seen with certainty: the stars of 1006, 1054, 1572 and 1604. However, supernovæ have often been found in other galaxies, and to astronomers they are of immense importance. When the next flare-up will occur in our Galaxy we do not know; it may be many centuries before we see anything to rival the magnificence of Tycho's Star.

January 26

Northern and Southern Lights

Auroræ, or polar lights (Aurora Borealis in the northern hemisphere, Aurora Australis in the southern) are due to electrified particles sent out from 'storms' on the Sun. These particles cross the 93 000 000 mile gap between the Sun and the Earth, and enter our upper air, admittedly in a rather roundabout fashion; the result is that the air begins to glow, rather in the manner of a spark coil. Because the particles are electrically charged, they tend to make for the Earth's magnetic poles, which is why auroræ are best seen from high latitudes. Go to Tromsø in northern Norway, and you may expect to see auroræ on at least 240 nights per year, but the frequency drops to only 25 nights per year in central Scotland, and no more than one or two nights per year in southern England – which is why the auroral display of 1938 caused such interest. From equatorial latitudes there are

Anniversary

1938: Brilliant display of Aurora Borealis (Northern Lights) seen over England.

almost no auroræ to be seen, though it is claimed that one display was detected from Singapore. Southern Lights are fairly common in places such as the Falkland Islands and the southernmost tip of New Zealand.

Auroræ may take many forms. There are glows, arcs, with or without rays; bands, draperies and 'curtains'. There are often vivid colours to be seen, and quick movement.

The heights of auroræ vary, but in general maximum activity is thought to occur at around 68 miles above sea-level. Very low auroræ have been reported now and then, but not with any certainty. There have been reports of sounds associated with auroræ, and even odours, but there is no proof; it would be difficult to explain associated sounds, and it must be said that smelly auroræ appear even less plausible than noisy auroræ!

The frequency of auroræ depends upon what is happening in the Sun. Every eleven years or so the Sun is at its most energetic, with many of the dark patches known as sunspots, and violent outbreaks which are termed flares; when the Sun is in its quietest mood there may be no spots for many consecutive days. Solar minimum last fell in 1996–7, so that the next maximum may be expected around the turn of the century – and it is quite likely that the new millennium will be ushered in by a number of brilliant displays of Polar Lights.

January 27

King Cepheus

Of the important constellations of the far north, we have yet to mention Cepheus, which lies more or less between Cassiopeia and the Little Bear. It was one of Ptolemy's original 48 constellations. The leading stars are given in the table below.

Cepheus is, of course, circumpolar. The north celestial pole shifts very slowly in the sky, and it is interesting to note

Cepheus

Greek letter	Name	Magnitude	Luminosity (Sun=1)	Distance (light-years)	Spectrum
α Alpha	Alderamin	2.4	14	46	A7
β Beta	Alphirk	3.2	2200	750	B2
γ Gamma	Alrai	3.2	10	59	K1
δ Delta	–	3.5–4.4 var.	6000 var.	1340	F8
ϵ Epsilon	–	4.2	15	98	F0
ζ Zeta	–	3.3	5000	700	K1
η Eta	–	3.5	4.5	46	K0
ι Iota	–	3.5	82	130	K1
μ Mu	–	3.4–5.1 var.	53 000	1560	M2

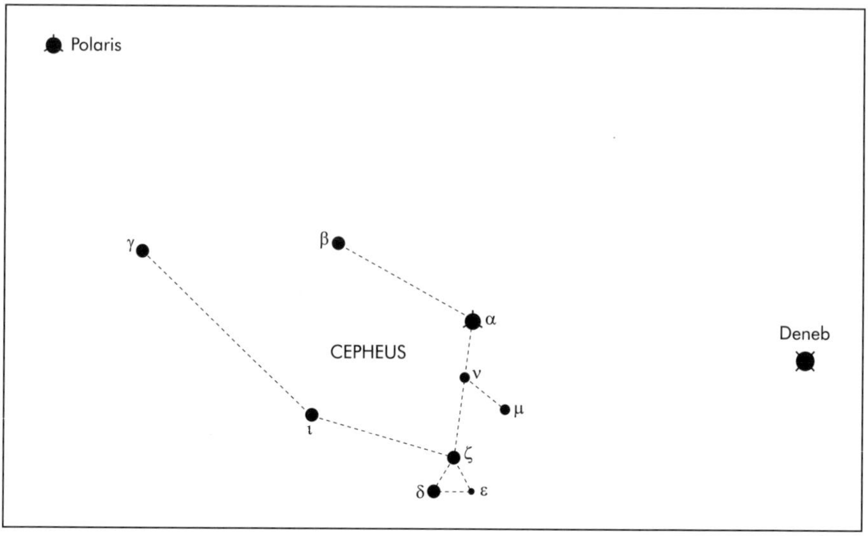

that in a few thousand years' time the pole star will be not our present Polaris, but Gamma Cephei.

Cepheus is not hard to identify; you should have no trouble in locating the rather distorted square formed by Alpha, Beta, Iota and Zeta. Pay special attention to the neat little triangle of Delta, Epsilon and Zeta. To astronomers, Delta Cephei is one of the most important stars in the sky; I will say more about it tomorrow.

January 28

Variable Stars in Cepheus

Let us turn straight back to Delta Cephei, which, as we have seen, makes up a small triangle with Epsilon (4.2) and Zeta (3.3). Delta is a variable star, with a range of magnitudes from 3.5 to 4.4. It is not hard to estimate its brilliancy by comparing it with Zeta and Epsilon; with a little practice the estimates can be made accurate to at least a tenth of a magnitude.

The period of Delta Cephei is 5.37 days. This is absolutely constant; the star is as regular as clockwork, so that we can always predict how bright it will be at any particular moment. It is one of many stars to behave in this way, and as it was the first of its type to be identified (by John Goodricke, in 1783), these variables are always known as Cepheids.

In 1912 a famous American astronomer, Henrietta Leavitt, was studying photographs of Cepheids in a nearby galaxy, the Small Cloud of Magellan, when she noticed something very interesting indeed. The Cepheids with longer periods were always brighter than those with shorter periods. To all intents

Future Points of Interest

2001: Eastern elongation of Mercury.

and purposes the stars in the Small Cloud could be said to be at the same distance from us, just as for all practical purposes it is good enough to say that Victoria and Charing Cross are the same distance from New York. It followed that the longer-period Cepheids really were the more powerful; there was a definite link between period and real luminosity. This meant that once the period had been measured, the luminosity could be found – and from this, the distance of the star.

It is difficult to over emphasize the importance of Miss Leavitt's discovery. Cepheids act as 'standard candles' in space, and since they are very powerful giants they can be seen over vast distances. Without them, we would be much less confident about our measurements of distances of galaxies far beyond our own Milky Way.

But by no means are all variables as regular in behaviour as Delta Cephei. One which is not is Mu Cephei, which can be located some way from the centre of a line linking the 'triangle' with Alpha. Mu is so intensely red that it has been nicknamed the Garnet Star; it fluctuates between magnitudes 3.4 and 5.1, but there is no well-marked period. Usually the magnitude is comparable with that of the adjacent star Nu (4.3). Mu Cephei is a red supergiant of the same type as Betelgeux in Orion; it is actually much more luminous than Betelgeux, but it is also much further away. Binoculars bring out the colour beautifully; it has been said that Mu Cephei looks rather like a glowing coal in the sky.

January 29

The Celestial Lizard

It is only too obvious that the constellation patterns which we follow are chaotic. The various groups are wildly unequal in both size and importance, and some of the smaller groups are so obscure that it is hard to justify their separate existence. We have already singled out some of these; Cælum, the Graving Tool, is a good example; for tonight's 'exercise' I recommend looking back at the notes for 19 January and re-identifying Cælum. True, it contains nothing of immediate interest, but it is always useful to know your way round the entire sky. Recall the words of Tycho Brahe: 'There is no great difficulty in attaining that knowledge.'

Another very dim constellation is Lacerta, the Lizard. It was created by Hevelius of Danzig in his star-maps of 1690, and has survived till the present day. It adjoins Cepheus, and may be found by using two of the stars in the famous triangle of Delta, Epsilon and Zeta. A line from Zeta, passing through Epsilon and prolonged, will come to the diamond pattern which marks the main part of Lacerta. Only the two brightest stars in the diamond have been given Greek letters: Alpha (magnitude 3.8) and the rather orange Beta (4.4). The other two members of the pattern are merely listed as 4 and 5 Lacertæ.

Lacerta is crossed by the Milky Way. It contains an open cluster, NGC 7243 (C16) which makes up an equilateral triangle with Alpha and Beta. However, it cannot be said to be at all conspicuous, and there is nothing special about it. It can be identified with binoculars, though not easily; a telescope resolves it without difficulty.

January 30

Introduction to Eridanus

One of the largest of all the constellations is Eridanus, the River. It covers a total area of 1138 square degrees (as against only 125 square degrees for Cælum), and is immensely long. It begins close to Rigel in Orion, and ends not far from the south celestial polar region. This means that only part of it is

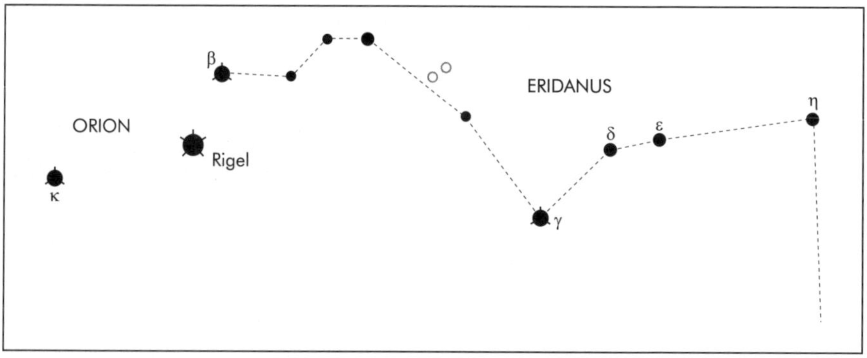

Anniversary

1910: Maximum brilliancy of the Daylight Comet. This was (so far!) the most brilliant comet of the twentieth century. It was discovered on 13 January by some diamond miners in South Africa, and soon become so bright that it could be seen in broad daylight. On 15 January the well-known astronomer R.T.A. Innes had a fine view of it when it was only 4.5° away from the Sun's limb; it was described as a snowy-white object with a tail 1° long. During January it moved into the evening sky, and was spectacular by any standards; on 30 January the tail reached its maximum length – at least 30°. It reached perihelion on 17 January, at magnitude –4. It then faded, and dropped below naked-eye visibility during the first week in February, but it was followed telescopically until 15 July, when the magnitude had fallen to 16.5. It has an elliptical orbit; as its period is around 4 000 000 years it will not be seen again in our time! It was much brighter than Halley's Comet, which was at its best a few weeks later, and it is fair to say that many people who claim to remember Halley's Comet in 1910 actually saw the Daylight Comet instead.

visible from British latitudes and its brightest star, Achernar, is permanently out of view, though to observers in the southern hemisphere it is very prominent in the south-west during January. From New Zealand Achernar is circumpolar, and from Australia and South Africa it sets only briefly. The brightest stars in Eridanus accessible from Britain are Beta (magnitude 2.8) and Gamma (2.9), both of which are well on view at the moment.

Eridanus has been linked with the legend of Phæthon's ride. Phæthon was a boy with a mortal mother but an immortal father – no less than Helios, the Sun-God. He persuaded Helios to allow him to drive the Sun-chariot across the sky for one day, but the results were disastrous; the white horses pulling the chariot bolted, and as Phæthon lost control the chariot swooped down, setting the Earth on fire. Jupiter, king of the gods, had no alternative but to strike the boy with a thunderbolt, toppling him down to his death in the waters of the river Eridanus.

This is one version. Others, however, identify the river with either the Nile or the Po. At any rate it is easy to track in the sky, though it contains few bright stars and only one – Achernar – which is outstandingly brilliant. The declination of Achernar is –57°, and that of Beta –5°, so that all in all Eridanus sprawls over a total of more than 50°.

January 31

Penumbral Eclipses

This is a mildly interesting evening for lunar observers, since there will be a 100% penumbral eclipse of the Moon in 1999.

We have seen that a lunar eclipse happens when the Moon passes into the Earth's shadow, and its supply of direct sunlight is temporarily cut off. However, the Sun is not a point source of light; it is a disk, and this means that the main cone of shadow cast by the Earth is flanked to either side by a zone of partial shadow, or penumbra (note that this has nothing to do with the penumbra of a sunspot). Of course, the Moon has to pass through the penumbra before entering the main cone or umbra, but there are also eclipses when the umbra is never entered at all. This is so on 31 January 1999. The Moon will be wholly immersed in the penumbra; the time of mid-eclipse is 16h 20m GMT.

With the naked eye, a slight dimming of the Moon will be evident; binoculars will show it well, but it will not be at all striking, and people who have not been told about it will probably not realize that anything unusual is happening.

It is obvious that eclipses, either umbral or penumbral, can take place only at full moon. But where will the Moon be on

Anniversary

1966: Launch of the Russian probe Luna 9 to the Moon. This was an attempt to achieve a soft landing, and obtain pictures direct from the lunar surface – which had not been done before. The mission was successful; on the 3 February Luna 9 came down in the Oceanus Procellarum, and excellent pictures were obtained.

Future Points of Interest

1999: Penumbral eclipse of the Moon.

31 January in the other years covered in this book? This is easy to work out from the table given on pages ix and x:

1998: A 4-day-old evening crescent.
1999: Full (penumbral eclipse).
2000: A waning crescent in the morning sky, 3 days after Last Quarter.
2001: Prominent – one day before half-phase.
2002: Gibbous, 3 days after full.
2003: Out of view – one day before new.

Professional astronomers are not fond of the nights close to full moon, when the sky is inconveniently light, but nobody can deny that the Moon itself, with its mountains, its craters and its waterless seas, is both fascinating and beautiful.

February

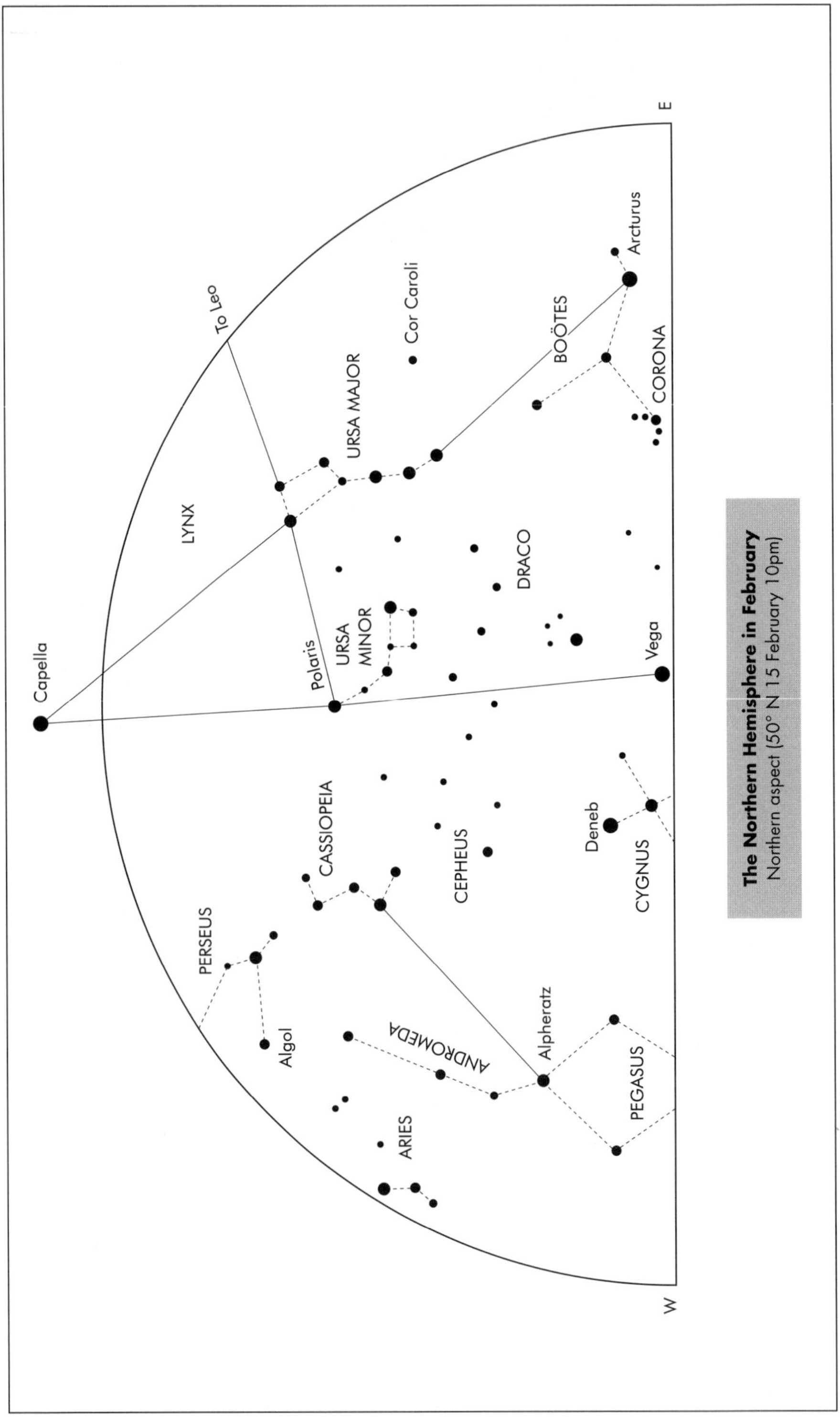

The Northern Hemisphere in February
Northern aspect (50° N 15 February 10pm)

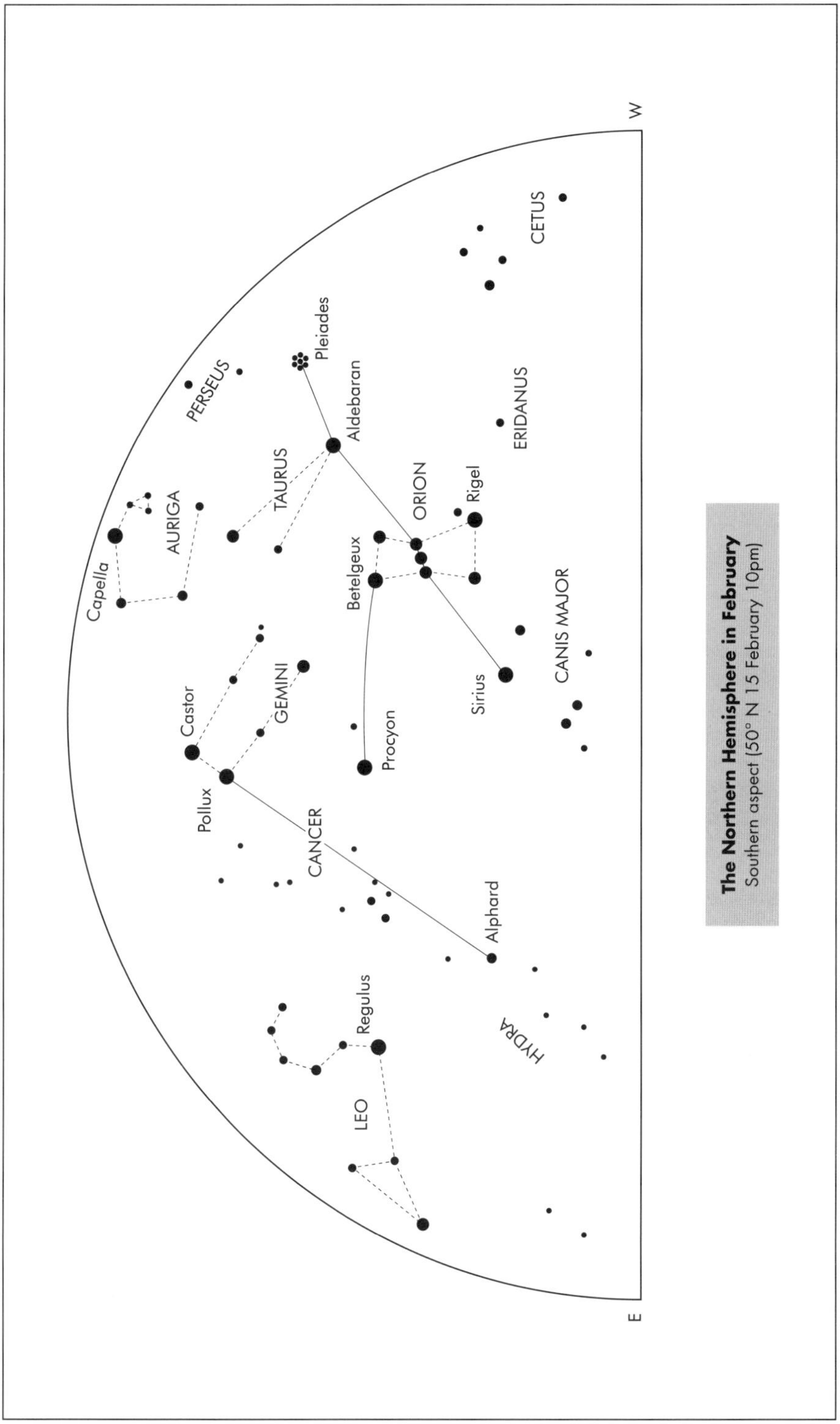

The Northern Hemisphere in February
Southern aspect (50° N 15 February 10pm)

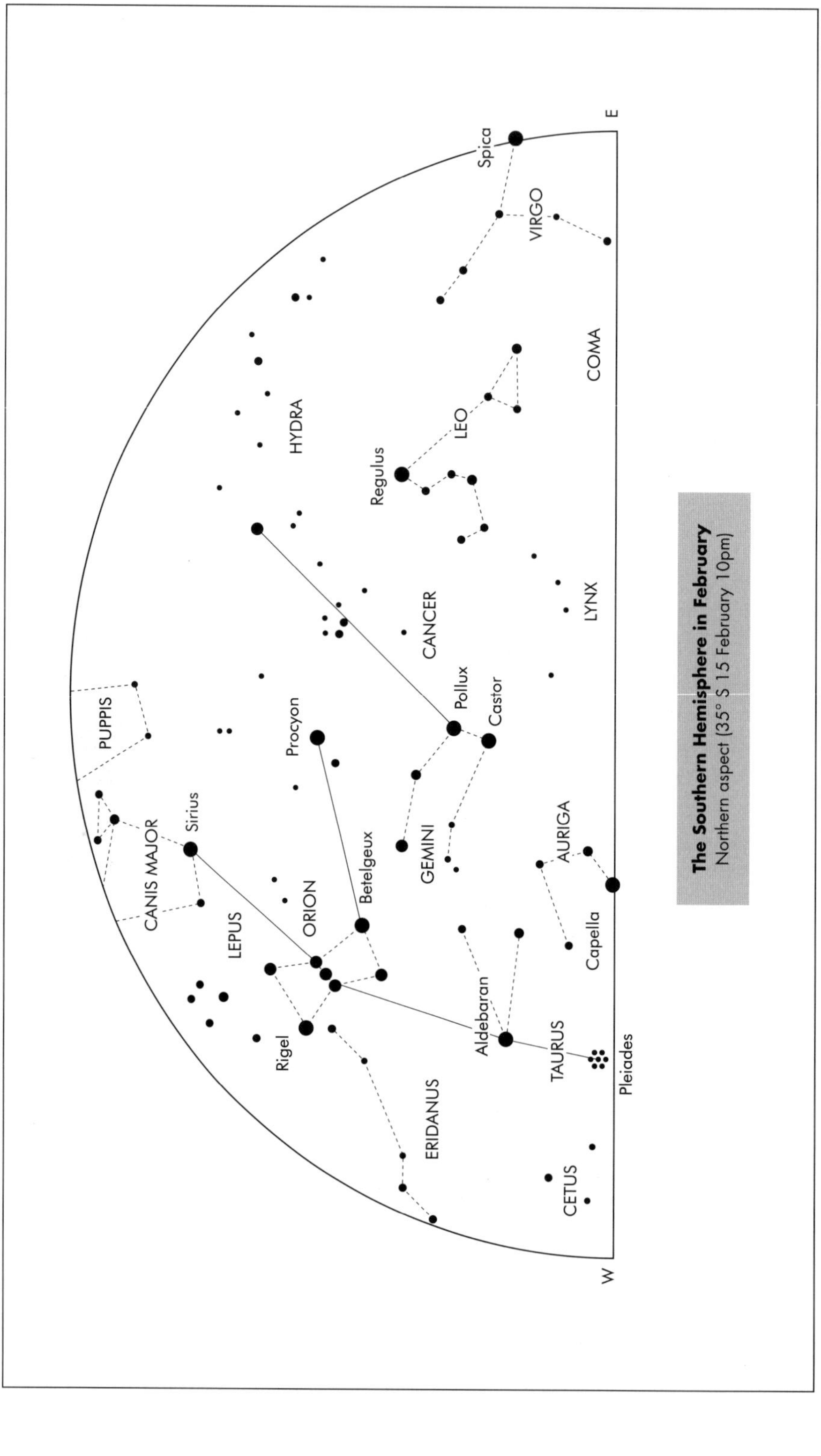

The Southern Hemisphere in February
Northern aspect (35° S 15 February 10pm)

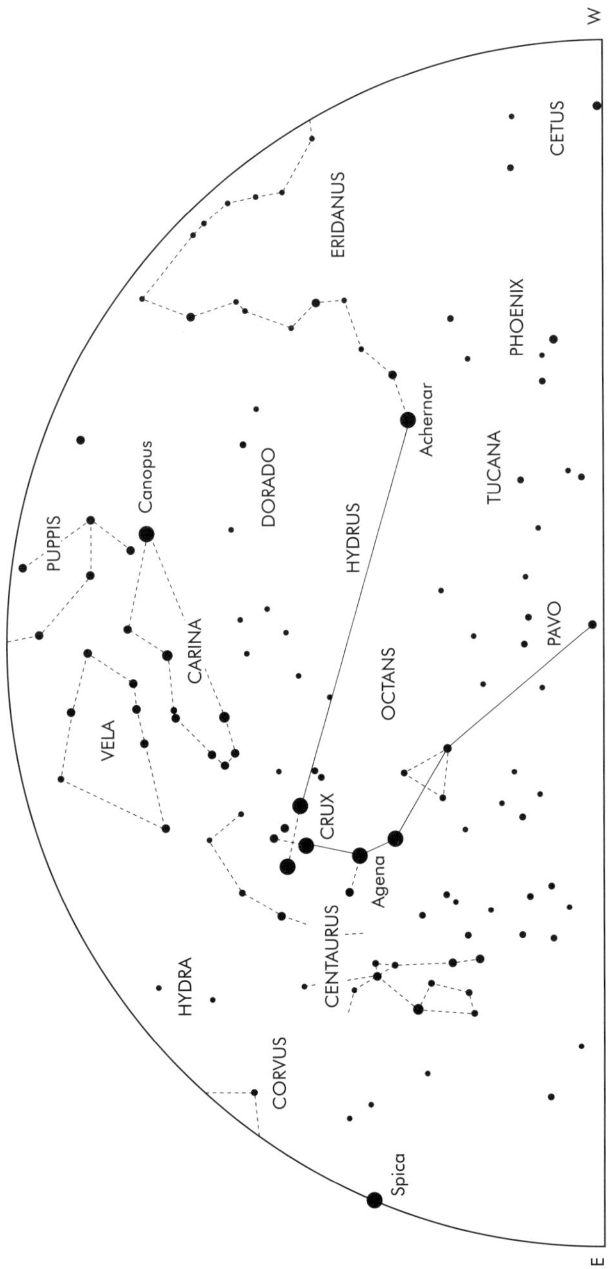

The Southern Hemisphere in February
Southern aspect (35° S 15 February 10pm)

February 1

The February Sky

As this is the start of the month, it will be useful to 'look round' and see which stars are on view. Fortunately, in the evening sky northern-hemisphere observers have both their main guides, Ursa Major and Orion. Ursa Major is in the north-east, with the Bear 'standing on its tail'; the W of Cassiopeia is in the north-west, about the same distance from the Pole Star. The Square of Pegasus is disappearing in the west, and will not be well seen again until autumn. The brilliant yellow Capella is almost overhead – which means that the equally brilliant Vega is very low over the northern horizon, even though from the latitude of the British Isles it never actually sets. Arcturus, actually the most brilliant star in the northern hemisphere of the sky, is starting to appear in the east. The Milky Way is beautifully displayed, running from the southern horizon past Orion, Cassiopeia and Capella down to the northern horizon.

Observers in southern latitudes – such as that of Australia – also have Orion, with Betelgeux lower down than Rigel. Canopus is near the zenith, with the Southern Cross in the east. Of course there is no sign of Ursa Major, but Leo, the Lion, is coming into view in the north-east. This is a good time to look for the two Clouds of Magellan, about which I will have much more to say later (24 February).

February 2

Oppositions of the Planets

When a planet reaches opposition – as Jupiter does on 2 February 2003 – it is exactly opposite to the Sun in the sky, and is best placed for observation. (Obviously the inferior planets, Mercury and Venus, can never come to opposition.) The mean interval between successive oppositions is termed the synodic period. The values for the various planets are:

Synodic Periods of the Planets

Planet	Synodic period (days)
Mars	780
Jupiter	399
Saturn	378
Uranus	370
Neptune	368
Pluto	367

Anniversary

1964: The US probe Ranger 6 landed on the Moon. It came down in the Mare Tranquillitatis, but owing to a TV failure no images were received.

Future Points of Interest

2003: Opposition of Jupiter.

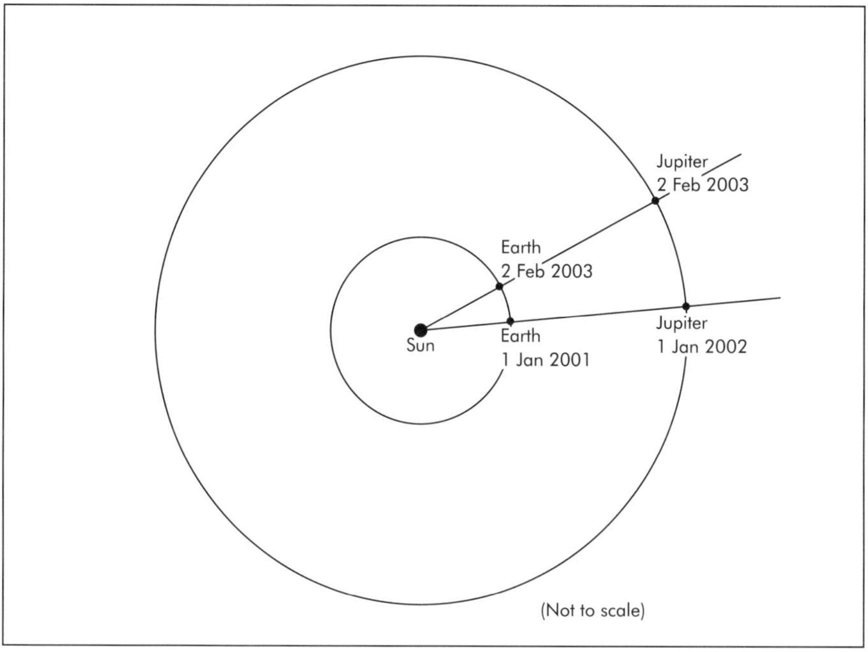

The diagram shows the situation with regard to Jupiter. An opposition fell on 1 January 2002. A year later the Earth had completed a full orbit, but Jupiter had not had time to do so – its orbital period is nearly 12 years! It takes the Earth some time to 'catch up', and on average Jupiter reaches opposition just over a month later each year; 2 February 2003 is followed by an opposition on 4 March 2004. Between 1999 and 2004 Jupiter is in the northern hemisphere of the sky, and its magnitude exceeds –2. It reaches maximum northern declination in 2002, in Gemini; by 2003 it has moved into Cancer, and in 2004 it will reach Leo.

Anniversary

1966: Luna 9 landed on the Moon.

1967: Launch of Lunar Orbiter 3.

February 3

Touchdown in the Ocean of Storms

Today marks an important anniversary. It was on 3 February 1966 that we obtained final proof that the Moon's surface is firm enough to bear the weight of a spacecraft.

The lunar maria are less crater-scarred than the bright uplands, and there is no doubt that they were once seas of molten lava. There had been a theory that they were filled with deep, soft dust, into which any spacecraft incautious enough to land would promptly sink. Observers had little faith in this idea, but it was definitely jettisoned when the Russian unmanned probe Luna 9, launched on 31 January 1966, made a controlled landing in the Oceanus Procellarum and sent back excellent images. Contact was maintained for four days after arrival. The landing position was: latitude 7°.1 N, longitude 64°.4 W.

The Oceanus Procellarum (Ocean of Storms) is rather irregular in outline; it contains the brilliant crater Aristarchus and the ray-crater Kepler. It is the largest of the lunar maria, and its area is greater than that of our Mediterranean Sea.

February 4

Future Points of Interest

2003: Western elongation of Mercury.

Gemini

The constellation of Gemini (the Twins) adjoins Orion, and is easy to locate, mainly because of the presence of its two bright leaders, Castor and Pollux. It lies in the Zodiac, so that it may contain planets; it is in fact the northernmost of the Zodiacal constellations, and it is also one of the largest and most conspicuous. The leading stars are listed in the table overleaf.

Though there is a bare half-magnitude difference between Pollux and Castor, Pollux is always included in the list of

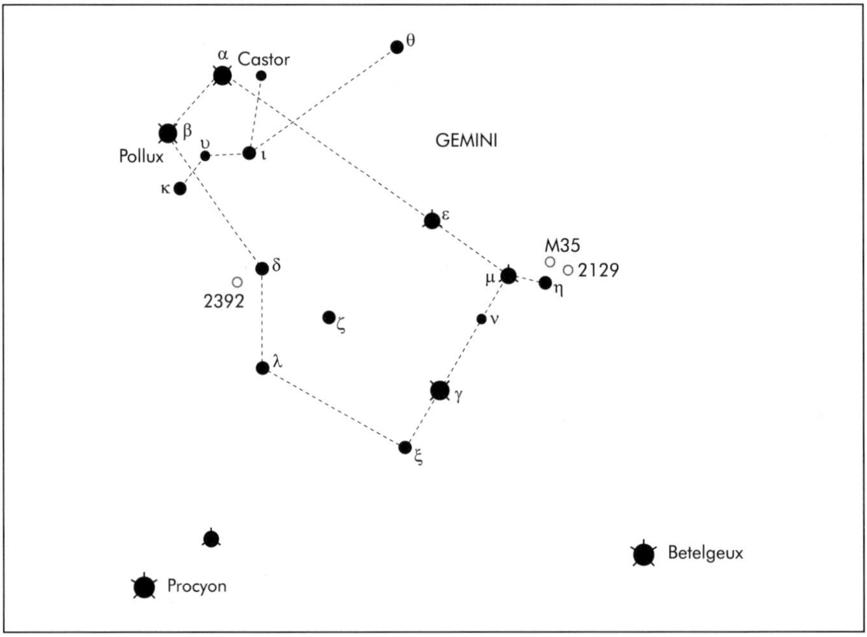

Gemini					
Greek letter	Name	Magnitude	Luminosity (Sun=1)	Distance (light-years)	Spectrum
α Alpha	Castor	1.6	36	46	A0
β Beta	Pollux	1.1	60	36	K0
γ Gamma	Alhena	1.9	82	35	A0
δ Delta	Wasat	3.5	14	59	F2
ϵ Epsilon	Mebsuta	3.0	5200	680	G8
ζ Zeta	Mekbuda	3.7 max.	5200	1400	G0
η Eta	Propus	3.1 max	120	186	M3
θ Theta	–	3.6	82	166	A3
ι Iota	–	3.8	60	163	K0
κ Kappa	–	3.6	58	147	G8
λ Lambda	–	3.6	17	81	A3
μ Mu	Tejat	2.9	120	230	M3
ξ Xi	Alzirr	3.4	46	75	F5

'first-magnitude stars', while Castor is not. The reasons for this distinction seem decidedly obscure!

Gemini is crossed by the Milky Way, and the whole area is very rich, so that it is well worth sweeping with binoculars or a wide-field telescope.

Anniversary

1971: Landing of Apollo 14 on the Moon. The astronauts (Alan Shepard and E. Mitchell) came down at latitude03° 40′ S, longitude 17° 28′ W, near the crater Fra Mauro. They were the first to take a 'lunar cart' to the lunar surface.

1974: Mariner 10 passed Venus at 3600 miles, en route for a rendezvous with Mercury.

Future Points of Interest

2000: Partial eclipse of the Sun, visible from the Antarctic area. At maximum (13h GMT) 56% of the solar disk will be obscured.

February 5

Fra Mauro

Fra Mauro is the largest member of a group of craters in the Mare Nubium (Sea of Clouds). It had been the target for the unsuccessful Apollo 13 mission. Fra Mauro is 50 miles in diameter, and has low, broken walls. Adjoining it is Bonpland (36 miles across) which is in a rather better state of preservation; Parry (26 miles across) is more regular. These formations have common boundaries, and are crossed by various rills. Further south is Guericke (33 miles across), with incomplete walls which are broken in the north and in places have almost been levelled by the Mare lava.

Groups of walled formations are common enough on the lunar maria; the overall distribution is not random. The low walls of the members of the Fra Mauro group mean that they become very obscure under high solar illumination.

The names given to them are of some interest. Fra Mauro (?–1459) was a Venetian monk who is best remembered for his excellent world map; Aimé Bonpland (1773–1858) was a French physician, explorer and naturalist; Otto von Guericke (1602–1686) was a German engineer and naturalist, and Sir William Edward Parry (1790–1855) was a British explorer of the Arctic, who once made a gallant if unsuccessful attempt to reach the North Pole by boat and sled.

February 6

The Non-Identical Twins

Though Castor and Pollux lie side by side in the sky, there is no genuine connection between them; Castor is a full ten light-years further away from us. However, there is a famous mythological legend about them. Castor and Pollux were twin boys, but there was one important difference between them; Pollux was immortal, while Castor was not. When Castor was killed in battle, Pollux pleaded to be allowed to share his immortality with his brother – so that both were transferred to the sky.

They differ also in colour; Castor is white, and is a multiple system, while Pollux is decidedly orange. The contrast between them is obvious even with the naked eye, and binoculars bring it out very well. This means, of course, that of the two Pollux has the cooler surface – a point about which I will have more to say when discussing stellar spectra.

As we have seen, Pollux is the brighter by half a magnitude, but there are suggestions that this may not always have been so. Both were listed by Ptolemy, the last great astronomer of Classical times, who rated them equal at magnitude 2. So did the Arabs of a thousand years later, whose magnitude estimates were in general very good. When Johann Bayer gave the stars their Greek letters, in 1603, he made Castor Alpha and Pollux Beta, while in 1700 John Flamsteed, the first British Astronomer Royal, rated Castor as being of the first magnitude and Pollux of the second. Since the early nineteenth century there has been no doubt about the superiority of Pollux.

The rest of the constellation Gemini is made up of lines of stars extending from Castor and Pollux in the general direction of Orion; Alhena or Gamma Geminorum, not very far below first magnitude, lies between Pollux and Betelgeux. There are several interesting telescopic objects in the constellation, as well as two variables, Eta and Zeta, whose fluctuations may be followed with the naked eye.

February 7

William Huggins and his Spectroscopes

This is a memorable 'anniversary day'. On 7 February 1824 William Huggins was born. He was never a professional astronomer; he had to enter the family drapery business, and it was only when he was able to retire from the firm that he was able to devote his whole time to astronomy.

Anniversary

1824: Birth of Sir William Huggins, the great astronomical spectroscopist. Huggins was a great pioneer. He was knighted in 1897, and served as President of the Royal Astronomical Society. He lived until 1910.

Anniversary

1936: The asteroid Adonis passed Earth at a little over a million miles. This was one of the earliest-observed 'NEA' or Near-Earth-Approach asteroids, though its record has been broken many times since.

He set up an observatory at Tulse Hill, in outer London, and concentrated on astronomical spectroscopy. As we have noted, a spectroscope splits up light, and a normal star produces a rainbow band crossed by dark lines, each of which is the trademark of some particular element or group of elements. Huggins found that the stars could be divided into definite types according to their colours; a red star showed a spectrum very different from that of a white star – so that, for example, the orange Pollux has a spectrum which is easily distinguished from that of a white star such as Castor. Similar work was being carried out in Italy by the Jesuit astronomer Angelo Secchi, and the stars were put into four distinct classes (the more detailed Harvard classification system came much later):

I White or bluish stars, with broad, dark spectral lines due to hydrogen. Example: Sirius.

II Yellow stars; hydrogen lines less prominent, but more evidence of lines due to metals. Examples: Capella, the Sun.

III Orange stars, with complicated banded spectra. Example: Betelgeux.

IV Red stars, with prominent lines due to carbon. All were below magnitude 5, and many of them were found to be variable. R Cygni in the Swan was a good example.

What about the nebulæ? Some, such as the Great Nebula in Andromeda, seemed to be made up of stars, but others, such as the nebula in Orion's Sword, looked more like patches of gas. Huggins knew that he could find out. The spectrum of a shining gas at low pressure does not yield a rainbow; instead, it produces isolated bright lines. When Huggins examined nebulæ such as Orion's Sword, he saw only bright lines – and this proved that the nebulæ themselves were indeed gaseous. On the other hand, the spectra of 'starry nebulæ' proved to be made up of the combined spectra of many millions of stars. We now know that they are galaxies in their own right, far beyond the Milky Way.

Anniversary

1992: The solar polar probe Ulysses passed Jupiter at 235 000 miles. Jupiter's powerful pull of gravity was used to put Ulysses into the required orbit.

February 8

The Castor Family

Look again at the Twins. With the naked eye, they seem like normal stars – one orange, one pure white. But with a telescope, things are different. Castor is seen to be double. Like Mizar in the Great Bear, it is made up of two components, one considerably brighter than the other; the magnitudes are respectively 1.9 and 2.9.

Castor was the first binary system to be recognized as such. In 1803 Sir William Herschel was busy trying to measure star distances, using the parallax method, and one of his targets was Castor. He did not succeed in measuring its distance (the first success in this field came later, in 1838, with Bessel's determination of the distance of the faint star 61 Cygni), but he did find that the two Castors were moving round their common centre of gravity. We now know that the orbital period is 420 years, and that on average the real distance between the two is around 8000 million miles. A three-inch telescope will show them well; the apparent separation is 2.5 seconds of arc.

This is only a beginning. Each component is again a close binary – too close to be split with ordinary telescopes, so that we have to use indirect methods. The brighter component (Castor A) is made up of two stars, each about twice the diameter of the Sun and rather over 4 000 000 miles apart; they orbit each other in 9.2 days. The secondary compoment (Castor B) has its two stars each about 1.5 times the diameter of the Sun, with a separation of a little over 3 000 000 miles and a period of 2.9 days. To complete the picture there is a third member of the group, Castor C, orbiting the main components in a period which may be as much as 10 000 years, at a distance of 100 000 million miles. It too is a close binary; each star is around 600 000 miles in diameter (as against 865 000 miles for our own Sun). Both components are dim and red, and below the 9th magnitude, so that they are none too easy to locate even with a telescope. Normal telescopes will not show them separately; they are less than 2 000 000 miles apart, and their orbital period is a mere 19.5 hours.

Interestingly, the two components of C seem to eclipse each other regularly as they move round their common centre of gravity, so that the brightness as seen from Earth changes, and Castor C has a variable star designation – YY Geminorum.

Castor, then, is a true stellar family, made up of six components – four hot and white, two faint and red. Multiple stars are by no means uncommon, but Castor is a particularly good example. Certainly it is quite unlike its solitary orange 'twin'.

February 9

Variable Stars in the Twins

Still with 'this week's constellation', Gemini, we have two more objects of interest. Both are variable stars, and both can be studied with the naked eye.

The first is Eta Geminorum, still sometimes called by its old name of Propus. It lies near Mu, which is a reddish M-type star of magnitude 2.9. Eta, also of type M, is variable over a magnitude range of from 3.1 to 3.9, so that it is never quite as bright as Mu.

Eta is what is termed a semiregular variable, because it has a rough period of 233 days – that is to say, there is an interval of 233 days between one maximum and the next – it is never predictable, and sometimes fluctuates quite erratically. In any case, the variations are slow, and you will have to watch it for a week at least before any definite change can be noticed. Apart from Mu, useful comparison stars are Xi Geminorum (magnitude 3.4) and Zeta Tauri (3.0).

The second variable is Zeta, which lies roughly between Alhena and Pollux. Here the range is from magnitude 3.7 to 4.1. The star is a Cepheid variable (see the notes for January 28) so that we always know how it is going to behave. Useful comparison stars are Delta (magnitude 3.5), Lambda (3.6), Iota (3.8) and Upsilon (4.1).

Follow these stars from night to night, and it is possible to draw up a light-curve. Variable stars are very common – it is lucky for us that our Sun is so sober and stable.

February 10

Messier's Catalogue – and M35

There is a great deal of interest in Gemini, which is still very well placed for observation during evenings – even from countries such as Australia and South Africa, where it is never very high above the horizon. Tonight let us turn to a cluster of stars, Messier 35 (M35 for short).

Charles Messier was a French astronomer who was interested mainly in hunting for comets. He kept on being deceived by star-clusters and nebulæ, and in the end he drew up a list of them as 'objects to avoid'. Ironically, we still use Messier's numbers, though few people remember the comets which he discovered.

M35 lies in the 'foot' of Gemini, near Mu and the variable Eta. The cluster is visible with the naked eye, though it is not too easy to sort out from the Milky Way. Binoculars show many stars in it, and a telescope reveals dozens; the total membership amounts to several hundreds, and the true diameter of the cluster is of the order of 30 light-years. It is thought to be about 2850 light-years away from us. It is a typical open or loose cluster, with no definite shape, but in a telescope it is a magnificent sight.

In the same region there is another open cluster; it was not listed by Messier, but is contained in the 'New General Catalogue' drawn up by J.L.E. Dreyer, so that we know it as NGC 2129. It contains about 40 stars, and is not a difficult telescopic object. Incidentally, Dreyer's catalogue is no longer 'new'; it was compiled more than a century ago.

In 1995 I drew up the Caldwell Catalogue, listing 109 objects which had not been listed by Messier; these C

Anniversary

1974: The Russian probe Mars 4 missed Mars by over 1200 miles. It had been launched on 21 July 1973.

1990: The Galileo probe, en route for Jupiter, passed Venus at just under 10 000 miles, and returned data.

numbers now seem to be coming into general use. (The 'Caldwell' comes from my surname, which is actually Caldwell-Moore. I could not use M numbers, for obvious reasons.)

February 11

The Eskimo Nebula

Before taking our leave of Gemini, let us try to locate a much more elusive object – C39 (NGC 2392), the so-called Eskimo Nebula. This time you will need a telescope of some size. The nebula lies about midway between Kappa and Lambda Geminorum, but its integrated magnitude is not much above 9, and I for one have never been able to see it in binoculars. Telescopically it appears as a small fuzzy blur; photographs taken with powerful instruments are needed to bring out its details. It is sometimes called the Clown-Face Nebula, though personally I prefer the more dignified Eskimo. It was discovered by Sir William Herschel in 1787.

It is known as a planetary nebula, but the name is misleading, since it has nothing whatsoever to do with a planet and is not truly a nebula. It is a very old star which has thrown off its outer layers; the star itself is of about the 10th magnitude, and is very small (by stellar standards), very dense and very hot. The distance from us is of the order of 3000 light-years.

It is something of a challenge; see if you can identify it! The position is: RA 07 h 29 m.2, declination +20° 55′.

February 12

Anniversary

1961: Launch of Venera 1, the first attempted interplanetary mission.

Messengers to Venus

Yet another important anniversary falls today. On 12 February 1961 the first of all interplanetary space-craft was launched: Venera 1. It was sent up by the Russians from their rocket base at Baikonur, in what is now Kazakhstan (then, of course, it was in the old USSR). It was aimed at Venus.

Venus is the nearest of the planets; it can approach us to within 24 000 000 miles, which is much closer than Mars can ever be. However, we cannot simply wait until the two worlds are closest together and then fire a rocket across the gap. This would mean using fuel for the whole journey, and no vehicle could possibly carry enough. What has to be done is to put the space-craft into a path which makes it 'coast' inward toward Venus without needing any extra power; this is what is officially termed free fall – and of course the Earth itself is in free fall around the Sun.

This is what the Russians did. Unfortunately, the mission was not a success. Contact with Venera 1 was lost at a distance of 4 600 000 miles, and was never regained. It is likely that the probe passed by Venus in mid-May 1961 at a distance of around 60 000 miles, but we will never know for certain; presumably it is still orbiting the Sun, but we have no hope of finding it again. It was not until, the following year that success came, with the fly-by of America's Mariner 2, which was launched on 27 August 1962 and passed Venus at 21 700 miles on the following 14 December. Since then there have been many missions to Venus, but we must not forget these pioneer efforts almost four decades ago.

February 13

Introducing the Milky Way

This is a good time of the year to start looking at the Milky Way. Ideally, you need a moonless night, and the Moon will not be obtrusive on 13 February in 1997, 1999 or 2002, though it will be brighter in 2000 and 2001 and will be near full in 1998 and 2003.

Whether you are observing from the north or south of the Earth's equator, the Milky Way runs right across the sky and passes near the zenith or overhead point; it cuts through Gemini, which has been our 'constellation of the week', though it is at its brightest in Cygnus and Sagittarius, neither of which are properly on view during evenings in February. The Milky Way is made up of stars, and there are so many of them, apparently so close together, that they may seem in danger of colliding with each other.

This is not true; as so often in astronomy, appearances are deceptive, and the Milky Way stars are not genuinely crowded together. We are dealing with a line-of-sight effect. The Galaxy is a flattened systen, and the Sun, with the Earth and the other planets, lies near the main plane – so that when we look along the main 'thickness', we see many stars almost one behind the other. This is what causes the Milky Way effect. The whole system is perhaps 100 000 light-years across, and the Sun is rather less than 30 000 light-years from the centre, which is beyond the lovely star-clouds in Sagittarius.

If it could be seen from 'above' or 'below', the Galaxy would show spiral arms, like a Catherine-wheel. The Sun lies near the edge of one of the arms.

February 14

The Solar Cycle – and Solar Max

So far we have said virtually nothing about the Sun, so it is time to turn to the daytime sky. The Sun may be a junior member of the Galaxy, but to us it is all-important.

The Sun is very large compared with the Earth; its diameter is 865 000 miles, and its huge globe could contain more than a million Earths. It is shining not because it is burning in the manner of a coal fire, but because of nuclear reactions going on deep inside it; as with other stars of its type, the main 'fuel' is hydrogen.

The Sun shows a more or less regular cycle of activity. Every eleven years it is energetic, with many sunspots and flares; it then dies down, until at minimum there may be many successive days with no spots at all (as happened, for instance, in April and May 1997), after which activity builds up once more. It would be idle to pretend that we yet have a full understanding of this cycle, and it was to gain extra information that the Solar Maximum Mission ('Solar Max') was launched in 1980. The Sun had been at minimum activity in 1976, so that the next maximum was due, and SMM was programmed to take full advantage of it. In the event there were serious operational problems, but a rescue mission by Shuttle astronauts put matters right, and Solar Max continued to operate almost up to the time of the next minimum. It came to the end of its career in December 1989, when it dropped back into the Earth's atmosphere.

The cycle is not completely regular, and 11 years is merely an average. The next maximum is timed for around the start of the new millennium, and solar workers will have an interesting time. Always remember, however, that the Sun is dangerous, and direct observation with any telescope or binoculars is emphatically not to be recommended, even with the addition of a dark filter. I will have much more to say about this later.

Anniversary

1980: Launch of the Solar Maximum Mission probe ('Solar Max'). This was a long-term mission designed to study the Sun. It was repaired in orbit in April 1984, when it was drawn into the Space Shuttle, overhauled and re-launched; it finally decayed in December 1989.

February 15

Lunar Insects?

Today we honour the memory of Galileo, the first great 'telescopic astronomer', who looked at the Moon in 1610 with his newly built instrument and described the mountains and craters. But for the moment let us turn to another astronomer, whose ideas about the Moon were, to put it mildly, unconventional.

Anniversary

1564: Birth of Galileo.

1858: Birth of W.H. Pickering.

Future Points of Interest

2000: Eastern elongation of Mercury.

William Henry Pickering worked at the Harvard Observatory, in America, where his brother Edward was Director. In 1904 he produced an excellent lunar photographic atlas, and then, between 1919 and 1924, went to Jamaica and carried out a detailed study of the lunar crater Eratosthenes, which lies at the southern end of the Apennine range; it is 38 miles across and is very deep, with high, terraced walls and a massive central peak. Inside it, Pickering saw strange dark patches which seemed to move during the course of each lunation. Pickering believed them to be due to low-type vegetation or even swarms of insects, and in a paper published in 1924 he went even further.

> While this suggestion of a round of lunar life may seem a little fanciful and the evidence upon which it is founded frail, yet it is based strictly on the analysis of the migration of the fur-bearing seals of the Pribiloff Islands ... The distance involved is about 20 miles, and is completed in 12 days. This involves an average speed of about 6 feet a minute, which, as we have seen, implies small animals.

Pickering believed that the Moon's atmosphere was dense enough to support low-type life, but we now know that this is wrong; the Moon's gravitational pull is so weak that it has been unable to hold on to any atmosphere it may once have had, and today the Moon is an airless world. The patches in Eratosthenes are obvious enough, and can be seen with a small telescope, but they do not move, and they are not due to any living thing. There has never been any life on the Moon.

Eratosthenes itself is one of the noblest of the lunar craters, and is conspicuous under almost any angle of solar illumination. The lunar Apennines form part of the border of the large, regular Mare Imbrium or Sea of Showers.

February 16

Anniversary

1996: Launch of NEAR (Near Earth Asteroid Rendezvous) to the minor planet Eros.

Future Points of Interest

1999: Annular eclipse of the Sun.

Annular Eclipses

Asteroid No. 433, Eros, is a small body, less than 20 miles long and shaped rather like a sausage. It is interesting because it was the first known asteroid to come well within the orbit of Mars; it was discovered in 1898. For today, however, let us concentrate on the annular eclipse, which reaches its maximum at 07h GMT and is visible from the area of the Indian Ocean, Australia and the Pacific.

As we have seen, a solar eclipse occurs when the Moon passes in front of the Sun, and temporarily blocks out the brilliant solar surface or photosphere. If the alignment is exact, the eclipse may be total – but not always. The Moon's path round the Earth is not circular; it is elliptical, and when the Moon is in the far part of its orbit it does not appear quite large enough to cover the Sun completely. The result is that at exact alignment, a ring of sunlight is left showing round the dark disk of the Moon. This is an annular eclipse (Latin *annulus*, a ring). At the 1999 eclipse, annularity will last for 1 minute 19 seconds.

Annular eclipses are interesting to watch, but the glorious phenomena of totality – the chromosphere, the prominences and the corona – cannot be seen. Because the Moon's shadow is only just long enough to reach the Earth at any time, annular eclipses are more common than totalities; the next will occur on 14 December 2001, 10 June 2002, 31 May 2003, 3 October 2005 and 22 September 2006.

February 17

Anniversary

1972: Landing of Luna 20 on the Moon.

Missions to the Sea of Crises

One of the most prominent of the lunar seas is the Mare Crisium (Sea of Crises), which is easily seen even with the naked eye. It lies not far from the Moon's limb, and is detached from the main Mare system. It measures 280 miles by 348 miles, so that in area it is about equal to Great Britain. It is fairly regular in outline; the longer diameter is east–west, not north–south as might be thought, because of the effects of foreshortening. Its floor contains a number of craters of which the largest are Picard (21 miles in diameter) and Peirce (12 miles). Some distance south of it is the 30-mile crater Apollonius, and it was near here, on 17 February 1972, that the unmanned Russian probe Luna 20 came down. The exact position was latitude 3°.5 north, longitude 56°.6 east.

Luna 20 was a new kind of space-craft. It landed, drilled into the lunar surface, collected samples, and then took off again, returning safely to Earth on 25 February. This was a technique perfected by the Russians, and was a notable 'first', though it certainly did not compensate for their failure to send a man to the Moon before the Americans were able to do so.

Since then there has been one more sample-and-return probe: Luna 24 of 1976, which landed right inside the Mare Crisium (latitude 12°.8 north, longitude 52°.2 east), drilled down to six feet below the surface, and landed back on Earth on 22 August.

February 18

Twinkling Stars

Let us start this evening by looking at Sirius, the leader of Canis Major – the Great Dog. It is easy to see why Sirius is often referred to as the Dog-Star. It lies in line with Orion's Belt, but it cannot be mistaken, because it is so brilliant. From British latitudes it is fairly low down in the south; from Australia, New Zealand or South Africa it is very high up, and indeed not far from the zenith.

Sirius is a pure white star, but a casual glance is enough to show that it is not shining steadily; it is twinkling, and as it does so it flashes various colours of the rainbow. This effect is known commonly as twinkling, officially as scintillation (who does not know the old nursery rhyme, which begins 'Twinkle, twinkle, little star'?). But twinkling has nothing directly to do with the stars themselves; it is purely an effect of the Earth's dirty, unsteady atmosphere, which, so to speak, 'shakes the starlight around' as it comes in from airless space.

A star which is low down in the sky will twinkle more than a star which is high up, because its light has had to pass through a thicker layer of atmosphere. All stars twinkle to some extent, but Sirius shows the effect most obviously, because it is so bright. Even from southern latitudes it will still twinkle.

It is often said that a planet will not twinkle. This is not completely true, but certainly a planet twinkles less than a star, because it shows up as a small disk rather than a mere dot of light.

Twinkling may be beautiful, but to an astronomer it is irritating; the steadier the image, the better. Of course, from space, and from the surface of the Moon, there is no twinkling at all.

Anniversary

1473: Birth of Copernicus.

February 19

De Revolutionibus

When the Moon is near full, the lunar scene is dominated by the systems of bright rays coming from a few of the craters. These rays are surface deposits, and cast no shadows, so that they are obscure when the Sun is low over them, but when the Sun is high above them they tend to drown much of the delicate surface detail. There are many ray systems, but two are particularly prominent; those from Tycho in the Moon's southern uplands, and from Copernicus in the Mare Nubium (Sea of Clouds). During the period covered in this book, the rays will be well seen on 19 February in 1997, 2000 and 2003, when the Moon will be only a few days from full.

The crater Copernicus is massive and terraced; it has been nicknamed 'the Monarch of the Moon'. It is 56 miles in diameter, with a central mountain group. Its walls are bright, and its rays extend far across the lunar surface. It is easy to identify with binoculars.

It is named after one of the most famous men in the history of science: Mikołaj Kopernik, always known to us by his Latinized name of Copernicus. He was born on 19 February 1473 in the little town of Torún, in Poland.

At an early stage he began to question the truth of the 'Ptolemaic theory', in which the Earth is regarded as the centre of the universe with everything else revolving round it. He realized that many of the clumsy features of the system could be removed by the simple expedient of removing the Earth from its proud central position and putting the Sun there instead. His theory was probably more or less complete by 1533, but he was reluctant to publish it, because he knew that the Church would regard it as heretical – and Copernicus was a Church official. So he kept quiet, but finally he was persuaded to allow the book to go forward, and it appeared in 1543, just before Copernicus died. Its title was *De Revolutionibus Orbium Cœlestium* – or in English, 'Concerning the Revolutions of the Celestial Orbs'.

His misgivings were amply justified. The Church was bitterly hostile, and its attitude was aptly summed up by Martin Luther, who thundered: 'This fool seeks to overturn the whole art of astronomy!' The book was placed on the Papal Index of forbidden works, where it remained for many years.

Copernicus made many errors, and his one real contribution was in replacing the central Earth with a central Sun, but it is fair to say that *De Revolutionibus* marked the beginning of what we may term the modern phase of astronomy, though the revolution in thought was not complete until the work of Isaac Newton, almost a hundred and fifty years after Copernicus' death.

February 20

The Great Dog

Canis Major, Orion's senior Dog, is of course dominated by Sirius, but it is a large and important constellation in its own right. Its brightest stars are given in the table below.

It seems strange to find that of these Sirius is much the least powerful! If Wezea lay at a mere 8.6 light-years from us,

Canis Major

Greek letter	Name	Magnitude	Luminosity (Sun=1)	Distance (light-years)	Spectrum
α Alpha	Sirius	−1.5	26	8.6	A1
β Beta	Mirzam	2.0	7200	720	B1
δ Delta	Wezea	1.9	132 000	3860	F8
ϵ Epsilon	Adhara	1.5	5000	490	B2
ζ Zeta	Phurad	3.0	450	290	B3
η Eta	Aludra	2.4	52 500	2500	B5
o^2 Omicron^2	–	3.0	43 000	2800	B3

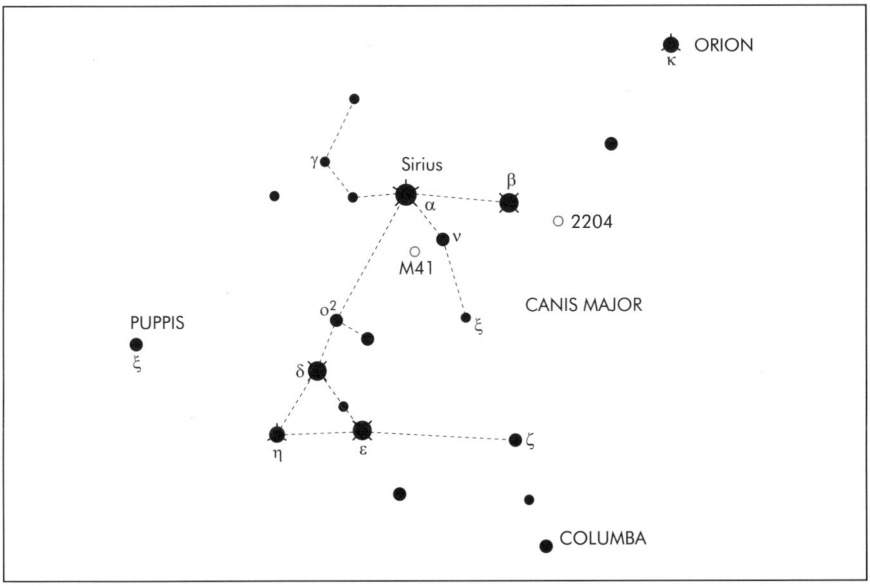

it would cast strong shadows. Adhara is only just below the official first magnitude, and is a particularly strong emitter of short-wave radiation. It is interesting to note that Gamma Canis Majoris is below fourth magnitude, so that the usual scheme of Greek lettering has been abandoned.

The declination of Sirius is around −16°, so that it can be seen from every inhabited country; even from places such as the Arctic towns of Tromsø in Norway or Fairbanks in Alaska it rises above the horizon, though admittedly not for long. The name Sirius comes from the Greek for 'scorching', and should really be pronounced with a long i – 'Syrius', not 'Sirrius'. To the Egyptians it was of special importance. It was worshipped as the Nile Star, So this, and its first appearance in the dawn sky every year ('heliacal rising') marked the imminence of the annual flooding of the Nile, upon which the whole Egyptian economy depended. Rather surprisingly, the Greeks and the Romans regarded it as an unlucky star, and in the *Æneid* Virgil refers to it as the star which 'brings drought and disease on sickly mortals', though later the Roman writer Geminus claimed that 'it is generally believed that Sirius produces the heat of the Dog Days, but this is an error, since the star merely marks the season of the year when the Sun's heat is greatest'.

Apart from Alpha Centauri, Sirius is the closest of the really brilliant stars. It is around 1 700 000 miles in diameter, larger than our own Sun; it is also hotter, with a surface temperature of 10 000°C. Its spectral type is A1. It is the equal of 26 Suns – and yet astronomers still class it as a stellar dwarf. Telescopically it is a magnificent sight, and it has been said to give the impression of a flashing diamond.

February 21

The Celestial Pup

Because Sirius is relatively so close to us, it has an appreciable individual or proper motion. In 1712 Edmond Halley realized that it had shifted in position against the background stars since Ptolemy of Alexandria drew up his catalogue in about AD 150; Sirius had crawled across the sky by a distance equal to one and a half times the apparent diameter of the full moon.

In 1838 the German astronomer F.W. Bessel, and found that Sirius was not moving in a direct line; it was 'weaving' its way along. Bessel saw that there could be only one explanation. Sirius is not a solitary traveller in space; it has a binary companion. Bessel looked for the companion, but could not see it, though from the motions of the bright star he calculated that the orbital period of the binary system must be 50 years. Then, in 1862, the American optician Alvan Clark was testing a new telescope when he saw a speck of light close to Sirius. It was the expected companion, almost precisely where Bessel had said it ought to be. Officially it is termed Sirius B, but since Sirius itself is the Dog-Star the companion is often referred to as the Pup. It has only 1/10 000 the brilliance of its primary, and it is not too easy to locate, simply because it is so close to Sirius; the angular separation can never be as much as 12 seconds of arc. Actually it is not really faint, and if it could be seen shining on its own it would be within the range of good binoculars.

It was naturally assumed that it must be large, cool and red, but in 1915 its spectrum was examined by W.S. Adams, using the 60-inch reflector at Mount Wilson in California. The result was surprising. Far from being cool, the Pup is extremely hot. It must therefore be small, and we now know that its diameter is a mere 26 000 miles, less than that of a planet such as Uranus or Neptune. Since it is as massive as the Sun, it must be very dense. If we could bring a cubic inch of its material down to Earth, the weight would be 2.5 tons; the density is 125 000 times that of water.

Sirius B is a white dwarf – a star which has used up all its nuclear 'fuel' and has collapsed; its atoms are crushed and broken, and packed together with almost no waste space. In the future it will simply go on shining feebly until all its light and heat leave it, and it becomes a dead star – a black dwarf, sending out no radiation at all.

Many other white dwarfs are now known, but Sirius B is the most famous of them. It is indeed a remarkable Pup.

Future Points of Interest

2000: Mercury at western elongation.

Anniversary

1824: Birth of Pierre Janssen. In 1876 Janssen became director of the Meudon Observatory, outside Paris, which has many powerful telescopes, including the great 33-inch refractor – one of the largest in the world. In 1904 he published an elaborate solar atlas, containing more than 800 photographs, and he continued his researches almost up to the time of his death in 1907. The square at the entrance of the Meudon Observatory is named the Place Janssen in his honour, and his statue is to be seen there.

February 22

The Sun's Surroundings

So far we have said little about the Sun, but this may be a good moment to start discussing it, because 22 February is the anniversary of the birth of a great French solar observer: Pierre Jules César Janssen.

Observing the Sun must be undertaken with caution. On no account look directly at it with any telescope or binoculars, even with the addition of a dark filter; to focus the Sun's light and (worse) heat on to your eye will certainly result in permanent blindness. The only sensible method is to use the telescope to project the Sun's image on to a white screen held or fixed behind the eyepiece. In this way it is easy to see the dark patches known as sunspots. Janssen, however, was more concerned with the Sun's surroundings, which cannot be seen in ordinary light except during the fleeting moments of a total solar eclipse, when the Moon passes in front of the Sun and blots it out.

Using a spectroscope, Janssen was able to isolate the light coming from hydrogen only, and he was able to study what was later known as the chromosphere, above the bright surface, as well as the masses of gas which used to be called red flames and are now termed prominences. This success, in 1868, may be said to mark the start of modern solar research.

February 23

Messier 41

Canis Major is so dominated by Sirius that we sometimes tend to forget that it contains anything else! In fact there are several very interesting objects, and one of these is an open star-cluster, No. 41 in Messier's list. Against a dark background it is easily visible with the naked eye.

To find it, begin at Sirius and then locate the fourth-magnitude star Nu^2, which is definitely reddish when seen through binoculars or a telescope (though not with the naked eye, because the star is too faint for its colour to show up without optical aid). M41 lies in the same binocular field with Nu^2, forming a triangle with it and Sirius. It has been known for a long time; it was first noted on the night of 16–17 February 1702 by the first Astronomer Royal, John Flamsteed.

The cluster is about 2400 light-years away, and is 20 light-years in diameter, though there are no sharp boundaries. It contains at least 80 stars, and probably more than 100. From its centre, marked by a reddish star, issue curved 'arms'. A low-power eyepiece on a telescope shows it well, and even good binoculars will resolve it into stars, so that it is well worth locating.

February 24

Anniversary

1987: Discovery of supernova SN 1987A in the Large Cloud of Magellan.

Supernovæ

One of the most important objects in the sky is the Large Cloud of Magellan. It is an independent galaxy, 169 000 light-years away, and ranks as a satellite of our Galaxy. It is prominent as seen with the naked eye, but unfortunately for British observers it lies in the far south, so that it is inaccessible from any part of Europe. Its declination is almost 70°S, so that to see it you have to go south of latitude 20°N. It lies mainly in the constellation of Dorado, the Swordfish, and during February evenings it is well placed for observation from countries such as Australia and New Zealand.

It contains objects of all kinds, including glant and dwarf stars, binaries, variable stars, open and globular clusters, novæ and gaseous nebulæ. There has also been one supernova, discovered on 24 February 1987 by Ian Shelton from the Las Campanas Observatory in Chile, and known officially as SN 1987A. At its peak it rose to magnitude 2.3, though it has now become very faint.

Supernovæ are the most colossal outbursts known in Nature. They are of two main types. A Type I supernova begins as a binary system, of which one component is a white dwarf; the dwarf collects material from its less dense companion, becomes unstable, and literally blows itself to pieces. A Type II supernova involves the collapse of a very massive star which runs out of nuclear 'fuel'; the collapse is followed by a rebound, and the shock-wave throws most of the star's material away into space, leaving only a very small, super-dense core made up of neutrons. SN 1987A was a Type II, and at maximum it shone as brightly as 250 000 000 Suns. Its progenitor star was not a red supergiant, as with most supernovæ, but a blue supergiant, classified as Sanduleak -69°202.

SN 1987A is the closest supernova to have been seen since the invention of the telescope, and is also the only one which has been visible with the naked eye since the outburst of a supernova in Ophiuchus in 1604. Many supernovæ have been found in distant galaxies, but all have been too remote to be studied in detail, so that SN 1987A proved to be of immense value to astronomers.

Anniversary

1842: Birth of Camille Flammarion, a French astronomer who was a noted observer as well as a populariser of astronomy. He founded the Société Astronomique de France. He died in 1924.

February 25

Fritz Zwicky and his Supernovæ

As we have seen, the supernova 1987A, in the Large Cloud of Magellan, was too far south to be seen from Europe. However, northern observers have one supernova remnant readily available: the Crab Nebula (Messier 1), near the third-magnitude star Zeta Tauri, in the Orion area. Binoculars show it as a dim patch, but photographs taken with large telescopes reveal its intricate structure. It is all that is left of the supernova of the year 1054, which became so brilliant that it could be seen with the naked eye even in broad daylight, and remained visible for months before fading away. The gas-patch contains a very small, very dense object known as a pulsar, made up of neutrons; it is spinning around 30 times per second, and it is this which is the Crab's 'power-source'.

The great pioneer of supernova research was a Swiss, Fritz Zwicky, who was born on 14 February 1898 (actually in Bulgaria; he spent most of his life in the United States, but retained his Swiss nationality). He worked out the theory of a Type I supernova, and linked supernovæ with neutron stars. He believed that he might be able to identify outbursts in far-away galaxies, and a photographic search carried out with the 48-inch Schmidt telescope at Mount Palomar, in California, showed him to be correct. He carried out much important work in other fields also, but it is as a supernova-hunter that he is chiefly remembered. He died on 8 February 1974.

Anniversary

1786: Birth of François Arago, a leading French astronomer, who was one of the first to realize the importance of photography in astronomy. From 1830 he was Director of the Paris Observatory. He died in 1853.

February 26

Procyon and the Little Dog

The second of Orion's dogs, Canis Minor, is a very small constellation. It contains only two fairly bright stars, listed below.

Procyon is easy to find, using Orion as a guide; it is the eighth brightest star in the sky. It is one of our nearest stellar neighbours; of the first-magnitude stars, only Alpha Centauri and Sirius are closer. Its diameter is around 3 500 000 miles,

Canis Minor

Greek letter	Name	Magnitude	Luminosity (Sun=1)	Distance (light-years)	Spectrum
α Alpha	Procyon	0.4	7	11.4	F5
β Beta	Gomeisa	2.9	106	137	B8

and, like Sirius, it has a white dwarf companion – though the companion is faint, and is a difficult object, well beyond the range of small telescopes. Procyon lies within 6° of the celestial equator, and during February evenings is prominent from both northern and southern countries. There is nothing else of immediate interest in Canis Minor.

February 27

The Little Snake

Just as the sky has a Little Dog, so it also has a Little Snake: Hydrus. But while the Dog is not far from the equator, the Snake is in the far south, and is never visible from any part of Europe. It has three main stars, listed below.

Hydrus is high up as seen from Australia or New Zealand, and one of its leaders, Alpha, is easy to find because it lies close to the brilliant Achernar in Eridanus. There is little of interest in the constellation, but it is worth noting that Beta is the nearest fairly bright star to the south celestial pole, though it is still 12° away from the pole itself. Southern observers never cease to regret that they have no pole star comparable with the northern Polaris.

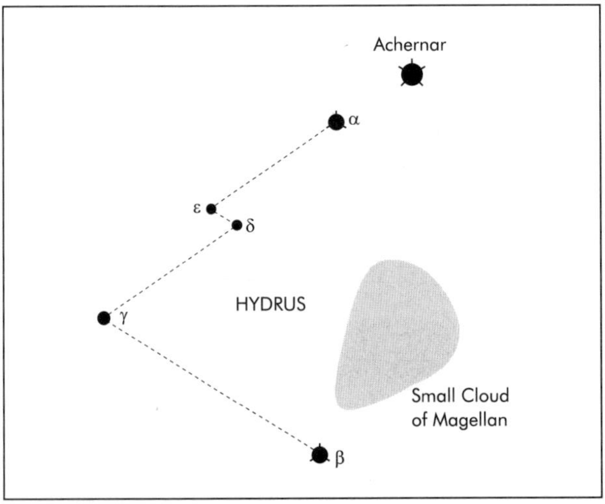

Anniversary

1826: Discovery of Biela's Comet, by W. von Biela – or rather rediscovery, since it had already been seen in 1772 and in 1806. It used to have a period of 6.75 years, but at the return of 1846 it split in two; the twins were seen for the last time in 1852. The comet has certainly disintegrated, but for some years afterwards a meteor shower was seen coming from the place where the comet ought to have been.

1897: Birth of the great French astronomer Bernard Lyot, who invented the coronagraph and became Director of the Meudon Observatory. He specialized in studies of the Sun, and made many important contributions to astronomy. He died in 1953 when returning from an expedition to observe a total solar eclipse.

Hydrus

Greek letter	Name	Magnitude	Luminosity (Sun=1)	Distance (light-years)	Spectrum
α Alpha	–	2.9	8	36	F0
β Beta	–	2.8	3	21	G1
γ Gamma	–	3.2	118	160	M0

Anniversary

1843: The brightest comet. The sky tonight looks dull when compared with that of 27 February 1843! What is often believed to have been the most brilliant comet of near-modern times was at its closest to the Sun, less than a million miles above the solar surface, and was visible close to the Sun in broad daylight. It had been discovered on 5 February; now it was at the peak of its glory. At night it cast strong shadows, and the tail stretched right across the sky. It was a 'sun-grazer', one of the group of comets known officially as the Kreutz family. Calculations indicate that it may be back in 540 years. It was last seen, as a dim speck, on 19 April 1843.

February 28

An Equatorial Star

The celestial equator divides the sky into two hemispheres. Not many stars above the third magnitude lie close to it, but one of these is Mintaka or Delta Orionis, in the Hunter's Belt. Its declination is –00° 18′ 27″.

It can be used to measure the angular diameter of a telescopic field. The method is to measure the time taken for the star to cross the field; this time, in minutes and seconds, when multiplied by 15, will give the angular diameter of the field in minutes and seconds of arc. Thus if Mintaka takes 1 m 3 s to cross the field, then the angular diameter is 1 m 3 s multiplied by 15 = 15′ 45″. Another star which can be used in the same way is Zeta Virginis (declination –00° 35′ 45″).

February 29

Leap Years

Nature is often very untidy. Thus the Earth does not go round the Sun in 365 days; it takes 365.256 days, or, in everyday language, $365\frac{1}{4}$ days. This means that the civil year of 365 days is actually a quarter of a day too short. If nothing were done, festivals such as Christmas would rotate round the calendar – and we would be celebrating Christmas in June, with summer in the northern hemisphere and winter in the southern.

The great Roman ruler Julius Cæsar took advice from an astronomer, Sosigenes, and revised the calendar (giving the year 46 BC no less than 445 days; not surprisingly, it was nicknamed the Year of Confusion). An extra day was tacked on to the shortest month, February, every four years; this made up for the fact that the calendar year is a quarter of a day too short. The 366-day years were called leap years. Any year which is exactly divisible by 4 is a leap year.

A small error remained; this was met by making the 'century' years (1700, 1800, 1900, 2000, etc.) leap years only if they could be divided by 400. Thus 1900 was not a leap year, but 2000 is. By the time this 'Gregorian calendar' was adopted in England, the error had risen to 11 days, so it was decreed in 1752 that 3 September should be followed immediately by 14 September. A favourite trick question is: 'What happened in England on 10 September 1752?' The answer – nothing, because there was no such day!

March

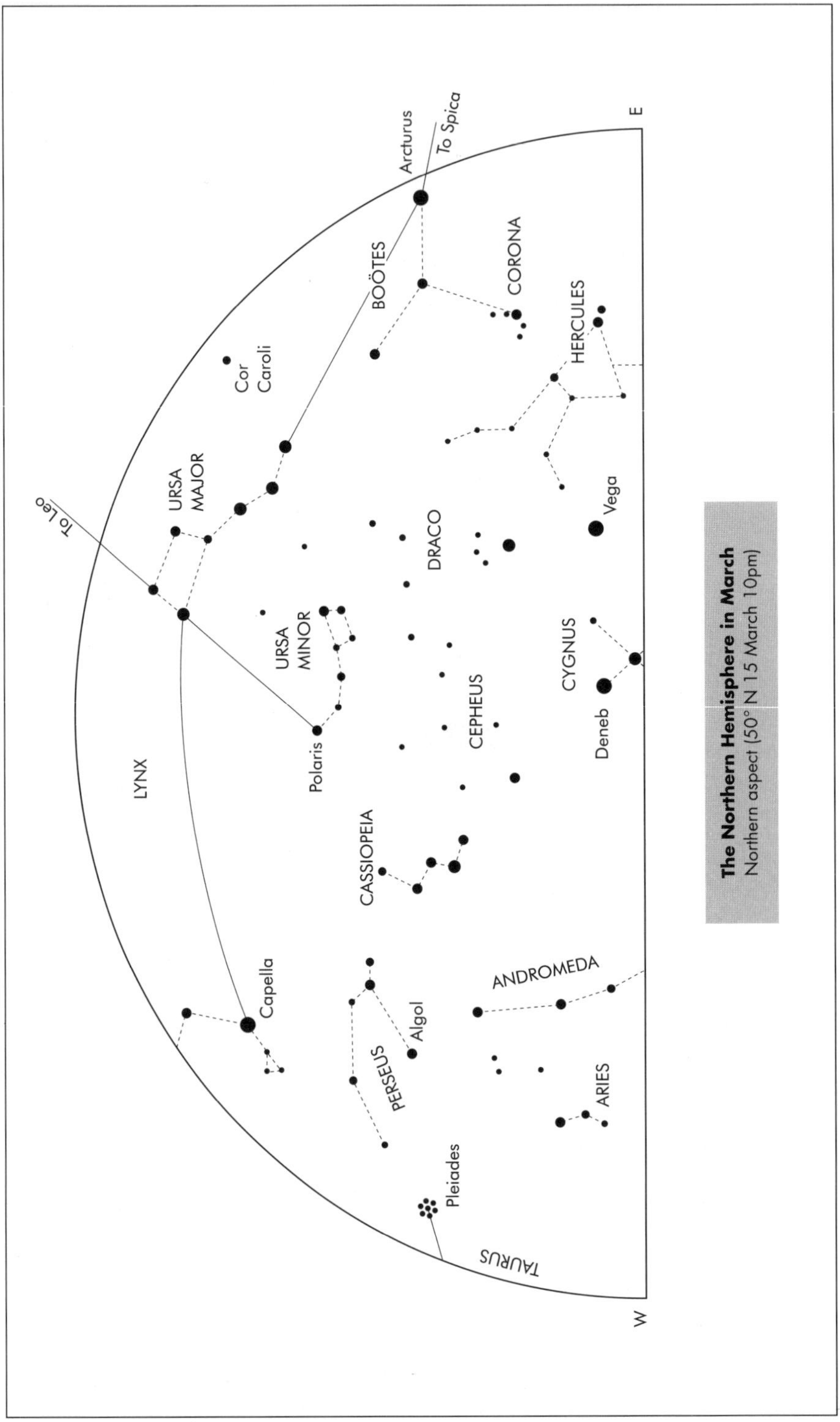

The Northern Hemisphere in March
Northern aspect (50° N 15 March 10pm)

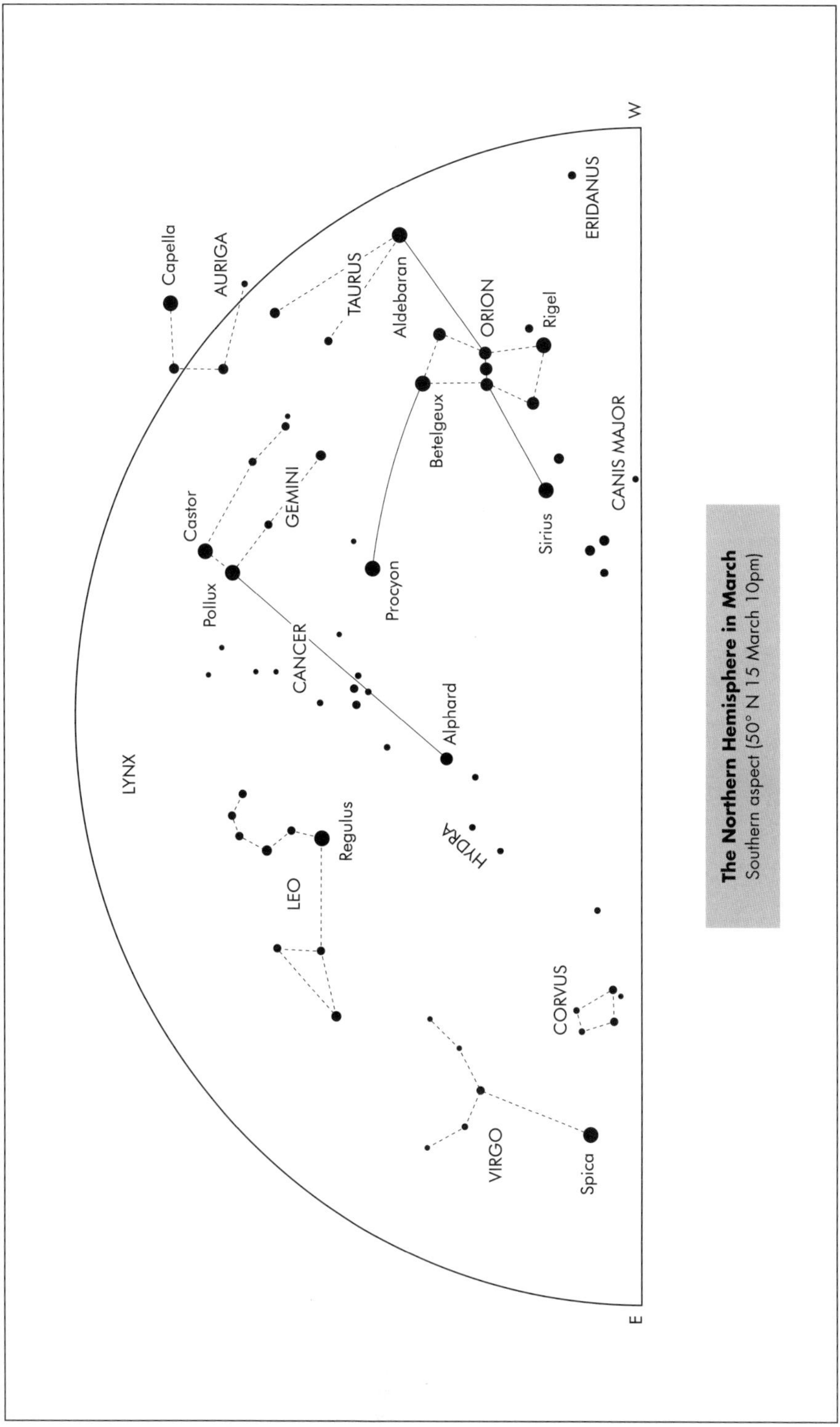

The Northern Hemisphere in March
Southern aspect (50° N 15 March 10pm)

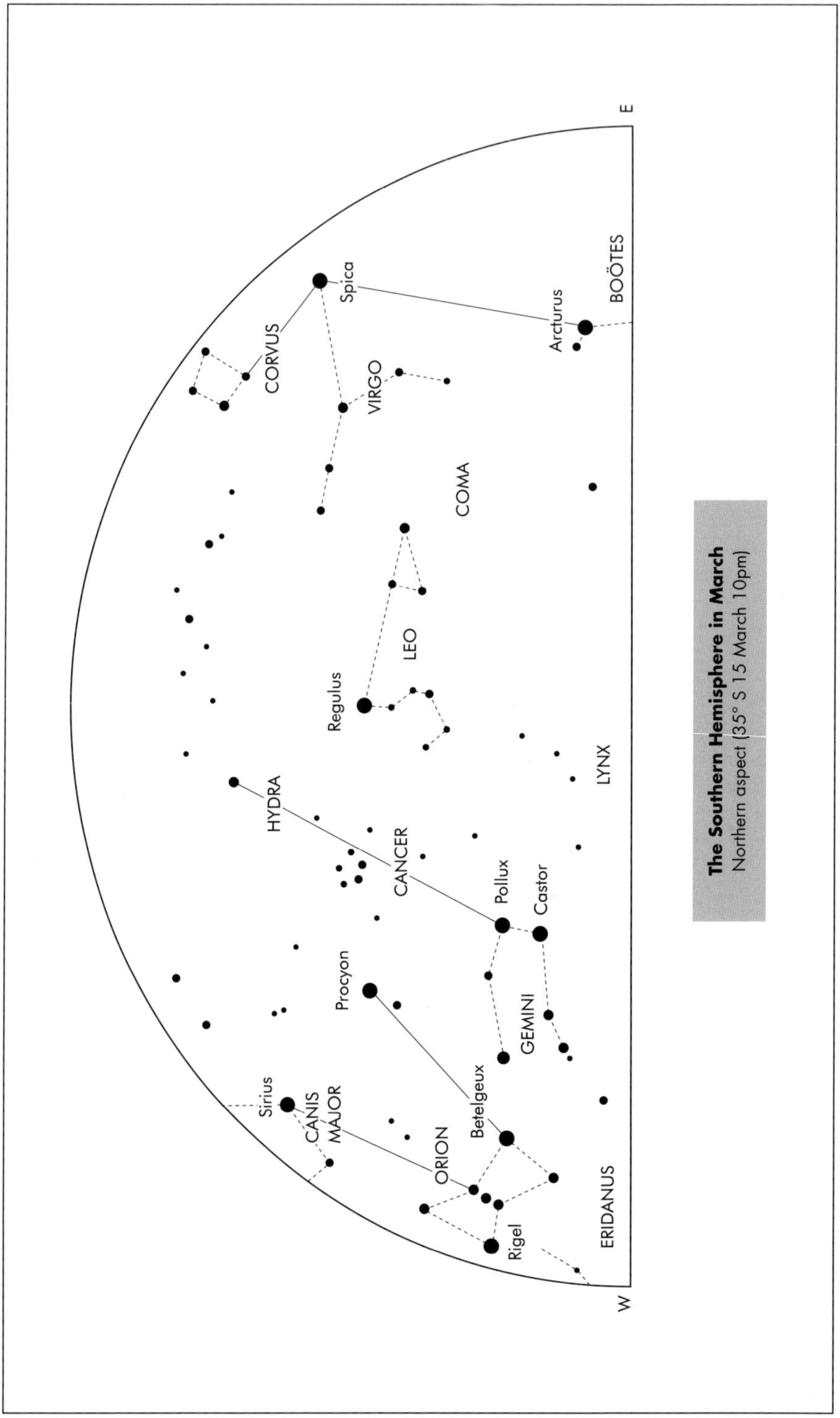

The Southern Hemisphere in March
Northern aspect (35° S 15 March 10pm)

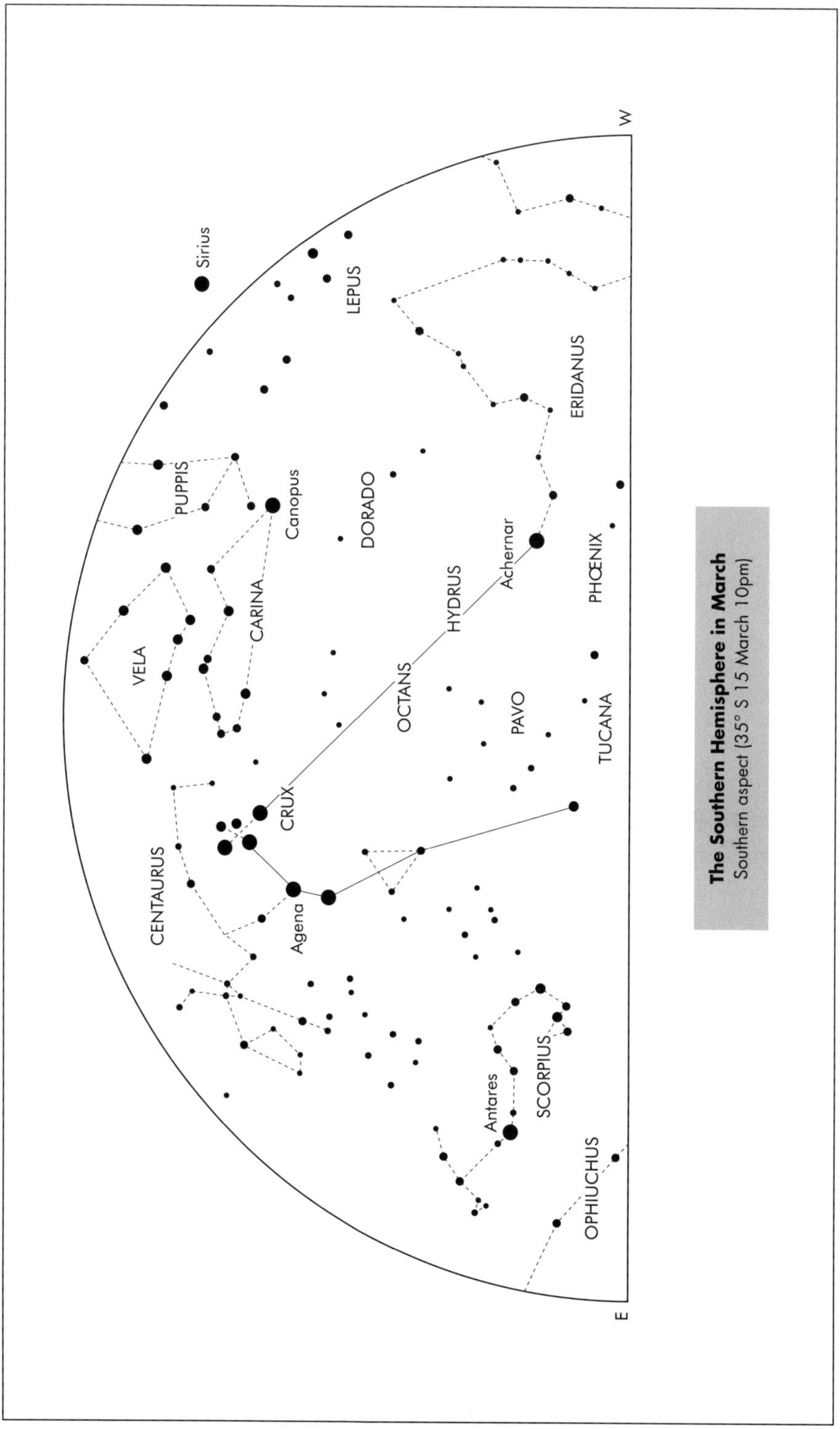

The Southern Hemisphere in March
Southern aspect (35° S 15 March 10pm)

March 1

The March Sky

By March we are beginning to see the start of a major 'change-over' in our view of the evening sky. Orion is still on view for the early part of the night, but is past its best, and by the end of the month it will start to fade into the twilight. From the northern hemisphere Ursa Major is very high; from the northern parts of Australia and South Africa (not New Zealand) a few of its stars may be glimpsed late in the evening, but they are of course very low down.

Look tonight for the Lynx, which is a large but faint constellation; it occupies most of the large triangle enclosed by Capella, Castor, and Dubhe in Ursa Major. Its only fairly bright star, Alpha Lyncis (magnitude 3.1), forms an equilateral triangle with Regulus and Pollux; binoculars show that it is decidedly red. It has been said that Lynx is so named because you need lynx-like eyes to see anything there at all.

Southern observers may care to search for Lupus, the Wolf, which lies near Centaurus (see 24 March). Here there are several reasonably bright stars, but there is no well-marked pattern, and very little of immediate interest. Lupus is one of the original constellations listed by Ptolemy, but it does not seem that there are any particular mythological legends attached to it.

March 2

The Celestial Crab

It is not easy to decide which is the most obscure of the constellations of the Zodiac. One candidate is certainly Pisces, the Fishes; another is Cancer, the Crab, which occupies most of the space enclosed by Procyon, Pollux and Regulus.

There is a legend attached to it. The great hero Hercules was doing battle with a particularly dangerous monster, the Hydra, when he was attacked by a crab sent by Juno, the Queen of Olympus, who had her own reasons for wishing Hercules no good at all. Not unnaturally, Hercules trod on the crab and squashed it, but as a reward for its efforts Juno placed it in the sky.

Cancer looks a little like a very dim and distorted version of Orion, but its brightest star, Beta Cancri or Altarf, is only of magnitude 3.5. At least Cancer is redeemed to some extent by the presence of two very fine open clusters, M44 or Praesepe (5 March) and M67 (28 March), and of course planets can pass through it.

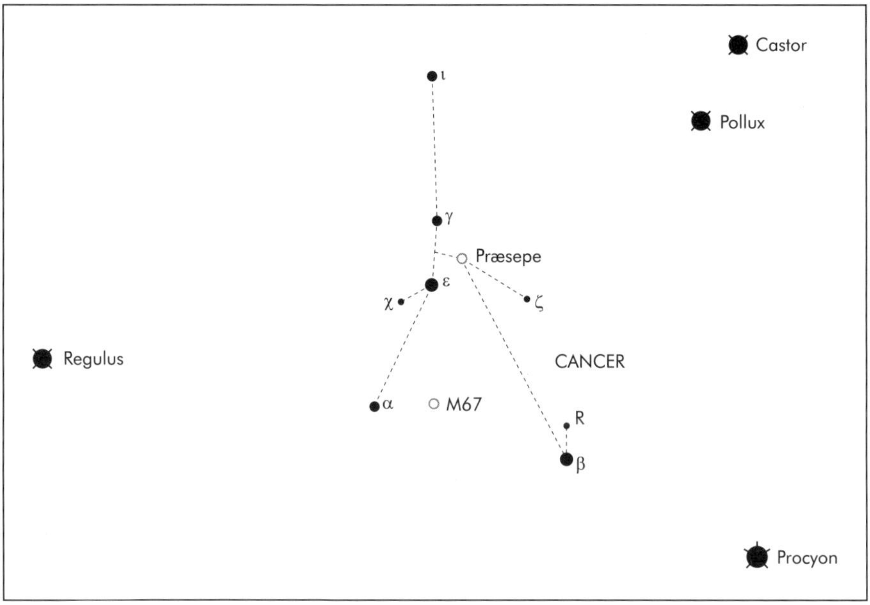

One interesting star is Zeta Cancri, or Tegmine. It is an easy telescopic double; the apparent separation is almost 6 seconds of arc, and the orbital period is 1150 years. The brighter component is itself a binary; magnitudes 5.7 and 6.2, separation 0.6 seconds of arc, orbital period 60 years. There is a third, very much fainter component at 288 seconds of arc, and there may be yet another star in the system. A small telescope gives a pleasing view of the main pair; the brightest member of the group is about 3 times as luminous as the Sun.

Anniversary

1969: Launch of Apollo 9. This was an Earth orbiter, used to test the Apollo techniques, and did not go to the Moon.

March 3

Back to the Moon

In view of the Apollo 9 anniversary, let us look back at the Moon, which, after all, is the most rewarding object in the sky for the owner of a small telescope. We can check on its position for 3 March during the period we are covering:

1998	Excellently placed in the evening; two days before half-phase.
1999	Past full, and still very bright.
2000	A very thin morning crescent, none too easy to see.

2001	Splendidly placed; exact half-phase.
2002	Gibbous, four days after full, well seen in the morning sky.
2003	New, and therefore out of view.

We have already noted that the border between the daylit and dark hemispheres of the Moon is called the terminator; it is here that the shadows are at their longest, and a crater right on the dividing line will be very prominent. Later, when the Sun is higher over the area, the shadows are less, and the crater may become hard to identify unless it has very bright walls or a very dark floor. When the Moon is near full, the entire scene is dominated by the bright streaks or rays which issue radially from a few craters, notably Tycho in the southern uplands and Copernicus in the waterless Sea of Clouds.

To learn your way around the Moon, the best procedure is to select a number of craters and draw them whenever they are in view. You will be surprised at the apparent changes in aspect over even a few hours. Some formations – such as the dark-floored, 60-mile Plato and the brilliant Aristarchus – can be identified whenever they are in sunlight, but others, very prominent under low light, may become very obscure indeed.

Future Points of Interest

1999: Eastern elongation of Mercury.

March 4

The Seas of the Moon

We are still concentrating on the Moon; yesterday's notes will show where it is at the moment, so let us now turn to the most prominent of all the features on the lunar surface – the 'seas' (Latin, *maria*).

These are great dark areas, covering much of the Earth-turned hemisphere of the Moon – remember that the Moon always keeps the same face turned toward us, because its orbital period is exactly the same as its rotation period (27.3 days). This means that to all intents and purposes the seas keep to the same apparent positions on the disk. The seas have been given romantic names which we still use, because not unnaturally, the early observers believed that the dark areas were true seas and that the bright regions were dry land. For a long time now it has been known that this is not so; the airless Moon can have no liquid water, and the 'seas' have never been watery, though certainly they must once have been seas of lava. Most of them make up a connected system, though one prominent sea, Mare Crisium (the Sea of Crises) is clearly separate.

The main seas are shown in the table opposite.

Future Points of Interest

2000: Conjunction of Venus and Uranus 0h 12m; separation 234 seconds of arc.

The Lunar Seas	
Mare Crisium	the Sea of Crises
Mare Fœcunditatis	the Sea of Fertility
Mare Frigoris	the Sea of Cold
Mare Humboldtianum	Humboldt's Sea
Mare Humorum	the Sea of Humours
Mare Imbrium	the Sea of Showers
Mare Nectaris	the Sea of Nectar
Mare Nubium	the Sea of Clouds
Mare Orientale	the Eastern Sea
Oceanus Procellarum	the Ocean of Storms
Mare Serenitatis	the Sea of Serenity
Mare Smythii	Smyth's Sea
Mare Tranquillitatis	the Sea of Tranquillity
Mare Vaporum	the Sea of Vapours
Lacus Somniorum	the Lake of the Dreamers
Palus Somnii	the Marsh of Sleep
Sinus Æstuum	the Bay of Heats
Sinus Iridum	the Bay of Rainbows
Sinus Medii	the Central Bay
Sinus Roris	the Bay of Dew

Some of these seas are fairly regular in outline, with mountainous borders; thus the vast Mare Imbrium is bounded in part by the lunar Alps and Apennines. Others are irregular in shape. They are easy to identify with the naked eye; they are not really smooth, but have fewer craters than the bright regions.

Rocks brought home by the lunar astronauts have confirmed that there has never been any water there. On the other hand, the seas were molten in the past, say around 4000 million years ago; there was tremendous volcanic activity, and often the lava has overwhelmed craters on the sea-floors, or levelled the seaward walls of craters and turned them into bays.

March 5

Anniversary

1979: Voyager 1 passed by Jupiter, at a distance of 217 500 miles. It had been launched in 1977, and sent back detailed information, notably the discovery of active volcanism on Jupiter's satellite Io. Voyager 1 went on to rendezvous with Saturn in 1980, and is now leaving the Solar System.

The Beehive

Præsepe, in Cancer – No. 44 in Messier's catalogue – is often said to be one of the finest open clusters visible from the northern hemisphere, and this is certainly true. It is inferior only to the Pleiades. It is a prominent naked-eye object at any time except when there is a misty sky, or the Moon is too bright. It was well known in ancient times, and around 280 BC two Greek astronomers, Aratus and Theophrastus, wrote that rain was on the way if Præsepe seemed dim or could not be seen in an apparently cloudless sky.

It lies almost in the middle of the large triangle formed by Procyon, Pollux and Regulus, and can be identified easily because it is flanked to either side by two stars – Delta Cancri (magnitude 3.9) and Gamma Cancri (4.7). Præsepe is often nicknamed 'the Manger', so that Delta and Gamma are the Aselli or Asses; but a more common name for Præsepe is 'the Beehive'.

It contains several dozens of stars, and is easily resolvable. In many ways binoculars give the best view of it, because it covers a wide area (70 minutes of arc) and cannot easily be fitted into the small field of an ordinary telescope. Its distance is about 525 light-years. It is not connected with the 'Asses', which are much closer to us.

Præsepe is a lovely cluster. It is therefore hard to see why the old Chinese gave it the unattractive name of 'the Exhalation of Piled-up Corpses'.

March 6

Binoculars and the Sky

We have noted that the most spectacular view of Præsepe is obtained with binoculars, and it is true that in some ways good binoculars are far more valuable astronomically than small telescopes. Their main drawback, of course, is lack of sheer magnification, and for astronomical work 'zoom' binoculars are never really satisfactory.

If you need only one pair, then 7×50 is a reasonable choice. (This means that the magnification is 7, and that each objective lens is 50 millimetres in diameter.) The magnification is high enough to be useful; the binoculars will be light, and they will have a wide field of view, so that they can be hand-held without excessive shaking. This is not the case with very strong binoculars, with magnifications of 12 or more; and above a magnification of 15, some sort of mounting is more or less essential.

One solution is to mount the binoculars on a camera tripod. This is straightforward enough, and it means that the binoculars can be kept steady – even if it is sometimes rather awkward to point them in the desired direction.

If you have only a limited amount of money to spend on equipment, it is wise to invest in binoculars rather than a small telescope. There is also the point that binoculars can be used for everyday activities, while small astronomical telescopes cannot.

Anniversary

1787: Birth of Josef von Fraunhofer, an outstanding German optical worker who was the first to map the dark lines in the spectrum of the Sun; these are still often referred to as the Fraunhofer lines. He was also the best lens-maker of his day. His comparatively early death, in 1826, was a tragedy for science.

1986: Russian probe Vega 1 bypassed Halley's Comet, sending back vital data which were of extreme importance to the Giotto probe, which reached the comet six days later.

March 7

X Cancri

Strong colours are not common among the stars – except for the reds. Of the really brilliant stars, Vega is always said to be

> *Anniversary*
>
> 1792: Birth of Sir John Herschel. He was the son of William Herschel, discoverer of the planet Uranus, and a brilliant astronomer in his own right; he was the first to undertake a really detailed survey of the far-southern stars – from 1832 to 1838 he was at the Cape of Good Hope. He died in 1871.
>
> 1973: Discovery of Kohoutek's Comet, by Dr Lubos Kohoutek at Hamburg Observatory. It was expected to become spectacular, but signally failed to do so.

'steely blue', though its hue is not pronounced; Capella is yellow, but not obviously so, and single green stars are almost non-existent – there is only one alleged example, Beta Libræ (see 28 June). But orange and red stars are much more in evidence.

There is an excellent example of a very red star in Cancer. It is known as X Cancri, and is not far from Delta Cancri or Asellus Australis, the southern 'Ass' flanking Præsepe. Like many red stars it is variable. The magnitude ranges between 5.6 and 7.5, so that it can always be seen in binoculars, and it is not hard to find, between Delta and a little pair of stars (Omicron and 63 Cancri). The spectral type is N, and there is a very rough period of around 195 days. However, this is a 'semi-regular' variable, so that its period is subject to marked irregularities.

When at its best, X Cancri can just be seen with the naked eye, but without optical aid you will not be able to see its colour, because the light-level is too low. In fact its colour is much the same as that of Mu Cephei, the Garnet Star (see 28 January), but Mu is much brighter, so that the colour is more evident.

If you have a suitable camera, aim it at Præsepe and give a time exposure. If all goes well, X Cancri should show up, and with sensitive colour film it really will look red.

March 8

'The Solitary One'

Some stars are easy to identify because of their brilliance; others because of their colour – usually red – and a few because of their particular positions in the sky. To this class belongs Alphard or Alpha Hydræ. It is not outstandingly bright – its magnitude is 2.0, so that it is equal to the Pole Star – but it is so much 'on its own' that it is always nicknamed 'the Solitary One'.

To find it, use the Twins, Castor and Pollux, as guides. A line from Castor, passed through Pollux and continued for some distance, will lead you straight to Alphard. There are no other bright stars anywhere in the region, and it is some way from the Zodiac, so that there is little danger of confusion with a planet.

As seen with the naked eye Alphard is clearly orange, and binoculars bring out the colour well. The spectrum is of type K, and the star is fairly luminous, 115 times as powerful as the Sun. Its distance is 88 light-years.

Orange or red stars of this kind are often variable, and Alphard has been suspected of changes in brightness. Sir John Herschel, on his way home from the Cape of Good Hope in 1838, made several observations of it from the deck of his ship, and was convinced that there were pronounced fluctuations, but these have never been confirmed, and Alphard is an awkward star to estimate with the naked eye because of the lack of suitable comparisons. If there are any variations, they cannot amount to more than a very few tenths of a magnitude. However, it is always worth finding 'the Solitary One', shining down with a subdued orange light in its lonely glory.

March 9

The Watersnake

We have already located Alphard, 'the Solitary One', the brightest star in Hydra, the Watersnake. There is, however, a different mythological legend; Hydra was said to be a monster with a hundred heads, living in the Lernæan marshes. It was rather difficult to kill, because each time a head was cut off another head appeared. It was during his battle with the Hydra that the hero Hercules had his encounter with a crab (see 2 March and 2 July). Needless to say, Hercules prevailed in the end!

Today Hydra is the largest accepted constellation, covering over 1300 square degrees. It sprawls all the way from the 'head', roughly between Procyon and Regulus, to the south of Virgo; the southernmost part of it is always very low from Britain, though none of it actually sets. Hydra, large though it is, contains very few bright stars. The leaders are listed below.

The snake's head, between the Twins and Alphard, is made up of a group of four stars of which Zeta is the brightest; the others are Delta (4.2), Eta (4.3) and Epsilon (3.4). Zeta is orange, and the others white; binoculars bring out the difference at once.

The most interesting of these stars is Epsilon, which is a multiple system. A small telescope will show that the main star has a 7th-magnitude companion, at a separation of 2.8 seconds of arc. The bright star is itself a binary; the components are rather unequal, and are very close together – their real separation is about 800 000 000 miles, which is 8.5 times the distance between the Earth and the Sun. The revolution period is 15 years, while the more distant star takes 890 years to complete one circuit of the close binary.

Anniversary

1870: Birth of E.M. Antoniadi, a Greek astronomer who spent much of his career on France and made use of Meudon 33-inch refractor. He was probably the best planetary observer of pre-space age days, and his nomenclature of features on Mars is still used. He died in Occupied France in 1944.

1986: Russia's probe Vega 2 passed Halley's Comet and sent back valuable data.

Hydra

Greek letter	Name	Magnitude	Luminosity (Sun=1)	Distance (light-years)	Spectrum
α Alpha	Alphard	2.0	115	85	K3
γ Gamma	–	3.0	58	104	G5
ζ Zeta	–	3.1	60	124	K0
ν Nu	–	3.1	96	127	K2

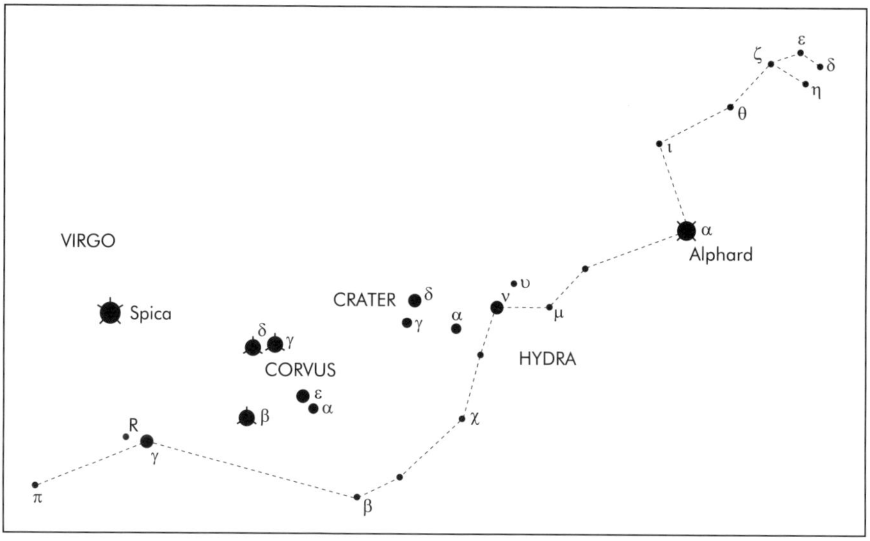

Eta Hydræ, the faintest of the main stars of the 'head', is actually much more luminous than Alphard, and could match 400 Suns, but it is well over 500 light-years away from us.

March 10

The Giraffe and the Southern Pointers

On March 1 we looked at Lynx, in the far north of the sky, which has at least one reasonably bright star. This is more than can be said of Camelopardalis, the Giraffe, which adjoins Lynx and the W of Cassiopeia. It is not an ancient constellation; it was added to the sky by Hevelius in 1690, and historians have claimed that it represents the camel which carried Rebecca to Isaac. In fact, it looks nothing like a camel, a giraffe or anything else, and consists of some dim, scattered stars, none of which is as bright as fourth magnitude. Probably the best way to locate the area is to look at the region between Capella and the Pole Star.

Most southern observers know Crux, the Southern Cross (see 26 April). It is almost surrounded by the much larger constellation of Centaurus, the Centaur, and two of the stars in the Centaur, Alpha and Beta Centauri, point to it. Once again appearances are deceptive. The two Pointers lie side by side in the sky, but they are nowhere near each other. Alpha Centauri is only just over 4 light-years away, and is the nearest bright star beyond the Sun; Beta Centauri

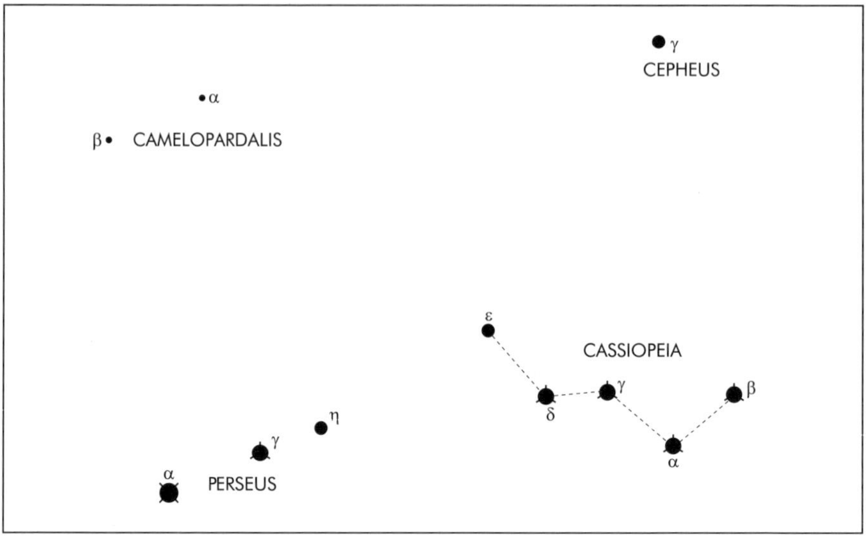

(Agena) is 480 light-years away, so that it lies far in the background. It is also very luminous, and could match 10 500 Suns, whereas Alpha Centauri is less than twice as powerful as the Sun.

To Australians and South Africans, Crux and the Pointers are now high in the east. Unfortunately, they are too far south to be seen from Britain or any part of Europe. From Sydney, Cape Town or any part of New Zealand they are circumpolar – that is to say, they never set.

March 11

The Planet-Hunter

This is an anniversary day. Urbain Jean Joseph Le Verrier, one of the greatest French astronomers of the nineteenth century, was born on 11 March 1811. It was he who made the calculations which led to the identification of the planet Neptune, in 1846.

Uranus, the first of the giant planets beyond the orbit of Saturn, had been discovered in 1781 by William Herschel (see 13 March). It was soon found that it was not moving as it had been expected to do. Something was pulling it out of position, and Le Verrier decided that this 'something' must be another planet, moving at a greater distance from the Sun. He was faced with a cosmic detective problem. He could see the victim – Uranus – and he had to work out where the 'attacker' must be. Finally he gave a position for the new

Anniversary

1811: Birth of U.J.J. Le Verrier. Le Verrier was not universally popular with his colleagues; indeed, he was compelled to resign as Director of the Paris Observatory, in 1870, because of his 'irritability', though he was reinstated when his successor, Delaunay, was drowned in a boating accident. Le Verrier died in 1877.

Future Points of Interest

2001: Mercury at western elongation.

planet to astronomers at the Berlin Observatory, and almost at once Johann Galle and Heinrich D' Arrest used a telescope there to find the planet, almost exactly where Le Verrier had predicted. After some discussion, it was named Neptune. (Similar calculations had been made by an Englishman, J.C. Adams, but Le Verrier did not hear about them until after the discovery had been made.)

Uranus and Neptune are not well placed this month, but will be better seen later in the year (see 24 August).

March 12

Star Magnitudes

Various bright stars are on view this evening – whether you are observing from the northern or the southern hemisphere – so this may be a good time to do a little checking upon apparent magnitudes. (Remember that this is a measure of how bright a star looks, not how luminous it really is. Absolute magnitude is different – see 21 May.)

Only four stars are above zero magnitude; Sirius (–1.5), Canopus (–0.7), Alpha Centauri (–0.3) and Arcturus (fractionally brighter than 0.0). Capella and Rigel are each of magnitude 0.1, but if you compare the two you will find a difference. In the northern hemisphere Capella will be higher up, and Rigel's light will be dimmed because it has to come to us through a thicker layer of the Earth's atmosphere – an effect known as extinction. To southern observers Rigel will be high and Capella low down, so that the reverse situation will apply. This is why it is always difficult to make reliable comparisons between stars which are at different altitudes.

Here are some useful magnitudes to act as guides:

0.1	Capella
1.1	Pollux
1.3	Regulus
1.6	Castor
2.0	Polaris
3.0	Gamma Boötis (see 7 June)
3.3	Megrez in Ursa Major
3.9	Asellus Australis (Delta Cancri), brighter of the 'Asses' (see 5 March)
4.7	Asellus Borealis (Gamma Cancri), fainter of the 'Asses'.

Stars below magnitude 5 can be seen only when the sky is dark, and below magnitude 6 the average observer will need optical aid. Sigma Octantis, the south pole star, is only of magnitude 5.5.

March 13

Solar System Anniversaries

This is certainly an 'anniversary day'. First comes the discovery of the first planet to be found in the telescopic era: Uranus. The man responsible was William Herschel.

Herschel was born in Hanover, and trained as a musician. He came to England while still a young man, and spent the rest of his life there; he became an organist in the city of Bath, and took up astronomy as a hobby. He made his own reflecting telescopes, and using one of these, from the garden of his home at 19 New King Street, Bath (now a Herschel museum) he found an unstarlike object which proved to be a new planet. Uranus is a giant, with about half the diameter of Saturn, and is only just visible with the naked eye.

Percival Lowell came of a rich American family; after some years as a diplomat he turned to astronomy, and established a major observatory at Flagstaff in Arizona, equipping it with a very powerful telescope (a 24-inch refractor). He was convinced that Mars was inhabited, and that the 'canals' were artificial waterways. Yet he was also an excellent mathematician, and from tiny irregularities in the movements of the outer planets he worked out a position for yet another member of the Sun's family. In 1930, fourteen years after Lowell's death, the new planet – Pluto – was discovered photographically by Clyde Tombaugh, close to the position that Lowell had given. Pluto has provided astronomers with puzzle after puzzle (see 4 June), but there is no doubt that its discovery was due to Lowell's persistent work.

The British-built spacecraft Giotto penetrated the head of Halley's Comet, and sent back close-range pictures of the nucleus. It survived the encounter, though the camera had been put out of action by the impact of a particle probably about the size of a grain of rice. Giotto then went on to encounter a second comet, Grigg-Skjellerup, on 10 July 1992.

Anniversary

1781: Discovery of Uranus, by William Herschel.

1855: Birth of Percival Lowell.

1930: Announcement of the discovery of Pluto, by Clyde Tombaugh at the Lowell Observatory in Arizona.

1986: The Giotto probe passed Halley's Comet, at 1040 miles from the nucleus.

Future Points of Interest

1998: Penumbral eclipse of the Moon. The Moon does not enter the main cone of shadow cast by the Earth; 72% of the surface is covered by the penumbra, or region of 'partial shadow'. The time of mid-eclipse is 04 h 42 m.

March 14

The Greatest Globular Cluster

Globular clusters form a sort of 'outer framework' to the main Galaxy. They are huge, symmetrical systems, some-

Anniversary

1853: Birth of G.V. Schiaparelli, the great Italian astronomer who is, unfortunately, best remembered for his spurious observations of the canals of Mars, but who carried out much valuable work, including the discovery of the close relationship between comets and meteor streams. He died in 1910.

1879: Birth of Albert Einstein, the greatest mathematician since Newton and founder of the theory of relativity. A German Jew, he left Germany in 1933 and spent the rest of his life in America, where he died in 1955.

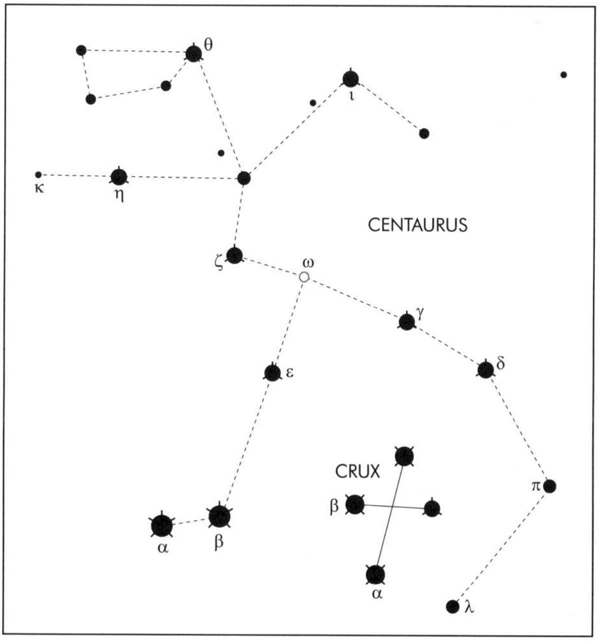

times containing over a million stars; over a hundred are known. The southern hemisphere of the sky contains more globulars than the northern, because the Sun lies well away from the centre of the Galaxy and we are having a lop-sided view; Sagittarius is particularly renowned for its globular clusters.

Only three globulars are clearly visible with the naked eye. These are Messier 13 in Hercules (4 July), 47 Tucanæ (29 November) and Omega Centauri. Of these, Omega Centauri is the brightest.

It lies in the northern part of the Centaur, and is now in the east as seen from Australia or South Africa, high above the horizon. Its apparent diameter is about the same as that of the full moon, and although the outer edges can easily be resolved into stars the central regions are very closely packed. Telescopically, it is difficult to show the whole cluster in the same field, because it is so large; you need a low magnification – and of course binoculars give a splendid view. It is prominent with the naked eye, looking like a hazy star. Obviously it is not in Messier's list, but it is No. 80 in the Caldwell catalogue.

The declination of Omega Centauri is approximately 48° S. Subtracting 48 from 90 gives us 42; to see the cluster, therefore, you must travel to a latitude south of 42°N. This means that the cluster cannot be seen from most of Europe, but it does actually rise over the southernmost part of the mainland; from Athens it skirts the horizon, but is clearly visible under good conditions. From Invercargill, in New Zealand, it is circumpolar.

March 15

Choosing a Telescope

As we have seen, binoculars are very useful in astronomy, and will give superb views of objects such as star-clusters. Yet they have low magnification, and for more detailed views a telescope is needed. With a telescope, for example, the rings of Saturn can be seen, as well as the belts and spots on Jupiter, the dark areas and the ice-caps on Mars and countless double stars, variables and nebular objects.

Unfortunately, modern telescopes are either good or cheap – not both! And in choosing a telescope, great care has to be taken. In particular, it is unwise to spend much money on a refractor with an object-glass less than three inches across, or a reflector with a mirror below six inches in aperture. Smaller telescopes are bound to be disappointing, particularly if they have unsteady mounts – as is usually the case.

Remember, the effectiveness of a telescope depends entirely upon its aperture; the actual magnification is carried out by the eyepiece. Therefore, do not buy any telescope which is advertised as 'magnifying so many times'. The maximum possible for good results is a power of 50 per inch of aperture, so that, for instance, a three-inch can never give good views with a power of over 150. Many manufacturers make exaggerated claims. Also, look for 'stops' inside the tube which reduce the actual aperture; this is a well-known method of concealing defective optics.

Lens-making is beyond the average amateur, but it is quite possible to make a good mirror for a Newtonian reflector. It takes time and patience, but all in all it is more laborious than difficult.

The golden rule is – do not rush into buying a telescope merely because it looks nice. If possible, obtain the help of an expert. Your local astronomical society will almost certainly be able to help; a full list of societies is given in the annual *Yearbook of Astronomy* (Macmillan, London).

Anniversary

1713: Birth of Nicolas Louis de Lacaille (1713–1762), the French astronomer who travelled to the Cape and drew up the first good catalogue of the far-southern stars. He also introduced a number of new constellations, including Sculptor, Fornax, Reticulum, Pictor and Octans.

March 16

Mars at Opposition

As we have seen (February 2) a planet is said to be at opposition when it is opposite to the Sun in the sky, and best placed for observation; the interval between one opposition and the next is termed the synodic period – in the case of Mars, 780

Anniversary

1750: Birth of Caroline Herschel. She was William Herschel's sister, who acted as her brother's assistant throughout his astronomical career. She was a considerable astronomer in her own right, and discovered eight comets. After William's death she returned to her native Hanover, and died there at the age of 98.

days. During the period covered here, there are oppositions of Mars on 24 April 1999 and 13 June 2001, followed by 28 August 2003. In 1999 Mars is in Virgo, and the minimum distance from Earth is 54 000 000 miles.

Mars is only 4219 miles in diameter, and the apparent diameter is never as much as 26 seconds of arc, so that a small telescope will show surface details only for a limited period to either side of opposition. Much of the disk is covered with the red 'deserts', which are not sandy but are coated with reddish minerals. The dark regions were once thought to be seas, but we now know that there can be no liquid water on Mars today, because the atmospheric pressure is too low (below 10 millibars everywhere). The dark regions are places where the reddish material has been blown away by winds in the thin Martian atmosphere, exposing the darker material underneath.

The polar caps are made of ice, but they are not the same as the caps of Earth, and are composed of a mixture of water ice and carbon dioxide ice. They wax and wane with the Martian seasons, but when at their best they can be very conspicuous. They are many feet thick, so that a great deal of H_2O is locked up in them – a fact which future colonists from Earth will no doubt find very useful.

Whether there is any life on Mars now is not certain. If there is, it must be very lowly. Certainly there can be nothing anything like so advanced as a blade of grass.

March 17

The Owl Nebula

This evening, for a change, let us look for a decidedly elusive object – Messier 97, the Owl Nebula. Its integrated magnitude is around 12, so that it is well beyond the range of binoculars, and probably you will need a telescope of at least six inches aperture.

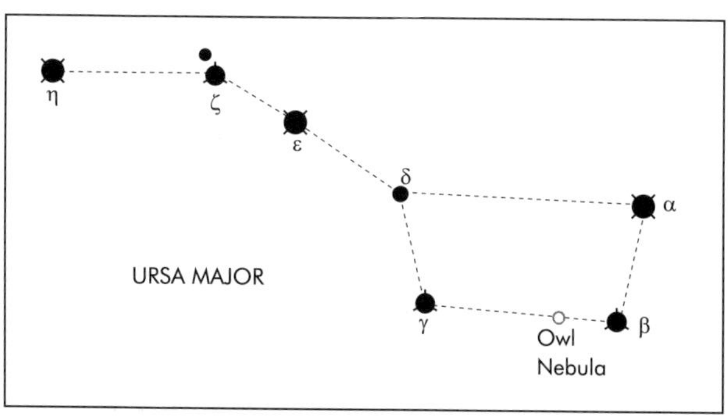

At least it is high up; it lies in Ursa Major, which at present is not far from the zenith or overhead point. It is not far from Beta Ursæ Majoris or Merak, the fainter of the two Pointers to the Pole Star. The position is: RA 11 h 15 m, declination +55° 01′.

The Owl is a planetary nebula. The name is misleading, because a planetary nebula has absolutely nothing to do with a planet and is not truly a nebula; it is a very old star which has thrown off its outer layers, and is surrounded by a shell of very tenuous expanding gas. Its nickname was given to it in 1848 by Lord Rosse, who observed it with his strange 72-inch reflector at Birr Castle in Ireland, and said that it bore 'a striking resemblance to the face of an owl', with 'a star in each cavity' giving the impression of an owl's eyes. Its real diameter is of the order of 3 light-years, and its distance seems to be around 3000 light-years.

M97 is one of the dimmest objects in the Messier catalogue. Look for it by all means, but do not be surprised if you fail to find it at the first attempt.

March 18

Anniversary

1965: Launch of Voskhod 2, carrying Alexei Leonov and Pavel Belyayev.

Space-Walking

In the early 1960s the space programme of the old USSR was in full swing. Voskhod 2 carried two cosmonauts, and it was during their trip that Alexei Leonov carried out the first of all 'space-walks'. Wearing a special suit, he went outside the space-craft, undertaking a series of manœuvres before going back through the airlock.

This has been done many times since, but Leonov's 'walk' was the very first. The obvious question is: Why did he not drift away from his space-craft? True, he was using a safety line, but in any case there was no force which would have separated him. He and the Voskhod were going round the Earth at the same speed, and at the same distance; therefore they simply stayed together. The best analogy is to picture two ants crawling on the rim of a bicycle wheel. If the wheel is made to spin, the ants will not drift apart, and this was the situation with Leonov and the space-craft; they represent the two ants, while the hub of the cycle wheel takes the place of the Earth.

Leonov's 'walk' was brief, but since then many complex manœuvres have been carried out, including the repair of the Hubble Space Telescope. In many cases the astronauts have not even been attached by safety cords. The first American to leave a space-craft in this way was Edward White, from the Gemini 4 vehicle in June 1965. This was successful, though, ironically, White later lost his life during a ground rehearsal for a space mission.

March 19

XI Ursæ Majoris: the First Computed Binary

We have seen that the Plough is only part of the large constellation of Ursa Major, the Great Bear. One of the much fainter stars in the Bear is of special interest. It has a name – Alula Australis, which means 'the First Spring' – but most astronomers refer to it simply as Xi Ursæ Majoris. It lies close by Alula Borealis or Nu Ursæ Majoris, a rather orange K-type star of magnitude 3.5. The pair is easy to locate; Nu is the brighter of the two by three-tenths of a magnitude.

Xi is an easy double. The components are of magnitudes 3.4 and 4.8, and the present separation is 1.3 seconds of arc. Its double nature was first seen by Sir William Herschel in 1780, during one of his 'reviews of the heavens'. At that time it was not realized that many (in fact, most) doubles are binary or physically associated systems; this discovery was due to Herschel himself, during his unsuccessful attempts to measure star distances. By 1804 it was seen that Xi Ursæ Majoris must be a binary, because the two components were moving significantly with respect to each other. The angle of a fainter object from a brighter one is known as the position angle (PA); in the case of Xi Ursæ Majoris, the PA had changed by almost 60 degrees since Herschel had first identified the pair.

The problem was taken up by a French astronomer, Félix Savery. In 1828 he announced that the orbital period of the binary was just under 60 years (actually it is 59.8 years), with a separation ranging between 0″.9 and 3″.1. This was the first binary to have its orbit computed, so that Alula Australis has its place in astronomical history.

The distance from Earth is only 25 light-years. It seems that each component is itself a spectroscopic binary, so that in reality Xi Ursæ Majoris is a quadruple system. If there are any planets there, they must have a fascinating view of the sky.

Future Points of Interest

1998: Mercury at eastern elongation.

March 20

The Equinox

The spring or Vernal Equinox always falls around this date. Owing to the idiosyncrasies of our calendar, the actual date is not constant, but it never changes much. During the period covered here, the times of the equinoxes are:

1998	March 20, 19 h 56 m
1999	March 21, 1 h 47 m
2000	March 20, 7 h 37 m
2001	March 20, 13 h 32 m
2002	March 20, 19 h 17 m
2003	March 21, 1 h 01 m

These times mark the moment when the Sun crosses the celestial equator, moving from south to north. For the next six months it will stay in the northern hemisphere of the sky, before again crossing the equator (autumnal equinox) and returning to the south.

The ecliptic – that is to say, the apparent yearly path of the Sun against the stars – is inclined to the equator at an angle of $23\frac{1}{2}°$. This is because the Earth's axis of rotation is tilted to the perpendicular to the orbit by this amount. The Vernal Equinox, where the ecliptic and the equator intersect, is also known as the First Point of Aries, because in early times it lay in the constellation of Aries, the Ram. However, the Earth's axis does not always point in precisely the same direction; it used to point to a position near the star Thuban in Draco (the Dragon), which was the pole star at the time when the Egyptian Pyramids were being built. The First Point has now moved from Aries into the adjacent constellation of Pisces (the Fishes), though we still use the old name. The wandering of the pole is due to the effect known as precession (see 18 October).

In our time, Polaris will remain the north pole star. Thuban is below the third magnitude; it lies between Alkaid in the Plough and the two Guardians of the Pole, Beta and Gamma Ursæ Minoris.

March 21

The Celestial Equator

Just as the Earth's equator divides the world into two hemispheres, so the equator of the sky divides the sky into two hemispheres. Since we are now at the Vernal Equinox, when the Sun is on the equator (either 20 or 21 March, depending on the year), this is a good time to see just where the equator runs.

Orion is still visible, low in the west, and we can start from there, because the equator passes just north of Mintaka or Delta Orionis, the faintest of the three stars in the Hunter's Belt; the declination of Mintaka is only 18 minutes of arc south (–00° 18′). The equator then runs across Monoceros, the Unicorn, where there are no bright

Anniversary

1965: Landing of the Ranger 9 probe on the Moon. It came down inside the crater Alphonsus, and sent back 5814 images. This was the last of the Ranger series of crash-landers.

stars (see 27 March) and thence into Canis Minor, the Little Dog, passing just under 6° south of Procyon, whose declination is +5° 14′. Next the equator runs across Hydra, south of the Watersnake's head (9 March); the brightest of the 'head' stars, Zeta Hydræ, is at declination +05° 56′. Between the Head and Alphard (8 March) there are two dim stars, Theta and Iota Hydræ, both of fourth magnitude; the equator runs between them. Next it crosses the faint constellation of Sextans (23 April) and then into Virgo (8 May), passing near two stars of the main pattern, Eta Virginis or Zaniah (−00° 40′) and Gamma Virginis or Arich (−01° 27′), after which it passes into the large constellations of Serpens and Ophiuchus.

Because the poles of the sky shift slightly, due to precession, the equator moves too, but the shift is much too slight to be noticed with the naked eye over a period of many lifetimes. In any case, there is no definite marker to show just where the equator is.

March 22

The Origin of the Planets

When making star maps, it is not possible to put in the positions of the planets, because they wander around the sky. Remember, they are very close compared with the stars, and are members of the Sun's family or Solar System.

The first man to propose a plausible theory to explain the origin of the planets was Laplace, whose anniversary falls today. He believed that the Solar System began as a shrinking cloud of gas. As it became smaller, because of the effect of gravity, it threw off various rings, each ring condensing into a planet. When he put forward this idea, in 1795, the two outermost planets, Neptune and Pluto, were not known (and in any case we can hardly regard Pluto as a proper planet), so that on Laplace's 'nebular hypothesis' Uranus would have been the oldest planet and Mercury the youngest.

The nebular hypothesis was accepted for many years. In its original form it was found to have fatal mathematical weaknesses, but modern theories are not too unlike it, and certainly it seems that the Earth and the other planets were produced from a 'solar nebula', a cloud of material associated with the youthful Sun. At least we are confident about the timescale; the Earth is 4.6 thousand million years old, and in all probability the other planets are of much the same age. There is no doubt that Laplace made very important contributions to astronomical theory.

Anniversary

1394: Birth of Ulugh Beigh, last of the great astronomers of the Arab school. He established an observatory at his capital, Samarkand; of course it had no telescopes, but it became an astronomical centre, where tables of the Moon and planets were drawn up. Ulugh Beigh was murdered in 1449 on the orders of his son, whom he had banished on astrological advice.

1749: Birth of Pierre Simon de Laplace, a French mathematician who drew up the 'nebular hypothesis' to explain the origin of the planets. He also made great advances in dynamical astronomy. He died in 1827.

March 23

The Hunting Dogs

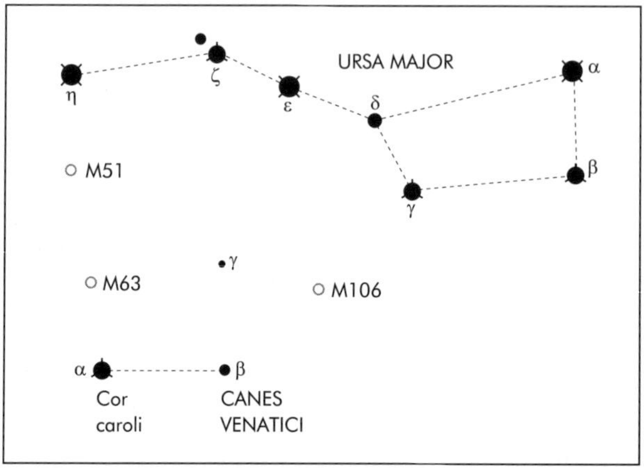

Tonight's constellation is Canes Venatici, the Hunting Dogs – Asterion and Chara. The constellation lies close to Ursa Major; it contains only one fairly bright star, Alpha (Cor Caroli). In old maps the Dogs are shown being held on leashes by Boötes, the Herdsman – possibly to stop them from chasing the Bears round the celestial pole.

Cor Caroli is of magnitude 2.9; it seems to have been given its name by Edmond Halley in honour of the memory of King Charles I, who had been executed in 1649 after the end of the Civil War between the Cavaliers and the Roundheads. It is 65 light-years away, and 75 times as luminous as the Sun. What makes it interesting is that it has a variable spectrum, and apparently a strong and variable magnetic field. It is also a fine double. The companion is of magnitude 5.5, and as the separation is over 19 seconds of arc this is a particularly easy telescopic pair.

The second star of the constellation is Beta or Chara, rather below the fourth-magnitude. It is only 29 light-years away, and is therefore one of our nearer stellar neighbours. Its spectrum is of type G.

If you have binoculars, look for the very red variable Y Canum Venaticorum, which lies about one-third of the way along a line joining Cor Caroli to Megrez in the Great Bear. Here we have one of the reddest of all stars; it is of type N, so that its surface is relatively cool. It is a semiregular variable; at maximum it is on the fringe of naked-eye visibility, but optical aid is needed to bring out its colour, which is so intense that the star has been christened La Superba. Its position is RA 12 h 45 m, declination +45° 26′. There is a very rough period of 157 days, which is however subject to marked fluctuations. La Superba is certainly worth finding, and it fully merits its nickname.

Anniversary

1893: Birth of Walter Baade, a German astronomer who spent most of his career in America. It was he who established that there had been a serious error in the Cepheid distance scale (see 28 January) and that the universe was twice as large as had been thought. He died in 1959.

1975: Contact finally lost with the spacecraft Mariner 10, which had been launched in 1973 and had bypassed Venus (3 February 1974) and Mercury three times (25 March, 21 September 1974 and 16 March 1975). It remains the only probe to have surveyed Mercury from close range. No doubt it is still in solar orbit.

March 24

The Whirlpool

While we are concentrating on Canes Venatici, let us search for one of the most famous galaxies in the sky: M51, nicknamed the Whirlpool for reasons which are obvious from a glance at any good photograph of it. Though it lies within the borders of Canes Venatici, the nearest bright star is Alkaid in the Plough; M51 lies about 3.5° from Alkaid. Its integrated magnitude is 8. It is usually said to be distinct in binoculars, though I admit that I have never been able to see it without a telescope (observers with keener eyesight will certainly do better). Telescopically, of course, it is easy enough. Small instruments show it as a fuzzy patch, but a good 12-inch will bring out the spiral form.

In fact the Whirlpool was the first galaxy to be recognized as a spiral – by Lord Rosse in 1845, during the initial tests with his strange 72-inch reflector at Birr Castle in Ireland. It was also Lord Rosse who gave the galaxy its nickname. His drawing of it was remarkably good, as is shown by comparing it with a modern photograph. He also shows the satellite galaxy, NGC 5195, which is joined to M51 by a bridge of luminous material.

M51 was discovered by Messier himself in 1773. It is about 37 000 000 light-years away, and is therefore one of the closer galaxies, though well beyond the limits of our Local Group. It is a favourite target for astronomical photographers, and is so striking because unlike the larger and closer Andromeda Spiral, M31, it is almost face-on to us.

Anniversary

1655: Discovery of Saturn's satellite Titan, by Christiaan Huygens.

March 25

Titan

For most of the period covered here Saturn is in the region of Cancer and Leo, so that it is well on view, and the rings have opened out (they were edgewise-on in 1995). This is therefore a good time to say something about its senior satellite, Titan, which was discovered by the Dutch astronomer Christiaan Huygens, using one of the small-aperture, long-focus refractors of the time. It moves round Saturn at a mean distance of 760 000 miles in a period of 16 days; it is of the 8th magnitude, so that almost any telescope will show it, and it is just within the range of powerful binoculars. It has a diameter of 3200 miles, so that it is larger than the planet Mercury; it is in fact the largest satellite in the Solar System apart from Ganymede in Jupiter's family.

What makes Titan so important is the fact that it has a dense atmosphere; the ground pressure is almost 1.5 times that of the Earth's air at sea level. Moreover, the bulk of the atmosphere is composed of nitrogen, though there is also a great deal of the unprepossessing gas methane (marshgas). Unfortunately we do not yet know much about the surface conditions, because of the constant cloud; the Voyager images showed only the top of a layer of what has been called 'orange smog', though recently the Hubble Space Telescope has done rather better. The temperature is low; about −168°C. This is near the triple point of methane, so that methane could exist as a solid, a liquid or a gas, just as H_2O can exist on Earth as solid ice, liquid water or water vapour. Titan may be covered, at least in part, by an ocean of ethane or methane.

The Cassini probe, due for launch in 1997, should drop a 'lander' on to Titan in 2004; this lander has been named in honour of Christiaan Huygens. Whether it will touch down on solid ground, or splash into a chemical sea, we do not know. The chances of life there are minimal, if only because of the low temperature, but certainly Titan is unlike any other world in the Solar System.

March 26

Messier Objects in Canes Venatici

We have not yet finished with the Hunting Dogs. Dim though it may be, the constellation contains a number of interesting objects, quite apart from the Whirlpool (24 March). One of these is the globular cluster M3, which has an integrated magnitude of between 6 and 7, and is not hard to find with binoculars.

M3 lies at the extreme edge of Canes Venatici, and probably the best guide to it is Beta Comæ in the adjacent constellation of Coma Berenices (18 May). M3 was discovered by Messier in 1764, and is around 48 000 light-years away. It is very symmetrical, and contains hundreds of thousands of stars. It is one of the three brightest globulars in the northern hemisphere of the sky; the others are M13 (4 July) and M5 (13 July), but in beauty M3 is the equal of any.

Canes Venatici is rich in galaxies. Apart from M51, the Whirlpool, there are three galaxies in the Messier catalogue (63, 94 and 106) and four in the Caldwell catalogue (21, 25, 29 and 32). Of these M63 is particularly notable. It is a tightly wound spiral, with an integrated magnitude of just below 7;

it is about 24 000 000 light-years away, and lies roughly between Cor Caroli and Alkaid. M94, of 8th magnitude, is face-on to us, but again the arms are tightly wound.

These and other galaxies in the Hunting Dogs are below binocular range, but the whole area is worth sweeping with a wide-field telescope, and with a little practice many of the galaxies may be identified.

March 27

Anniversary

1989: Phobos 2 lost. This was a Russian probe, intended to touch down on Phobos, the inner satellite of Mars. Unfortunately contact was lost before the main programme had begun, and was never regained.

Future Points of Interest

1998: Venus at western elongation.

Monoceros

Monoceros, the Unicorn, is by no means a prominent constellation. It adjoins Orion, and occupies most of the large triangle outlined by Betelgeux, Sirius and Procyon. The brightest star, Beta, is only of magnitude 3.7, but the Milky Way runs right through the constellation, so that there are plenty of rich star-fields. Beta is an easy double; the components are of magnitudes 4.7 and 5.2, and the separation is over 7 seconds of arc. There is a third component, of magnitude 6, at a separation of 10 seconds of arc, so that the whole field is decidedly attractive.

Though Monoceros has no brilliant stars, and no distinctive shape, it does contain some interesting objects. Pride of place must surely go to the lovely Rosette Nebula.

To locate it, first find the open cluster C50 (NGC 2244), round the dim star 12 Monocerotis. The cluster is an easy

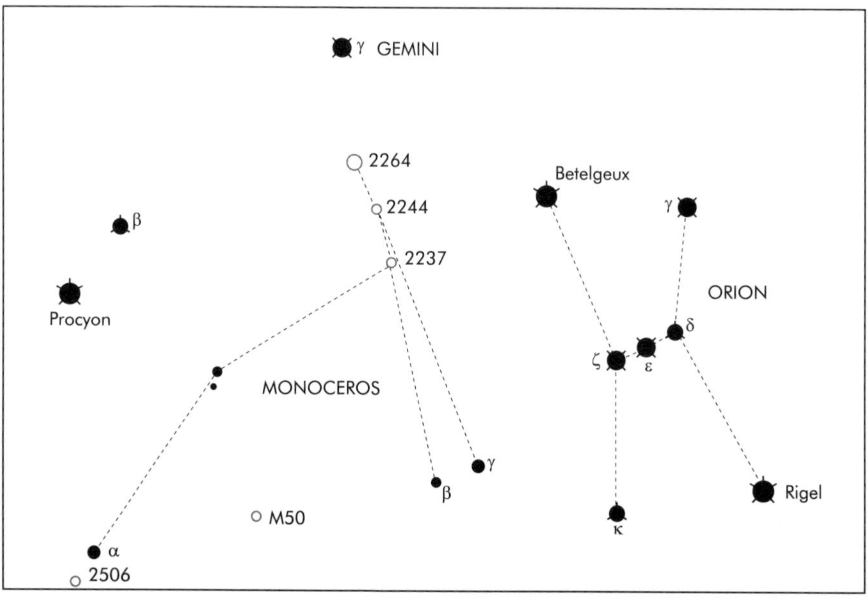

binocular object, and it is rather surprising not to find it in Messier's list; it lies slightly off the line joining Procyon to Betelgeux, considerably closer to Betelgeux. The cluster contains a small but distinctive quadrilateral of stars, which are plain with fairly strong binoculars.

The Rosette Nebula, C49 (NGC 2237), surrounds the cluster. It is none too easy to see visually – strong binoculars or a low power on a telescope show a soft glow round the cluster – but when photographed it is magnificent, and is a favourite target. The colours are superb; note also the central cavity, due to grains of material being 'pushed away' by the hot stars in the middle of the nebula. The real diameter of the nebula is about 55 light-years, almost seven times the distance between the Sun and Sirius.

Also in Monoceros is the open cluster M50, roughly between Sirius and Procyon. Strong binoculars show individual stars in it. Beyond it lies the 3.9-magnitude star Alpha Monocerotis, and in the same binocular field is the open cluster C54 (NGC 2506), which is by no means a difficult object even though it is unspectacular. It lies close to the border between Monoceros and Puppis, the Poop of the old Ship Argo (27 December).

March 28

The Oldest Open Cluster

This evening let us turn back to Cancer, which is a Zodiacal constellation and is so near the celestial equator that it can be seen from every inhabited country. We have already looked at the main cluster, Præsepe (see 5 March), but there is also another open cluster in the Crab which is within the range of binoculars, and is exceptionally interesting. This is M67.

It is easy to find, north of the head of Hydra and south or Præsepe; it is close to the fourth-magnitude star Acubens (Alpha Cancri), and is on the fringe of naked-eye visibility. It is an attractive cluster, more symmetrical than Præsepe, and is very rich, with hundreds of faint stars. Its distance is of the order of 2700 light-years.

What makes it so unusual is that it seems to be much older than most open clusters. It is thought that the stars in a cluster are born out of the same 'cloud' of material, and start their lives in the same area; but inevitably the cluster is perturbed by non-cluster stars, so that eventually it is scattered and loses its identity. M67, however, is well away from the main plane of the Galaxy, so that there are not many stars near it to disrupt it, and it has preserved its identity over a very long period. It is thought to be at least 4000 million years old, and probably older, whereas the ages of clusters such as the Pleiades can be measured in a few millions of years.

Anniversary

1802: Discovery of Pallas, the second asteroid, by Olbers. Pallas is one of the largest of the asteroids – it measures 360 by 292 miles – but is never visible with the naked eye. It is a 'main-belt' asteroid, moving around the Sun at a mean distance of 225 million miles in a period of 4.6 years.

Future Points of Interest

2000: Mercury at western elongation.

> *Anniversary*
>
> 1807: Discovery of the fourth asteroid, Vesta, by Olbers.
>
> 1890: Birth of Sir Harold Spencer Jones, who was Astronomer Royal from 1933 to 1955. He was an eminent astrophysicist, who carried out work of fundamental importance; during his régime as Astronomer Royal the Greenwich Observatory moved from its old site in Greenwich Park to Herstmonceux, in Sussex. Spencer Jones died in 1960.
>
> 1974: First close pass of Mercury by the spacecraft Mariner 10. It sent back 647 high-quality pictures, giving us our first close-range views of the Mercurian craters.

March 29

Miniature Worlds

This is an 'asteroidal anniversary' month. On 28 March the German amateur astronomer Dr Heinrich Wilhelm Matthias Olbers discovered the second asteroid, Pallas, and on 29 March 1807 he discovered Vesta, the brightest member of the whole swarm.

The asteroid story began on the first day of the nineteenth century, 1 January 1801, when Piazzi, at Palermo, discovered Ceres, which remains much the largest of all the asteroids. Ceres then vanished into the twilight but was recovered in the following year, probably by Olbers (though there have been a few other claimants!). During his search Olbers discovered Pallas. The searches continued; asteroid No. 3, Juno, was found by Karl Harding in 1804, and then came Olbers' detection of Vesta in 1808. Vesta is the only asteroid ever visible with the naked eye.

Vesta is 358 miles in diameter. Surface details on it have been recorded from the Hubble Space Telescope, and a rough map has been drawn up; there seems to be a large basin, over 120 miles across, which has been provisionally named 'Olbers'. The two hemispheres of the asteroid are different; one seems to be covered with quenched lava flows, while the other has the characteristics of molten rock that cooled and solidified underground, and was later exposed when Vesta was struck by wandering bodies, of which there must be many in the main asteroid zone. It has even been suggested that some particular meteorites, known as eucrites, come from Vesta, though the evidence is very uncertain.

Olbers, who found the asteroid, was a medical doctor who made notable contributions to astronomy. He also had the reputation of being an exceptionally pleasant person. He was born in 1758, and died in 1840.

March 30

Nu Hydræ

One of the brighter stars in the Watersnake is Nu Hydræ. It is now well placed, in the south for British observers, high up for Australians and New Zealanders; its declination is –16°. It is one member of the long line of stars leading from the Watersnake's head through Alphard and along to the neighbourhood of Corvus, the Crow (23 May); it is also the best guide to the inconspicuous but ancient constellation of

Crater, the Cup (25 May). Nu Hydræ is of magnitude 3.1, slightly brighter than Megrez, the dimmest of the seven stars in the Plough. What sort of star is it?

It is of spectral type K, which means that it is rather orange in hue. The colour is not evident with the naked eye, because the star is not bright enough, but binoculars will show it. This means that the surface temperature is only about 4000 °C, as against over 5000 °C for our yellow Sun. It is much larger than the Sun, and 90 times as luminous; it is 127 light-years away, so that we now see it as it used to be 127 years ago.

The absolute magnitude is −0.1 (see 2 and 6 January), so that as seen from our standard distance of 32.6 light-years it would shine more brightly than Capella does to us. From a planet orbiting Nu Hydræ, our Sun would be faint, and you would need a telescope of some size to see it at all. But Nu Hydræ does not seem to be the sort of star which would be expected to have a planetary system, and it is well advanced in its life-story, so that it has used up much of its store of nuclear 'fuel'.

Near it is U Hydræ, which is of type N and is one of the reddest stars in the sky; it rivals X Cancri (7 March) and R Leporis (18 January). Like many of its kind, it is a semiregular variable with a rough period of 450 days; the magnitude range is between 4.8 and 5.8, so that it is always an easy binocular object. Certainly it is worth finding, and with binoculars or a telescope it is a beautiful sight.

March 31

Van Maanen and the Galaxies

Today is the anniversary of a well-known Dutch astronomer, Adriaan van Maanen. He was born on 31 March 1884, and died in 1947. He achieved much useful work, but some of his results proved to be wrong, and held up progress for some time.

In the early part of the twentieth century it was still widely believed that the 'spiral nebulæ' were minor features of our Galaxy, and that our Galaxy was the only one. This point of view was championed by astronomers such as Harlow Shapley, the first man to measure the size of the Galaxy (in 1918) and also by van Maanen. It was strongly opposed by Edwin Hubble, who was convinced that the spirals were independent galaxies millions of light-years away; and this, of course, included M51, the Whirlpool in Canes Venatici, which we have already discussed (24 March).

At Mount Wilson, van Maanen took photographs of galaxies, notably the Whirlpool. He then claimed that his photographs, taken over a period of years, showed changes in the structure of the spiral arms. If this had been correct, then

Anniversary

1884: Birth of Adriaan van Maanen.

1966: Launch of the Russian unmanned probe Luna 10. It approached the Moon to within 220 miles and entered lunar orbit, carrying out gamma-ray studies of the lunar surface layer. Contact was lost on 30 May, after 460 orbits.

M51 could not be millions of light-years from us; it would indeed have to be a member of the Milky Way Galaxy. Hubble's results, also at Mount Wilson, were quite different; he was sure that there had been no changes in the spiral arms. The argument raged for some time, and things were not helped by the fact that Hubble and van Maanen disliked each other intensely! In the end it was shown that van Maanen had made an honest mistake; the spirals had not changed, and they really were as remote as Hubble believed. By 1923 Hubble was able to measure the distances of the nearer galaxies, using Cepheid variables as 'standard candles', and the problem was finally solved.

April

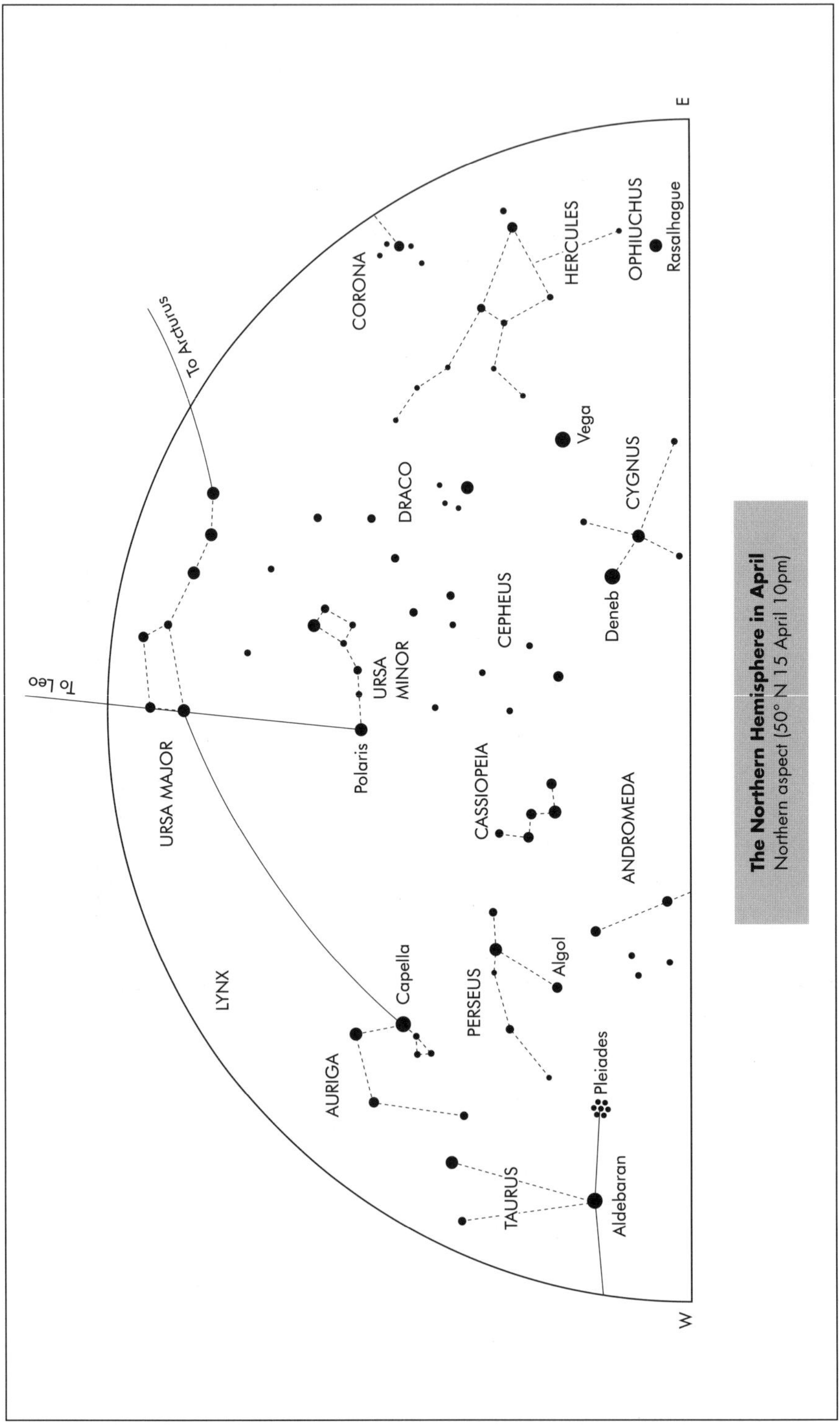

The Northern Hemisphere in April
Northern aspect (50° N 15 April 10pm)

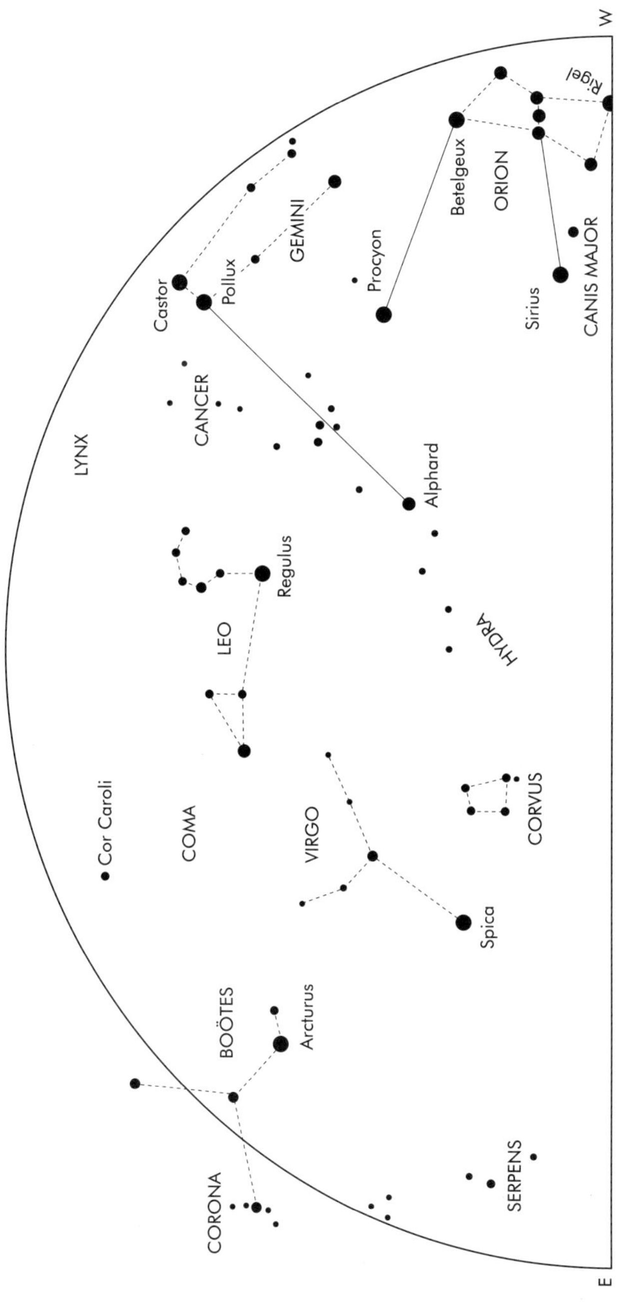

The Northern Hemisphere in April
Southern aspect (50° N 15 April 10pm)

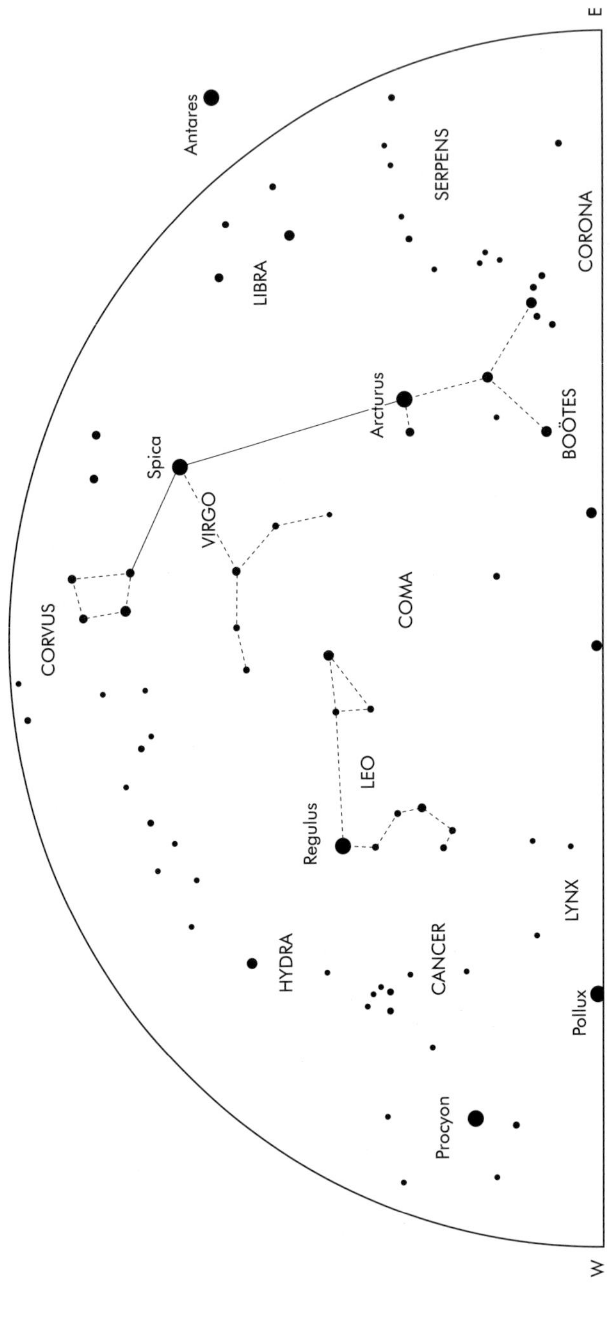

The Southern Hemisphere in April
Northern aspect (35° S 15 April 10pm)

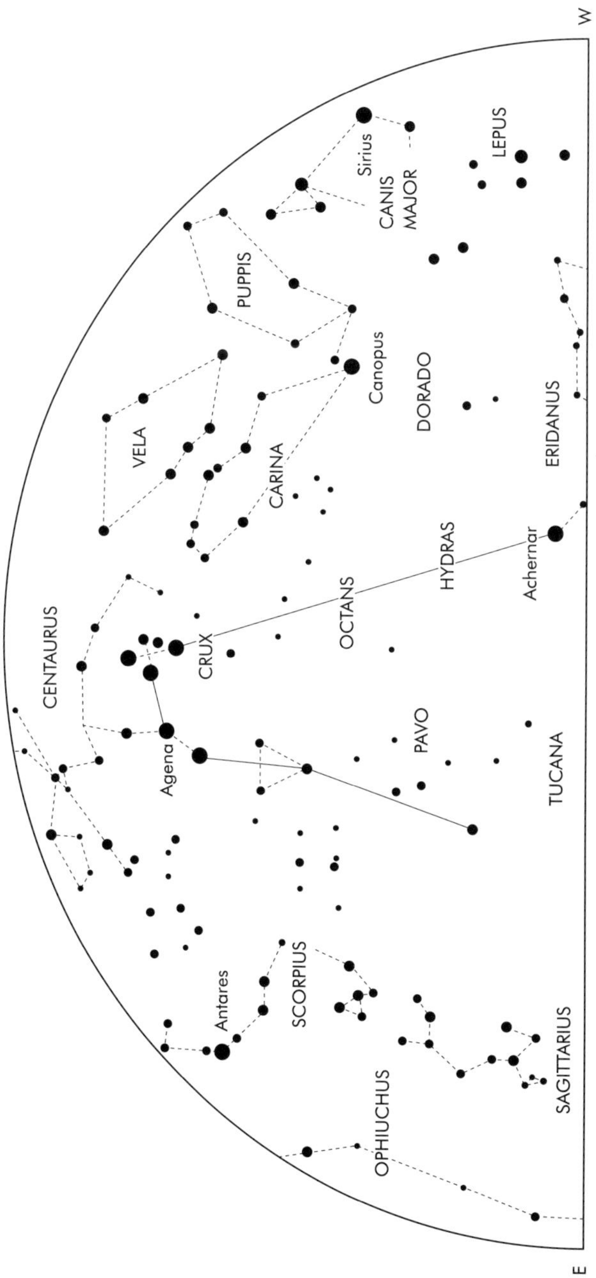

The Southern Hemisphere in April
Southern aspect (35° S 15 April 10pm)

April 1

The April Sky

By April evenings we have to all intents and purposes lost Orion, so that we are deprived of one of our most valuable guides. Fortunately there are means of identifying the less obvious groups.

To northern-hemisphere observers, the Great Bear is now almost overhead, which means that Cassiopeia is at its lowest, though from Britain neither sets; remember, the Bear and Cassiopeia are on opposite sides of the celestial pole, and about equally distant from it. Parts of Orion's retinue can still be seen; the Twins are there, and so is Procyon, but not Sirius. Capella is in the west, and on the opposite side of the pole, gaining altitude in the east, is the steely-blue Vega, in Lyra (the Lyre), which will occupy the near-zenith position during evenings in summer. In the east the brilliant orange Arcturus, in Boötes (the Herdsman) is high, and due south we find Leo, the main constellation of spring.

To Australians and New Zealanders, the overhead position is now occupied by the Southern Cross, with the two brilliant Pointers and the rest of Centaurus. Sirius can still be seen in the west, and so can Canopus; Leo is in the north, and Arcturus can be seen rising rather low in the north-east. Also rising is Scorpius, the Scorpion, with its red leader Antares; it is the Scorpion which will dominate the evening scene throughout southern winter.

Anniversary

1997: Perihelion passage of Comet Hale–Bopp. This was the second bright comet of very recent times; the first was Hyakutake (1996). Both these comets have periods of thousands of years, so that we will not see them again.

April 2

Leo, the Lion

Leo is one of the most brilliant of the Zodiacal constellations. In mythology it represents the Nemæan Lion which was one of Hercules' victims, but in the sky Leo is much more imposing than his conqueror. The main pattern is known as the Sickle, which rather resembles the mirror image of a question-mark; the brightest member is Regulus, of the first magnitude. The rest of Leo consists of at fairly prominent triangle of stars, one of which, Denebola, is of the second magnitude.

The leading stars of Leo are listed opposite.

There should be little difficulty in finding Leo, but northern observers have an extra advantage in as much as they can use Ursa Major. Simply trace the line of the Pointers 'the wrong way', i.e. away from the Pole Star. This line will cross a rather blank area and then reach Leo; the alignment is not perfect, but it is good enough.

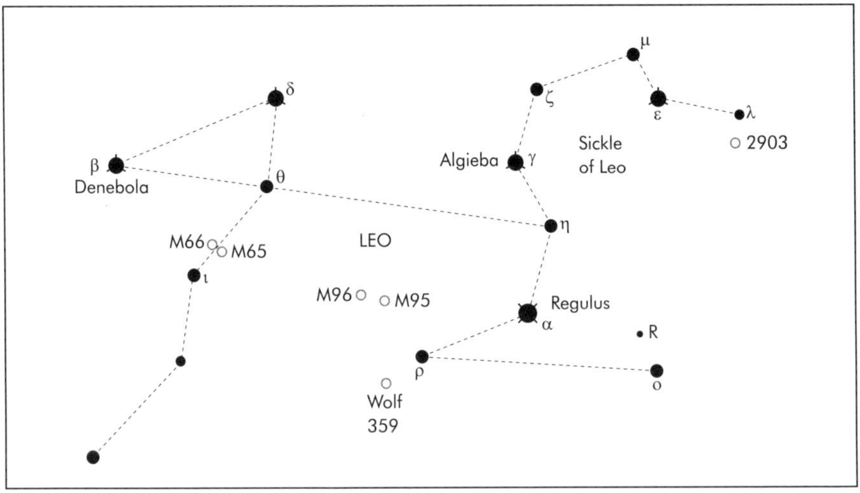

Leo					
Greek letter	Name	Magnitude	Luminosity (Sun=1)	Distance (light-years)	Spectrum
α Alpha	Regulus	1.4	130	85	B7
β Beta	Denebola	2.1	17	39	A3
γ Gamma	Algieba	2.0	60	90	K0+G7
δ Delta	Zosma	2.6	13	52	A4
ϵ Epsilon	Asad Australis	3.0	520	310	G0
ζ Zeta	Adhafera	3.4	50	117	F0
η Eta	–	3.5	9500	1800	A0
θ Theta	Chort	3.3	26	78	A2

Southern observers lack Ursa Major; the stars of the Plough are so low down that they skirt the horizon, and any mist or artificial lighting will hide them (from New Zealand they do not rise at all). However, Leo is not too far north of the celestial equator, and is sufficiently prominent to be recognized at once, particularly as there are no other bright groups close to it. The region between Leo and the Southern Cross is indeed very barren, and is occupied largely by Hydra; the only bright star in the area is Alphard, 'the Solitary One' (see 8 March).

April 3

The Royal Star

Now that we have found Leo, let us take a closer look at Regulus. It has its place in ancient lore; it was the leader of the four Royal Stars of the Persian monarchy, and was one of the 'Guardians of the Heavens' (the others were Aldebaran in

Taurus, Fomalhaut in Piscis Australis, and Antares in the Scorpion). Early English astrologers regarded it as a lucky star, and in 1552 William of Salisbury wrote that 'they that are born under it, are thought to have a royal nativity'.

Regulus, of magnitude 1.38, is the faintest of the stars usually ranked as being of first magnitude; it is not clear how the 'cut-off' point was chosen – Adhara in Canis Major, next in order, is of magnitude 1.50, so that the difference between it and Regulus is very slight indeed.

Regulus is a B-type star, and is therefore white or slightly bluish; it is 130 times as luminous as the Sun, and its diameter is 5 times that of the Sun. In the sky it lies less than 12° from the celestial equator, and only about half a degree from the ecliptic, so that it is one of the very few bright stars which can occasionally be hidden or occulted by the Moon. In 1959 Regulus itself was occulted by the planet Venus – an extremely rare event.

Regulus has a companion of magnitude 7.7, at a separation of 4.3 seconds of arc; the companion is about half as luminous as the Sun, and is an easy telescopic object. Its real distance from Regulus is around 12 000 million miles, and it is a close binary. There is also a 13th-magnitude star at a separation of 217 seconds of arc, but this has no genuine connection with Regulus.

Anniversary

1842: Birth of Hermann Carl Vogel, a German astronomer who was one of the pioneers of stellar spectroscopy, and was among the first to recognise the existence of spectroscopic binaries. He spent much of his career at the Potsdam Observatory. He died in 1907.

April 4

Algieba, the Double Star in the Sickle

Continuing with Leo, our 'constellation of the week', we come next to Algieba or Gamma Leonis, the second brightest star in the Sickle. At magnitude 2.0 it looks much fainter than Regulus, and as it is only five light-years further away from us it is clearly less luminous. It is also cooler, and even with the naked eye it looks somewhat orange.

Its proper name has caused comment in the past; the famous nineteenth-century astronomer Admiral W.G. Smyth wrote that 'the star has been improperly called Algieba, from *Al jeb-bah*, the forehead; for no representation of the Lion which I have examined will justify that position'. However, proper names are seldom used except for stars of the first magnitude and a few special cases, so that in general astronomers simply prefer to call the star Gamma Leonis.

It is a beautiful double. The primary is of type K; the companion, at magnitude 3.5, has a G-type spectrum. Smyth called the primary orange and the companion greenish-yellow, but most observers can see little colour in the fainter star. The separation is over 4 seconds of arc, and is increasing; the pair is a binary system with an orbital period of 619 years, and we are seeing it at a more and more favourable angle. The pair is one of the best doubles in the sky for the user of a small telescope.

There are two other stars nearby; one of magnitude 9.2 at 260 seconds of arc, and the other of magnitude 9.6 at 333 seconds of arc. Neither of these seem to be genuinely associated with the main pair.

Gamma Leonis lies very close to the radiant point of the Leonid meteors of November (see 17 November), and this is useful to remember when setting out to observe the Leonids, which are notoriously unreliable.

April 5

Anniversary

1973: Launch of Pioneer 11. This was a Jupiter Mission, and on 2 December 1974 it by-passed the Giant Planet at the range of 28 800 miles, sending back excellent data. It then went on to a rendezvous with Saturn on 1 September 1979. It is now on its way out of the Solar System.

Features of the Sickle

Leo is nowhere near the Milky Way, so that in general it is not particularly rich, but it contains many interesting objects, and for the owner of binoculars there are some coloured stars well worth seeking out. Three of them are in the Sickle; Gamma or Algieba, which we have already discussed, and Mu (Rassalas) and Lambda (Alterf). Between them is the brighter star Epsilon Leonis (Asad Australis), which is white.

Both Mu and Lambda have K-type spectra, so that in colour they are very similar to Algieba, but they are so much fainter that optical aid is needed to bring out their hues; their magnitudes are 3.9 and 4.3, respectively. Lambda is the more luminous of the two, and is over 100 times as powerful as the Sun; Mu could match 90 Suns. Neither has any special feature.

Also in this part of the sky is a star known only by its catalogue number of Wolf 359. Its position is RA 1 h 54 m, declination +07° 20′, but its visual magnitude is below 13, so that a telescope of some size is needed to show it – and certainly it is not easy to identify. Its interest lies in the fact that it is exceptionally feeble; it is a red dwarf, with only 0.000 02 the luminosity of the Sun. Apart from Barnard's Star and the members of the Alpha Centauri group, it is our nearest stellar neighbour, at 7.8 light-years – closer even than Sirius. For some years it was ranked as the least luminous normal star known, but this record has now been broken by a star known as MH18, discovered by M. Hawkins in 1990 and which is considerably feebler than Wolf 359.

April 6

Denebola: a Fading Star?

We will stay with Leo this evening; after all, it is well placed for both northern and southern observers. The next star to engage our attention is Beta Leonis or Denebola, the brightest star in the triangle some way from the Sickle which makes up the rest of the main pattern of the constellation.

Outwardly there is nothing unusual about Denebola. Its name comes from the Arabic Al Dhanab al Asad, the Lion's Tail; it is of magnitude 2.1, and at 39 light-years is one of our nearer neighbours. It is white, and 17 times as luminous as the Sun.

The main interest stems from the fact that almost all observers up to the seventeenth century ranked it as of magnitude 1, equal to Regulus and considerably brighter than Algieba. Today there is no doubt about the situation – Denebola and Algieba are virtually equal, and Regulus is brighter by more than half a magnitude. It is always wise to be sceptical about alleged changes of star magnitudes based on ancient observations, and this applies to Denebola, though it is fair to say that the evidence is slightly stronger than in most other cases of secular variation. Yet Denebola is not the sort of star expected to show changes over a period of a few thousand years, and it is best to conclude that there has been no real fading; all the same, an uneasy doubt remains. The two other stars of the triangle are quite normal; Delta Leonis or Zosma (magnitude 2.6) and Theta or Chort (3.3).

It may be interesting to compare Denebola with Algieba. They should look virtually equal; extinction must always be allowed for (see 15 January) but Denebola and Algieba will probably be at about the same altitude above the horizon.

April 7

R Leonis and the Purkinje Effect

Close to Regulus in the sky is a very interesting red variable, R Leonis; it lies between Regulus and the star Omicron Leonis, of magnitude 3.5. Using binoculars this is sometimes easy enough, because when at its best R Leonis can reach magnitude 4.4, so that it is then a naked-eye object and is very prominent in binoculars. At other times things are more difficult. Like so many red stars of type M, R Leonis is a Mira variable; at minimum it sinks to magnitude 11, and to see it you will have to use a telescope. The average period between one maximum and the next is 312 days. Approximate dates of maxima for the period covered here are:

1998 August 22	2001 March 17
1999 July 2	2002 January 22
2000 May 8	2003 December 1

but it is not possible to be precise, because, as with all Mira stars, both the period and the amplitude change slightly from one cycle to another.

Anniversary

1968: Launch of the Russian probe Luna 14, which approached the Moon to a distance of 99 miles, and sent back valuable data.

R Leonis is a true supergiant. Earth-based telescopes show stars as virtual point sources of light, but the Hubble Space Telescope, operating from high above the atmosphere, can do better, and can measure amazingly small angles. It has been able to measure the apparent diameter of R Leonis, which proves to be egg-shaped; the values are 78 × 70 micro-arc seconds, corresponding to real dimensions of 800 000 000 × 900 000 000 miles. This means that the huge globe could swallow up the orbits of all the planets in the Solar System out to and including that of Jupiter. Compared with this, even Betelgeux in Orion looks decidedly puny.

Telescopically, the magnitude of R Leonis can be estimated by comparing it with the stars in the same field, of which the two brightest are 18 Leonis (magnitude 5.8) and 19 Leonis (6.4). However, there is an extra problem. Both these comparison stars are white, whereas R Leonis is fiery red. If two objects of different colours appear equally bright, and then both are dimmed by the same amount, the red object will appear the fainter of the two. This is called the Purkinje effect, and must always taken into account when estimating red variables. However, R Leonis and the comparison stars are in the same low-power field, and their altitudes above the horizon are almost the same, so that at least the observer does not have to worry about extinction.

April 8

Chort and Zosma: Radial Velocities of Stars

We have looked at Denebola or Beta Leonis, the brightest member of the 'triangle' in the Lion. The other two members are Delta or Zosma (magnitude 2.6) and Theta or Chort (3.3).

Neither star has any really special features. Zosma is 52 light-years away, and 14 times as luminous as the Sun; Chort lies at 78 light-years and is the equal of 26 Suns. Both are white, with A-type spectra. They are moving across the sky in different directions – though their shifts are so slight that they cannot be noticed with the naked eye even over periods of thousands of years. However, we can tell that Zosma is approaching us at a rate of 13 miles per second, while Chort is receding at 5 miles per second.

This can be found by examination of their spectra. As we have seen, a stellar spectrum is made up of a rainbow background crossed by dark lines. If these lines are shifted over to the short-wave or blue end of the rainbow, the star is approaching us; if the shift is to the red, the star is receding. This is the well-known Doppler effect. The toward-or-away speeds are known as radial velocities; positive if the body is receding, negative if it is approaching. Thus in technical terms, the radial velocity of Chort is +8 miles per second and of Zosma –21 miles per second.

However, it must not be thought that in the far future we are in any danger of being hit by Zosma, or that Chort will eventually fade into the distance. Our Galaxy is rotating, and all its stars (including the Sun) are taking part in this rotation, so that the movements we observe are simply parts of the general picture. We have now measured the radial velocities of a great many stars, and we know that collisions, or even close encounters, must be very rare indeed.

April 9

Galaxies in Leo

For the last time – at any rate, for the moment – we will spend this evening with Leo, which will remain prominent in the evening sky all through the rest of this month and May. There are several galaxies worth seeking out, and five of these are in Messier's list (and listed below).

All these are on the fringe of binocular range, though a telescope is needed to show them properly.

M65 and M66 make up a pair not far from the fourth-magnitude star Iota Leonis, which lies south of Chort, the third member of the Leo triangle. Both are spiral systems, and both are around 30 000 000 light-years away from us; both were discovered by the French astronomer Pierre Méchain in 1780. M65 is almost edgewise-on to us, so that the full beauty of the spiral shape is lost; M66 is only about half a degree from it, and is rather larger, though less massive. A small telescope will show both without difficulty, though the spiral forms need a much larger aperture and are best brought out photographically. M105 is a small but fairly bright elliptical system close to M96. It was not in Messier's original catalogue, but it too was discovered by Méchain and was tacked on to the catalogue later. Telescopically it looks rather like a globular cluster.

M95 and M96 lie side by side, roughly between Regulus and Iota Leonis; M96, a normal spiral, is definitely the more conspicuous of the two, though it is very difficult to see with

Messier Galaxies in Leo

M number	NGC number	RA h	RA m	Dec. °	Dec. ′	Magnitude	Type
95	3351	10	44.0	+11	42	9.7	Barred spiral
96	3368	10	46.8	+11	49	9.2	Spiral
105	3379	10	47.8	+12	35	9.3	Elliptical
65	3623	11	18.9	+13	05	9.3	Spiral
66	3627	11	20.2	+12	59	9.0	Spiral

binoculars even as a faint patch. M95 is what is termed a barred spiral, so that the arms seem to issue from the ends of a 'bar' across the main plane of the system.

All these, and several others, form a definite group or cluster of galaxies. Indeed, galaxies do tend to congregate in clusters, and our own Milky Way system is a member of what we call the Local Group, other members of which include the two Clouds of Magellan and the Andromeda and Triangulum spirals.

April 10

The Little Lion

It does not take very long to learn the main constellations. Identifying the smaller ones needs more time and patience, but it can be done, and so let us turn now to one of these obscure groups, Leo Minor or the Little Lion. It is not an original constellation; it was added to the sky in 1690 by the Polish astronomer Hevelius, and there are no mythological legends attached to it. Neither does it contain much of immediate interest. However, let us find out where it is.

It lies between the Plough and the Sickle of Leo, so that at present it is very high up as seen from British latitudes; to southern observers it is very low, but still well above the horizon except from the southern part of New Zealand. Its three main stars, all of about the fourth magnitude, are 46,

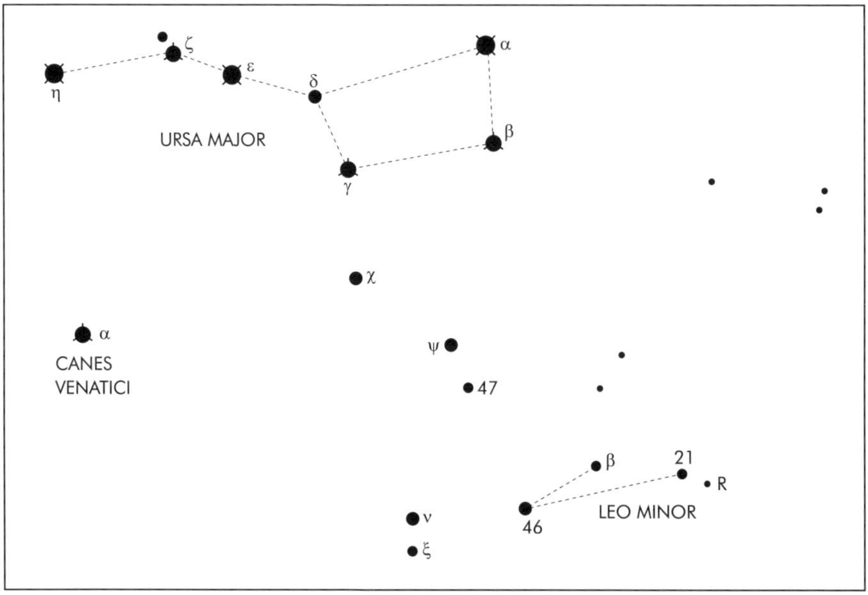

Beta and 21. For some curious reason Beta is the only star in the constellation to have been given a Greek letter; there is no Alpha Leonis Minoris.

Locate the triangle of stars below the Pointers; Psi Ursæ Majoris (magnitude 3.0), Lambda or Tania Borealis (3.4) and Mu or Tania Australis (3.0) The two Tanias lie close together, and make a good contrasting pair – Tania Australis is very red, as binoculars will show, while its neighbour is white. The fainter triangle of Leo Minor is close by, between Alpha Lyncis in the Lynx (11 April) and two fainter stars in Ursa Major, Nu (3.5) and the famous binary Xi (3.8).

46 Leonis Minoris, the brightest star in the Little Lion, is of magnitude 3.8; it has a proper name – Præcipua; it is of spectral type K. The red Mira variable R Leonis Minoris can reach magnitude 6.3 at maximum; its period is 372 days, so that it reaches maximum a week later each year. The 1995 maximum fell on September 5, so that it needs only a little mental arithmetic to work out the maxima for the rest of the period covered here – though, of course, all Mira stars have cycles which are themselves variable to some extent.

April 11

Lynx

Lynx, the celestial Lynx, covers a fairly wide area of the sky (almost 450 square degrees) but is very obscure, with only one reasonably bright star. This is Alpha, of magnitude 3.1,

Anniversary

1970: Launch of Apollo 13. This was the mission which so nearly ended in tragedy. On the outward journey to the Moon, an explosion in

(continued opposite)

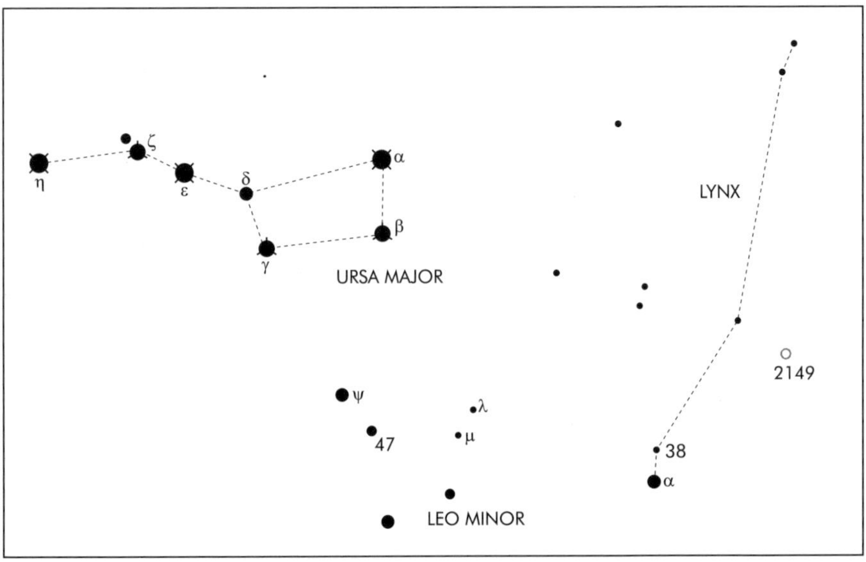

the Command Module put the main motors out of action, and it was only by a combination of courage, skill and improvisation that the astronauts returned safely, touching down on April 17 after having been round the Moon.

which is easy to locate because it makes up a triangle with Regulus in Leo and Pollux, the brighter Twin. It is of type M, and binoculars show it to be very red; it is 166 light-years away, and 115 times as luminous as the Sun.

The only other star in Lynx above the fourth magnitude is 38 (magnitude 3.9). Again there is a strange lack of Greek lettering; only Alpha has been so honoured, though 38 does have a little-used proper name, Alsciaukat.

Anniversary

1961: First manned space-flight, by Yuri Gagarin of what was then the USSR. He was launched from the Baikonur cosmodrome at seven minutes past six GMT in his vehicle Vostok 1. He reached peak velocity at 17 500 mph., and entered free fall; he made one full circuit of the Earth, at a height ranging from 180 to 327 miles, and remained aloft for 1 h 48 m before landing safely in the prearranged area. Sadly, this was Gagarin's only space-flight; he lost his life in an ordinary aircraft crash in 1968.

April 12

The Intergalactic Tramp

Lynx may be a dim constellation, but it does contain one object of real note: a distant globular cluster NGC 2149. It was not listed by Messier, but is No. 25 in the Caldwell catalogue. It was discovered by William Herschel in 1788; in 1861 the Earl of Rosse surveyed it with his great reflector and suggested that it might be a globular cluster, though it was not until 1922 that photographs taken at the Lowell Observatory in Arizona showed that this suggestion was correct.

It is not too easy to find, because its integrated magnitude is only 10.4 and there are no obvious guides to it; the position is RA 7 h 38 m, declination +38° 53′. It looks small, with an apparent diameter of 4 minutes of arc, and its individual stars are faint and closely packed. It is interesting because its distance has been given as 180 000 light-years, further away than the Large Cloud of Magellan. It is receding at 12 miles per second, and may be a true 'intergalactic tramp', not now a genuine member of any galaxy. Its real diameter is not far short of 400 light-years.

It is really a 'tramp'? Probably – and if so there may be many other globulars wandering between the galaxies, though they will not be easy to identify. Meantime it is well worth looking for C25. Its appearance as a small fuzzy patch belies its true significance.

Future Points of Interest

Maximum of the annual Virginid meteor shower. This is one of the minor showers, and is never rich; it begins around 7 April and ends on 18 April. In general the meteors are rather slow, with long paths. The radiant is not well defined.

April 13

47 Ursæ Majoris: a Planetary Centre?

Let us go back to the Great Bear for a few moments, and look at one of its very inconspicuous stars – 47 Ursæ Majoris, near the triangle formed by Psi, Lambda and Mu which we discussed when locating Leo Minor (10 April). The position of 47 is RA 10 h 59 m, declination +40° 25′.

It is of the fifth magnitude, and is 42 light-years away. It has a G-type spectrum, and is not too unlike the Sun, though it is rather more luminous. What marks it out is the fact that in 1995 studies of it indicated that there was a companion – not an ordinary star, but a body only about 2.3 times the mass of Jupiter, making it a planet rather than a star. If the calculations are correct, it moves round its primary at a distance of just under 200 000 000 miles in a period of 3 years.

True, the evidence is indirect, and depends entirely upon very delicate spectroscopic measurements, and other explanations have been offered. But if the system does contain a Jupiter-type planet, why should there not be smaller planets also – other Earths, in fact?

Logically there seems no reason why this should not be so. We cannot see them, even if they exist; not even the Hubble Space Telescope can do that, but if you look at 47 Ursæ Majoris it is sobering to reflect upon the possibility that some alien astronomer in that system may, at this very moment, be turning a telescope toward the yellow star which we call the Sun.

April 14

The Mountains of the Moon

It is high time that we looked back at the Moon. We have already said something about the main surface features, so now let us turn our particular attention to the mountains.

The Moon is a highly mountainous world. There are great ranges, such as the Lunar Apennines, which contain peaks higher than those of our own Apennines. It is difficult to give precise comparisons, because on Earth we reckon the height relative to sea level, and there is no water on the Moon, so that we have to calculate from an agreed mean radius for the lunar globe.

The most important ranges are as follows are given in the list opposite.

Anniversary

1624: Birth of Christiaan Huygens. He was Dutch, and probably the best observer of his period; he discovered Saturn's satellite Titan, in 1655, and was the first to realize the true nature of Saturn's rings. He was also the first to see markings on Mars. He is probably best remembered today as being the inventor of the pendulum clock. He died in 1695.

Lunar Mountain Ranges			
Name	Latitude	Longitude	
Alps	45° N	1° E	Northern border of Mare Imbrium
Altai	24° S	23° E	In the Mare Nectaris basin; better called the Altai Scarp
Apennines	20° N	3° W	Border of Mare Imbrium; length 380 miles; high peaks
Carpathians	15° N	25° W	250-mile range along the southern border of Mare Imbrium
Caucasus	39° N	9° E	Continuation of the Apennines
Hæmus	17° N	13° E	Border of Mare Serenitatis; 250 miles long
Harbingers	27° N	41° W	Mountain clumps near Aristarchus
Juras	47° N	37° W	Border of Sinus Iridum
Riphæans	7° S	28° W	In Mare Nubium, near Euclides
Rook	20° D	83° W	Bordering Mare Orientale
Spitzbergen	35° N	5° W	Chain of peaks near Archimedes
Straight Range	48° N	20° W	Very regular range in Mare Imbrium, near Plato
Taurus	26° N	36° E	Mountain clumps near Rømer; Mare Crisium area

Most of the great ranges form parts of the borders of the regular seas; thus the Mare Imbrium (Sea of Showers) is in part bordered by the Apennines and the Carpathians. (In many cases the lunar mountains are named after terrestrial ranges.) It seems that these ranges were formed at the same time as the sea basins, and are not in the least like our Himalayas. Isolated peaks and groups of peaks abound.

The heights are measured by the shadows which the peaks cast on to the adjacent landscape. The altitude of the Sun above the peak can be worked out; knowing this, and the shadow length, gives the height of the peak itself.

Anniversary

1793: Birth of F.G.W. Struve. The first of a family of major astronomers. Friedrich Georg Wilhelm Struve was German, but went to Dorpat in Estonia and became Director of the observatory there. Using a 9-inch refractor (one of the first to be clock-driven) he concentrated upon stars; he moved to the Pulkovo Observatory in 1839, and produced a catalogue of over 3000 stars. He also measured the parallax of Vega. He died in 1864 and was succeeded at Pulkovo by his son Otto.

April 15

The Spin of the Moon

Look at the Moon tonight, and you should be able to see the Mare Crisium or Sea of Crises, which is well marked and is separate from the main 'sea complex'. It seems to be elongated in a north–south direction. In fact the north–south diameter is only 280 miles, while the length east–west is 348 miles; appearances are misleading, because the Mare Crisium is not far from the edge or limb of the Moon, and is foreshortened.

The Moon's orbital period is 27.3 days. It spins on its axis in exactly the same time: 27.3 days. This means that it keeps the same face turned toward the Earth all the time, and there is part of the Moon which we can never see from here, because it is always turned away from us. Moreover, the visible features always keep to the same positions on the disk – for instance Mare Crisium is always to the upper right, as seen from the northern hemisphere of the Earth.

There is no mystery about this behaviour; it is due to the effects of tidal friction over the ages. Originally both Earth and Moon were molten, and raised tides in each other. Those due to the Earth were much the more powerful, because the Earth has 81 times the mass of the Moon. The effect was to keep a 'bulge' in the Moon turned Earthward, and as the Moon rotated it had to fight against this force; the situation may be likened to that of a spinning cycle wheel between two brake shoes. The Moon's rotation was slowed down, until relative to the Earth – not relative to the Sun – it had stopped. Most planetary satellites have similarly 'captured' or synchronous rotation.

There is, however, a qualification. The Moon spins on its axis at a constant speed, but it does not move round the Earth at constant speed, because its orbit is appreciably eccentric, and it moves quickest when at its nearest to Earth (perigee). Each month, therefore, the amount of spin and the position in orbit become out of step, and we can see for a little way beyond first one mean limb and then the other. This rocking movement is termed 'libration in longitude'. Together with various other effects, it means that all in all we can examine 59% of the total surface, though of course no more than 50% at any one time. The remaining 41% is permanently out of view, and until the round trip flight of the Russian probe Lunik 3, in 1959, we had no direct information about it.

Near the limb, foreshortening is extreme, and it is often difficult to distinguish between a crater and a ridge. Moreover, some features are carried alternately in and out of view. The effects are easily detectable with the naked eye. The Mare Crisium, in the eastern part of the Moon, almost touches the limb at the least favourable libration, while when best placed it is well clear; the same applies to the dark-floored crater Grimaldi, on the opposite or western limb.

April 16

Méchain and the Nebulæ

Pierre François André Méchain, of France, was one of the best observers of the eighteenth century. He discovered eight comets, and one of these, which he found in 1790, has a period of 13.75 years; it was seen again by Horace Tuttle in 1858 and is now known officially as Tuttle's Comet, though to be fair it really ought to be Méchain–Tuttle. It has been seen at every return to perihelion since 1858 except in 1953, when it was hopelessly badly placed in the sky.

Méchain's father was an architect. Pierre took up a career in mathematics, and in 1774 he obtained a post as calculator at the Navy Depôt; it was here that he met Charles Messier,

Anniversary

1744: Birth of Pierre Méchain.

Future Points of Interest

1999: Mercury at western elongation.

2003: Mercury at eastern elongation.

who was with the same department. Méchain searched energetically for comets, and in the process he discovered a number of clusters and nebulæ. He passed on all his observations to Messier, who used them for his famous catalogue. It is pleasant to record that although Méchain and Messier were in a sense observational rivals, they remained on terms of close friendship throughout their lives.

Méchain was away from Paris during the French Revolution, and did not return there until 1795. He served for a while as Director of the Paris Observatory, and finally retired in 1803. He died of yellow fever in Spain on 28 September 1804.

Altogether he discovered 22 of the objects in Messier's catalogue: Nos 63, 78, 65, 66, 68, 75, 72, 76, 74, 79, 77, 85, 98, 99, 100, 95, 96, 94, 101 and 103. Only Messier himself, with 42 discoveries, found more.

April 17

Seas at the Moon's Limb

Of the major seas on the Earth-turned hemisphere of the Moon, only the Mare Crisium is separate from the main system. There are however various seas close to the limb which are worth finding, and which can be seen with a small telescope even though they are so foreshortened.

There is, for example, the Mare Humboldtianum, or Humboldt's Sea (latitude 57° N, longitude 80° E), near the north-east limb. It is fairly regular, and is never actually carried out of view even at the most unfavourable libration. The Mare Marginis (Marginal Sea), east of Mare Crisium, is of the same type, but is rather smaller and less well placed (12° N, 88° E). On the eastern limb is the Mare Smythii, Smyth's Sea (2° S, 87° E), named after the famous astronomer–admiral of the nineteenth century. It is well defined, with an area about half that of the Mare Crisium. Further along the limb, not far from the large walled plain Furnerius, is the Mare Australe (Southern Sea) (46° S, 91° E) which is really an irregular, patchy area rather than a single well-defined sea, and extends on to the Moon's far side.

On the opposite limb is the Mare Orientale or Eastern Sea (20° S, 96° E). It is widely believed to be the youngest of all the maria. It is a vast ringed structure extending well on to the Moon's far side, and only the extreme eastern edge is ever visible from Earth. In fact I discovered it myself, years before the start of the Space Age, when I was mapping the libration areas with the modest 12.5-inch reflector in my private observatory, then at East Grinstead in Sussex. I recognized it as being new, but I had no idea of its true nature or its importance. I suggested the name of 'the Eastern Sea', which was accepted. However, later on, the International Astronomical Union reversed lunar east and west, with the result that my *Eastern* Sea is now on the Moon's *western* limb! No part of it can be seen except under favourable libration, but it is interesting to find.

April 18

Constellation Shapes – and the Southern Triangle

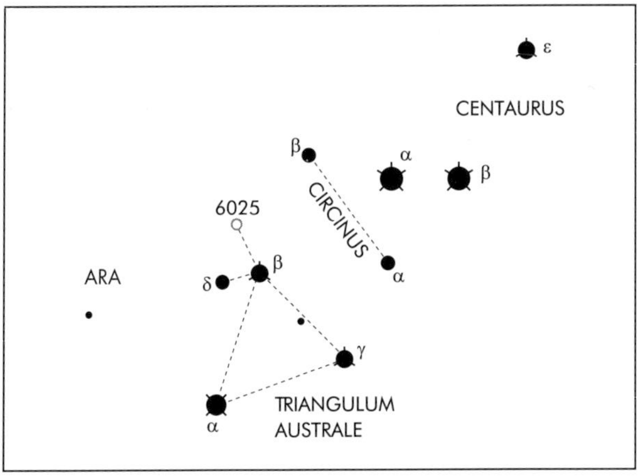

Triangulum Australe is too far south to be seen from Britain or Europe. Its chief star, Alpha, lies at declination –69°, so that to see it you must go south of latitude 21° N. (90 – 69 = 21.) The three leaders, making up the triangle, are listed below.

Next in order comes Delta (magnitude 3.8), which lies close to Beta. In fact Delta is much more powerful than the other leaders, and could match 600 Suns, but at 360 light-years it is much further away. Triangulum Australe is easy to locate, because it lies close to the two Pointers to the Southern Cross, Alpha and Beta Centauri. The orange-red hue of Alpha is evident with the naked eye, and very obvious with binoculars. Probably the most interesting object in the

Triangulum Australe

Greek letter	Name	Magnitude	Luminosity (Sun=1)	Distance (light-years)	Spectrum
α Alpha	Atria	1.9	96	55	K2
β Beta	–	2.8	5	33	F5
γ Gamma	–	2.9	50	91	A0

constellation is the open cluster C95 (NGC 6025), which is on the fringe of naked eye visibility and is easy with binoculars.

Between Triangulum Australe and the Pointers lies the constellation of Circinus (the Compasses). Its only fairly bright star is Alpha, of magnitude 3.2, which has a companion of magnitude 8.6 at a separation of almost 16 seconds of arc. Since the two share a common motion through space, they must be physically associated.

Anniversary

1967: Landing of the US unmanned probe Surveyor 3 on the Moon.

April 19

The Lunar Surveyors

It has long been known that the lunar 'seas' have never had any water in them, but until quite recently there was a widely held theory that they were filled with deep, treacherous dust. In particular Dr Thomas Gold, of Cambridge, maintained that any vehicle unwise enough to land there would simply sink out of sight with devastating permanence. The theory was not widely supported by lunar students, because it did not seem to fit the observational facts, but officially it was taken very seriously indeed, particularly in the United States. It was finally disproved on 3 February 1966, when the automatic Russian probe Luna 9 made a controlled descent on to the Oceanus Procellarum and showed no inclination to sink.

The first American soft-landers were the Surveyors of 1966–68. There were seven in all, of which only Nos 2 and 4 were unsuccessful. Surveyor 3 landed in the Oceanus Procellarum, at latitude 2.9° S, longitude 23.3° W, and returned 6315 images, as well as carrying out analysis of the lunar soil. More than two years later the astronauts of Apollo 12 walked over to it, removed some pieces and brought them home for analysis.

Surveyor 3 is still standing in the waterless Ocean of Storms. There is nothing to make it deteriorate, and there it will remain until a future expedition collects it and removes it to a lunar museum.

Future Points of Interest

Maximum of the annual Lyrid meteor shower (April 20–21).

April 20

The Lyrids – and Thatcher's Comet

Tonight marks the maximum of the Lyrid meteor shower, with its radiant not far from the brilliant Vega. Generally the shower is not very striking, with a ZHR (zenithal hourly rate) of about 10, but occasionally it can produce good displays; there seems to have been a real 'Lyrid storm' in 1803, and in modern times there were rich displays in 1922 and again in 1982, so that it is well worth keeping a careful watch tonight and tomorrow.

There seems to be little doubt that the parent of the Lyrid stream is Thatcher's Comet of 1861. The comet was discovered on 5 April of that year by the American observer A.E. Thatcher (no relation of any modern politician, let it be added!) when it was below the seventh magnitude. It brightened slowly, and at one time in May had reached magnitude 2.5, with a tail extending for at least a degree. It reached perihelion on 3 June, and was followed telescopically until 7 September. The orbit is elliptical, but the period is very long – 415 years – so that it will not be seen again in our time.

April 21

Anniversary

1972: Landing of Apollo 16 on the Moon.

Apollo 16 and the Highlands of the Moon

On 21 April 1972 the lunar module of Apollo 16 touched down gently on the surface of the Moon. Astronauts John Young and Charles Duke began their exploration of the area, driving around in the LRV or Lunar Roving Vehicle that they had brought with them; the third member of the crew, Thomas Mattingly, remained in lunar orbit in the command module of the vehicle.

Apollo 16 landed in the region of the crater Descartes (latitude 8° 36′ S, longitude 15° 31′ E), in the southern uplands of the Moon. This was in fact the only mission to come down in an area of this type; all the rest were targeted at the waterless seas, so that the programme carried out by Young and Duke was of special significance.

The lunar uplands are much rougher than the seas. They are also older. There can be no doubt that around the period from 3200 to 3000 million years ago there was widespread volcanic activity on the Moon, with lava pouring out from below the crust and flooding the 'sea' basins; most of the craters which were already there were destroyed, whereas the upland regions were able to preserve the craters which had been formed at an earlier epoch. Young and Duke brought back samples of the surface; the rocks were of essentially the same type as those recovered by the other Apollo missions, as was only to be expected. When the astronauts left the Moon, the 'moon car' remained behind. We know exactly where it is, and one day, no doubt, it will be given a new battery and driven off to its museum.

April 22

Antlia

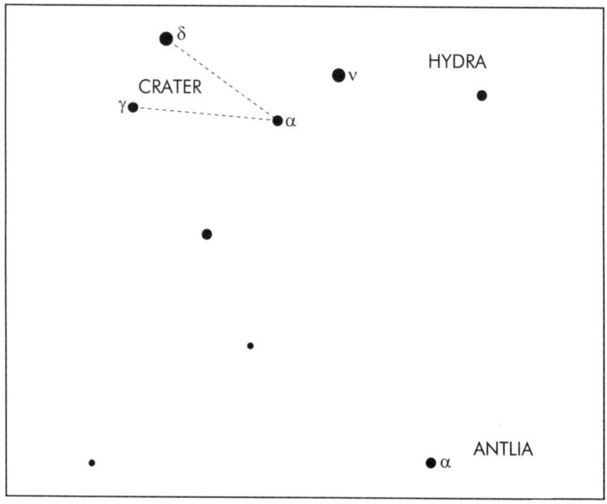

All the original constellations were named either after mythological characters, or else after living creatures or familiar objects. However, the far-southern groups were not included, for obvious reasons, and the constellations here were added by later astronomers. One of these was Nicolas Louis de Lacaille, of France, who went to the Cape between 1750 and 1754 and compiled an important star catalogue. During this work he introduced fourteen new constellations, all of which are still to be found on our maps and some of which have decidedly modern sounding names; in fact Lacaille's original names have been modified, so that Antlia Pneumatica, the Air-Pump, is today known simply as Antlia.

Southern observers will find it high tonight; it lies between Hydra and Carina (9 March). Northern observers will have more trouble. Antlia is very low down in the south, below the long, straggly line of stars marking Hydra, and its stars are faint; even the brightest of them, the red M-type Alpha Antliæ, is only of magnitude 4.2, and there are no interesting objects in the group, so that identifying it is mainly a matter of pride.

Antlia covers 239 square degrees, so that its area is about the same as that of Leo Minor. Whether either constellation has a valid claim to separate identity is questionable.

April 23

The Sextant

Sextans is one of the groups added to the sky by Hevelius in 1690; its original name was Sextans Uraniæ, Urania's Sextant. It lies between Leo and Hydra, and is cut by the celestial equator; it also contains the pole of the ecliptic. It is a very barren group, and has no definite outline. Its brightest star, Alpha Sextantis, is only of magnitude 4.5.

However, it contains one very interesting object: the Spindle Galaxy, C53 (NGC 3115). Unfortunately it is rather faint, with an integrated magnitude of below 9, and well beyond the reach of normal binoculars, but its actual surface brightness is higher than for many galaxies, so that it will bear magnification well. It seems to be about 25 000 000 light-years away, and to have a real diameter of the order of 30 000 light-years.

Is it a spiral? This is uncertain. When seen visually it looks like an elongated blur; photographs taken with larger telescopes show it as being rather lens-shaped, but there are indications of structure. No spiral arms have been seen, but of course much depends on the angle of observation; a true spiral might not betray its nature if seen edgewise-on. At the moment the Spindle Galaxy is classed as an elliptical system, but this may have to be revised in the future.

If you have an adequate telescope, search for the Spindle by all means, but do not be too disappointed if you fail to find it. If your telescope has good setting circles, things are easier; the position is RA 10 h 05m.2, declination –07° 43′.

April 24

The Centaur and the Southern Cross

So far we have been concentrating on the view from the northern hemisphere of the Earth, but it so happens that some of the most important objects in the stellar sky are in the far south, I hope, therefore, you will forgive me if for the next five nights we 'go south' and discuss some of the objects which Britons never see.

First and foremost, of course, are Centaurus (the Centaur) and Crux Australis (the Southern Cross), which during April evenings are almost overhead to Australians or New Zealanders. As Centaurus almost surrounds the Cross, it will be best to list their principal stars together (see table opposite).

A small part of Centaurus protrudes above the British horizon, but not much, and the two brilliant Pointers to the

Future Points of Interest

1999: Mars at opposition. The planet is in Virgo; magnitude –1.5; apparent diameter 16″.2; minimum distance from Earth 54 000 000 miles.

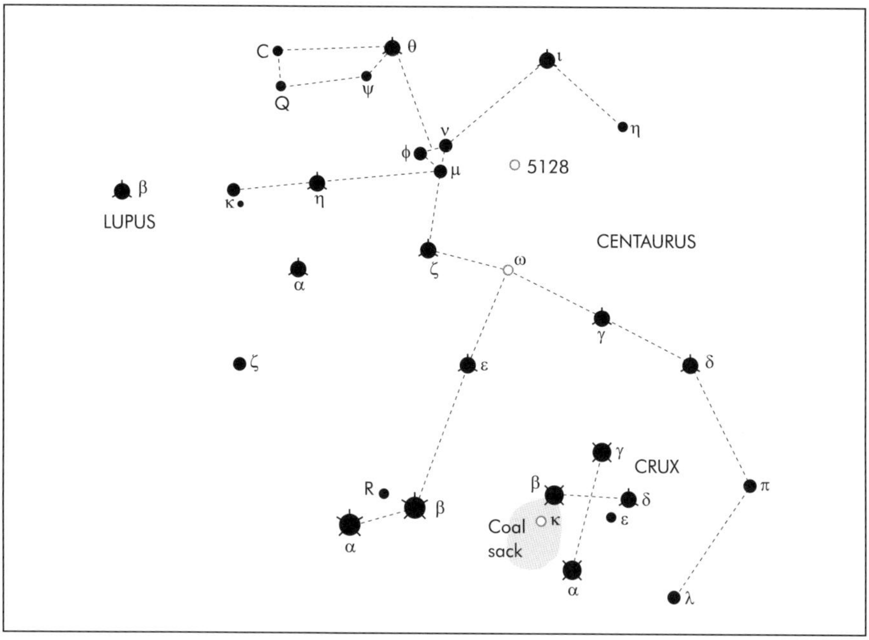

Centaurus and Crux					
Greek letter	Name	Magnitude	Luminosity (Sun=1)	Distance (light-years)	Spectrum
Centaurus					
α Alpha	–	–0.3	1.7+0.45	4.3	G2+K1
β Beta	Agena	0.6	10 000	460	B1
γ Gamma	Menkent	2.2	130	110	A0
δ Delta	–	2.6	830	325	B2
ϵ Epsilon	–	2.3	2100	490	B1
ζ Zeta	Al Nair al Kentaurus	2.6	1300	360	B2
η Eta	–	2.3	1200	360	B3
θ Theta	Haratan	2.1	17	46	K0
ι Iota	–	2.7	26	52	A2
κ Kappa	Ke Kwan	3.1	830	420	B2
λ Lambda	–	3.1	180	190	B9
μ Mu	–	3.0	450	300	B3
Crux					
α Alpha	Acrux	0.8	3200+2000	360	B1+B3
β Beta	Mimosa	1.2	8200	425	B0
γ Gamma	–	1.6	120	88	M3
δ Delta	–	2.8	1320	260	B2

Cross, Alpha and Beta Centauri, are well out of view. They are not true neighbours. Alpha is the nearest of all the bright stars, at just over 4 light-years; its faint red dwarf companion, Proxima, is slightly closer, but is too faint to see without a powerful telescope and is none too easy to identify. Beta is a distant, powerful giant. Alpha Centauri is a magnificent binary; the components are of magnitudes 0.0 and 1.2, and the separation is great enough for the pair to be split with a very small telescope. It is a binary system, with an orbital period of 79.9 years, so that separation and position angle change quite quickly. Oddly enough, Alpha Centauri has never had a universally accepted proper name; Toliman, Bundula and Rigel Kent have all been used, but astronomers prefer to call it simply Alpha Centauri.

April 25

Anniversary

1990: Launch of the Hubble Space Telescope.

Features of the Centaur

Centaurus contains a wealth of interesting objects. For example Gamma is a fine double with equal components (magnitude 2.9), though it is rather close; it is a binary with a period of 84.5 years. Between Alpha and Beta lies the red Mira variable R Centauri, with a range of magnitudes from 5.3 to 11.8 and the unusually long period of 546 days. (The 1997 maximum fell on 11 May.) But pride of place must go to the globular cluster Omega Centauri, which is easily visible with the naked eye in the guise of a hazy star. It is much the brightest of all the globulars; it contains well over a million stars, and even from its distance of 17 000 light-years it is a superb sight in a telescope. At its centre, the average distance between stars is only about a tenth of a light-year.

Much further north in the constellation, at RA 13 h 25 m, declination −43° 01′, lies the remarkable galaxy C77 or NGC 5128, known to radio astronomers as Centaurus A because it is a powerful radio emitter. Telescopically it is seen to be crossed by a dark band; it is of integrated magnitude 7, so that it is easy to locate near the little triangle made up of Mu (magnitude 3.0), Nu (4.3) and Phi (3.8). Centaurus A is no more than 10 000 000 light-years away at most, and is the nearest of the major radio sources. Find it if you can; it is one of the most remarkable objects in the sky.

April 26

The Southern Cross

Crux, the most famous of all southern constellations, used to be included in Centaurus; it was first separated out in 1679 by an otherwise obscure astronomer named Augustin Royer. It is more or less surrounded by Centaurus; its only other common boundary is with Musca.

A casual glance is enough to show that of the four main stars, three are bluish-white while the fourth, Gamma Crucis, is orange-red. Alpha is a fine double; the components are of magnitudes 1.4 and 1.8, separated by 4.4 seconds of arc. No doubt they are physically connected, but they are a long way apart. A third nearby star, magnitude 4.9, is also a remote member of the system. Gamma Crucis has a 6.7-magnitude companion at a separation of 111 seconds of arc, but this is an optical pair, not a binary.

We have to admit that Crux is nothing like an X; it more nearly resembles a kite. It is a pity that there is no central star to complete the pattern – and moreover Epsilon, of magnitude 3.6, rather spoils the effect. However, it is in a class of its own, and the southern sky would seem drab without it.

April 27

The Jewel Box and the Coal-Sack

Though Crux is so small, it is crammed with spectacular objects. Close beside Beta is the superb open cluster C94 (NGC 4755) round Kappa Crucis. Most of the stars in it are white, with a prominent triangular arrangement, but there is one red supergiant which stands out; the cluster has been nicknamed the Jewel Box, and is as beautiful as any in the sky. It is around 7700 light-years away, and is probably no more than a few million years old.

Adjoining Alpha and Beta is something very different: an apparently starless area, known as the Coal Sack (C99). It is simply a dark nebula, no more than 500 light-years away, blotting out the light from stars beyond; only a few stars are seen in the foreground. There are plenty of dark nebulæ in the sky, but none to rival the Coal Sack.

It is worth adding that the only difference between a bright nebula and a dark one is the presence or absence of any nearby stars to provide illumination. For all we know, there may be suitable stars on the far side of the Coal Sack, so that if we could observe it from a different vantage point in the Galaxy it might well look bright instead of inky black.

April 28

Lupus

Before returning to the northern sky, we must pause to look briefly at Lupus, the Wolf, which adjoins Centaurus and contains some reasonably bright stars, listed overleaf.

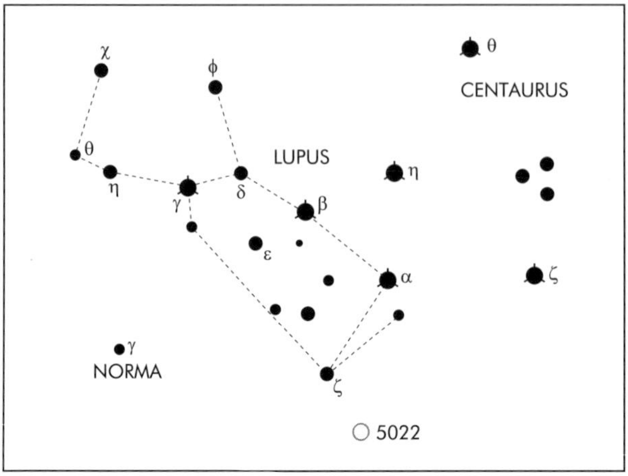

> ## *Anniversary*
>
> 1900: Birth of Jan Hendrik Oort. Oort was one of the greatest of all Dutch astronomers; he was Director of the Leiden Observatory from 1945 to 1970, and President of the International Astronomical Union from 1958 to 1961. With Bertil Lindblad, he was the first to recognize the rotation of the Galaxy; he proposed the 'dirty iceball' model for a comet, and he put forward the idea of what is now called the Oort Cloud, a swarm of cometary bodies orbiting the Sun far beyond the path of Neptune. He died in 1992.

Lupus

Greek letter	Name	Magnitude	Luminosity (Sun=1)	Distance (light-years)	Spectrum
α Alpha	Men	2.3	5000	6800	B1
β Beta	KeKouan	2.7	830	360	B2
γ Gamma	–	2.8	450	258	B3
δ Delta	–	3.2	1320	587	B2
ϵ Epsilon	–	3.4	700	456	B3
ζ Zeta	–	3.4	58	137	G8
η Eta	–	3.4	830	490	B2

Lupus has no really distinctive shape, and there is nothing much of immediate interest here, but it may be worth locating the open cluster NGC 5822, near Zeta (RA 15 h 05 m, declination –54° 21′), which has an integrated magnitude of 6.5, and is an easy binocular object.

April 29

Ara

Ara, the Altar, is an original constellation, though it does not seem to have any definite mythological legends attached to it. It is in the southern part of the sky, between Scorpius and Triangulum Australe, and so cannot be seen from Britain or most of Europe; part of it rises briefly from the latitude of Athens. Its leading stars are listed opposite.

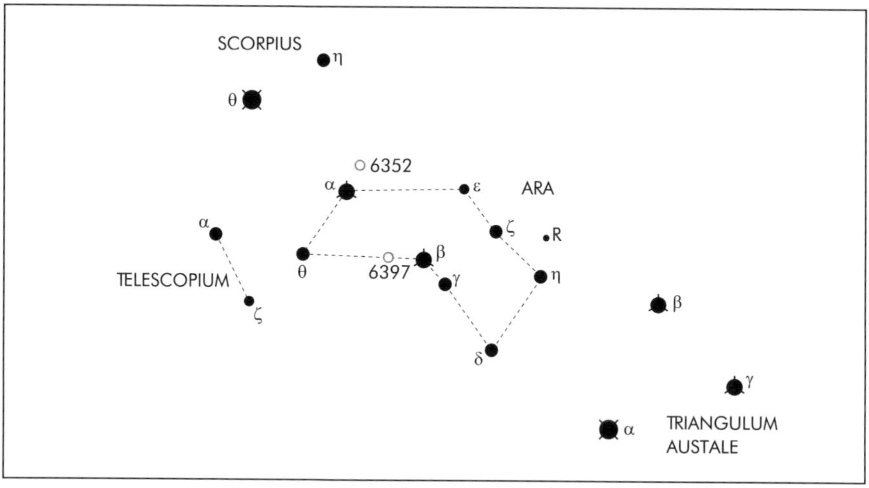

Ara					
Greek letter	Name	Magnitude	Luminosity (Sun=1)	Distance (light-years)	Spectrum
α Alpha	Choo	2.9	450	190	B3
β Beta	–	2.8	5000	780	K3
γ Gamma	–	3.6	5000	1075	B1
δ Delta	–	3.6	105	95	B8
ζ Zeta	–	3.1	110	137	K5
η Eta	–	3.8	110	190	K5
θ Theta	–	3.8	9000	1570	B1

Three of these (Zeta, Beta and Eta) are obviously orange in colour. In the same binocular field with Zeta is a variable star, R Aræ, which is not intrinsically variable at all, but is an eclipsing binary of the Algol type (see 13 November). Its period is 4.4 days, and as the magnitude range is from 6.0 to 6.9 it can always be seen with binoculars. It makes up a small triangle with Zeta and Eta.

Ara contains several clusters within binocular range. The most notable is C86 (NGC 6397), close to the Beta–Gamma pair. It is an easy binocular object, and is less condensed than most globulars, so that it is comparatively easy to resolve except near its centre. It is also much smaller than most clusters of its type, and at its estimated distance of 7000 light-years it is probably the closest to us of all the globulars. Its leading stars are old red giants, well advanced in their life-stories, and it seems that C86 must be very ancient even by globular cluster standards. C81 (NGC 6352) is a fainter globular cluster, near Alpha, and there are also several open clusters. Part of Ara is crossed by the Milky Way, and the whole area is decidedly rich.

April 30

The Moon and Easter

Easter is one of the public holidays which is not fixed on a definite date in the year, as most are; it depends on the Moon. This was decided in the year 325 at a major religious conference, the Council of Nicæa, after a great deal of squabbling between the various Church leaders.

The rule is: Easter Sunday is the first Sunday after the full moon which happens upon or next after the Vernal Equinox (see 22 January). If the full moon occurs on a Sunday, then Easter day is the following Sunday. It can be shown that Easter can therefore fall on any date between March 22 and April 25 inclusive.

There are, of course, many popular customs associated with Easter quite apart from the religious aspect. For example, in nineteenth-century Britain it was widely believed that the Sun dances for joy at dawn on Easter Sunday, and this led to the custom of going to the hills at dawn to take part in the celebration.

Here are the dates of full moon around equinox time for the period covered in this book. From these, it is easy to work out the date of Easter.

1998:	March 13, April 11
1999:	March 31, April 30
2000:	March 20, April 18
2001:	March 9, April 8
2002:	March 28, April 27
2003:	March 18, April 16

There have been various attempts to make Easter a fixed day in the calendar year, but so far no action has been taken.

May

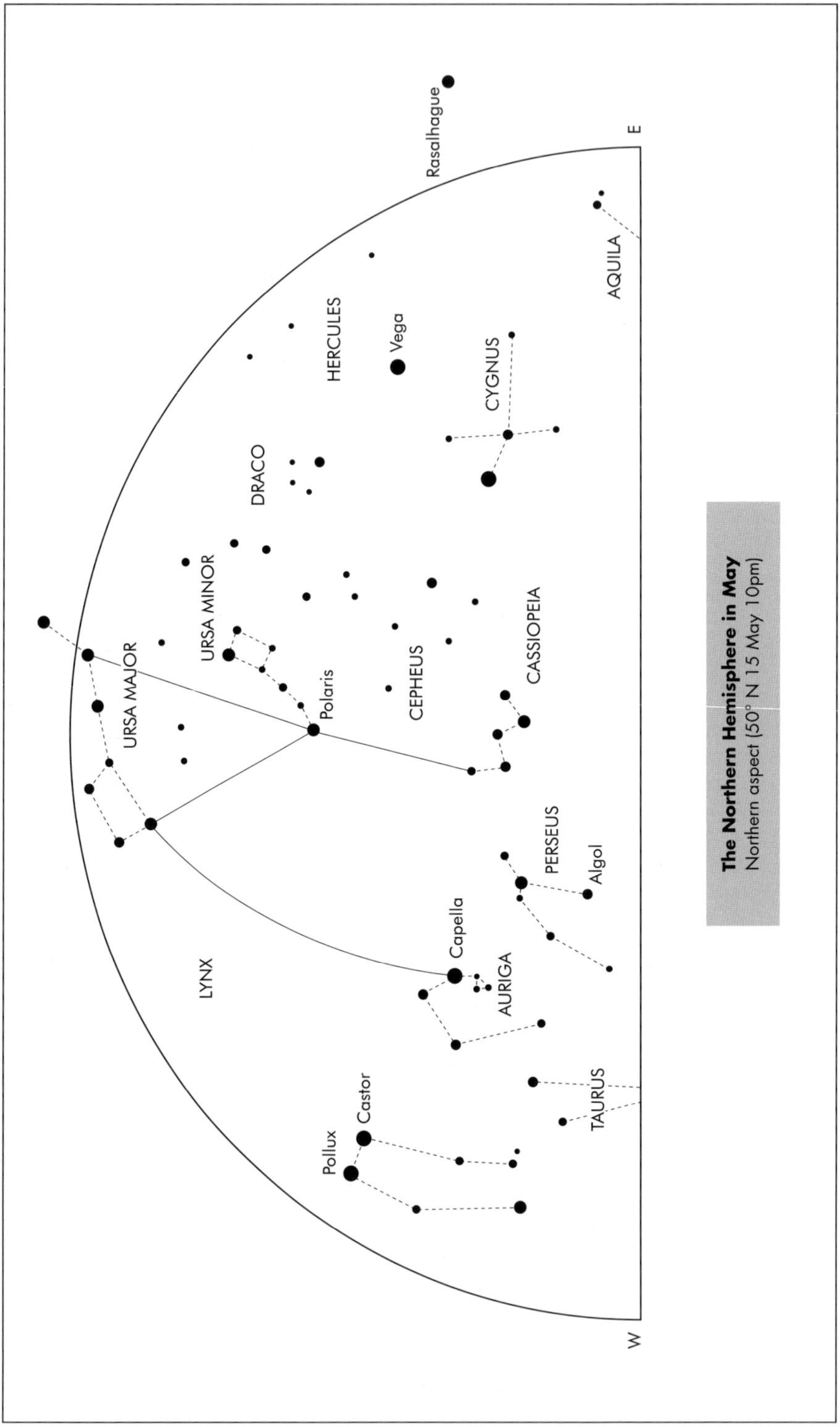

The Northern Hemisphere in May
Northern aspect (50° N 15 May 10pm)

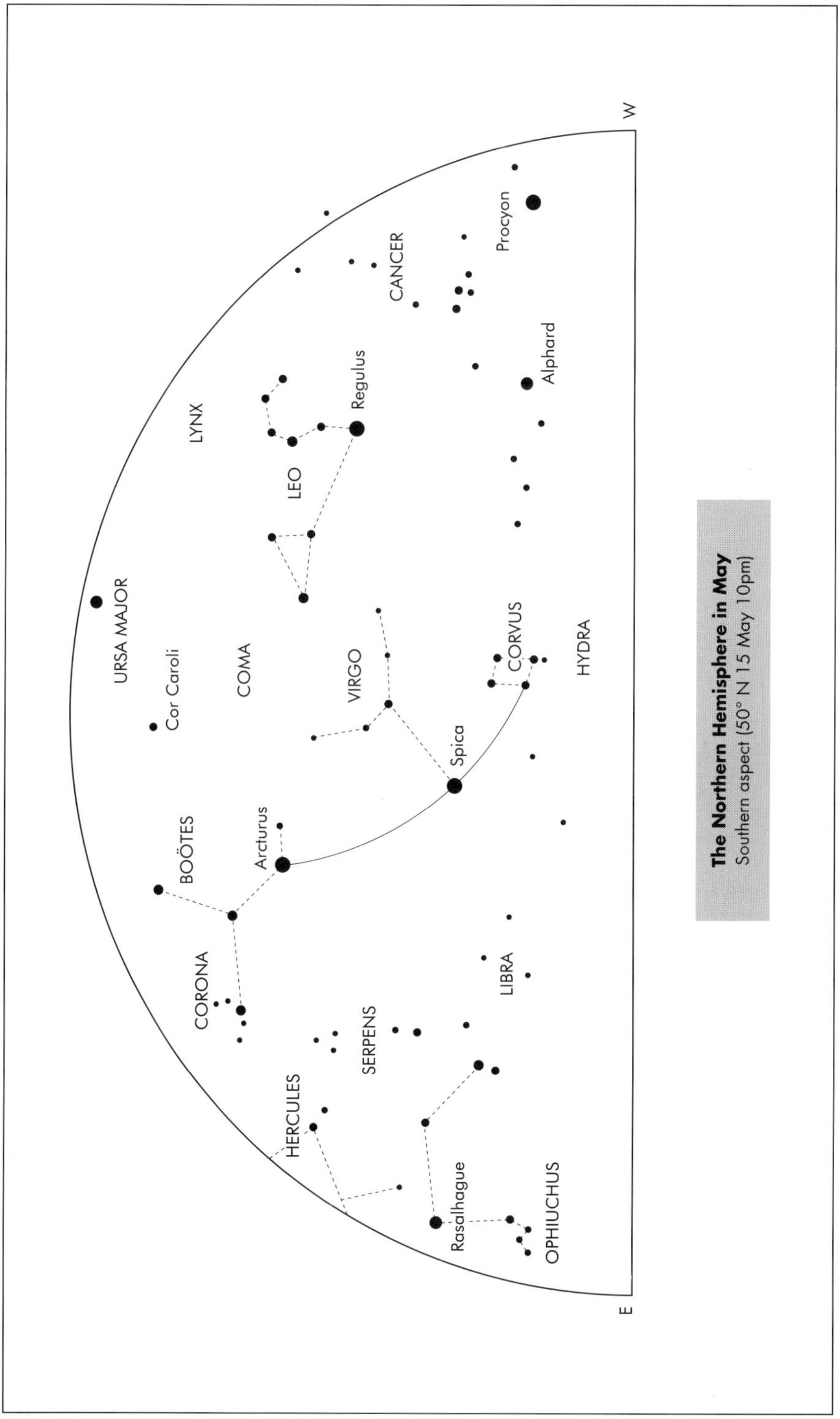

The Northern Hemisphere in May
Southern aspect (50° N 15 May 10pm)

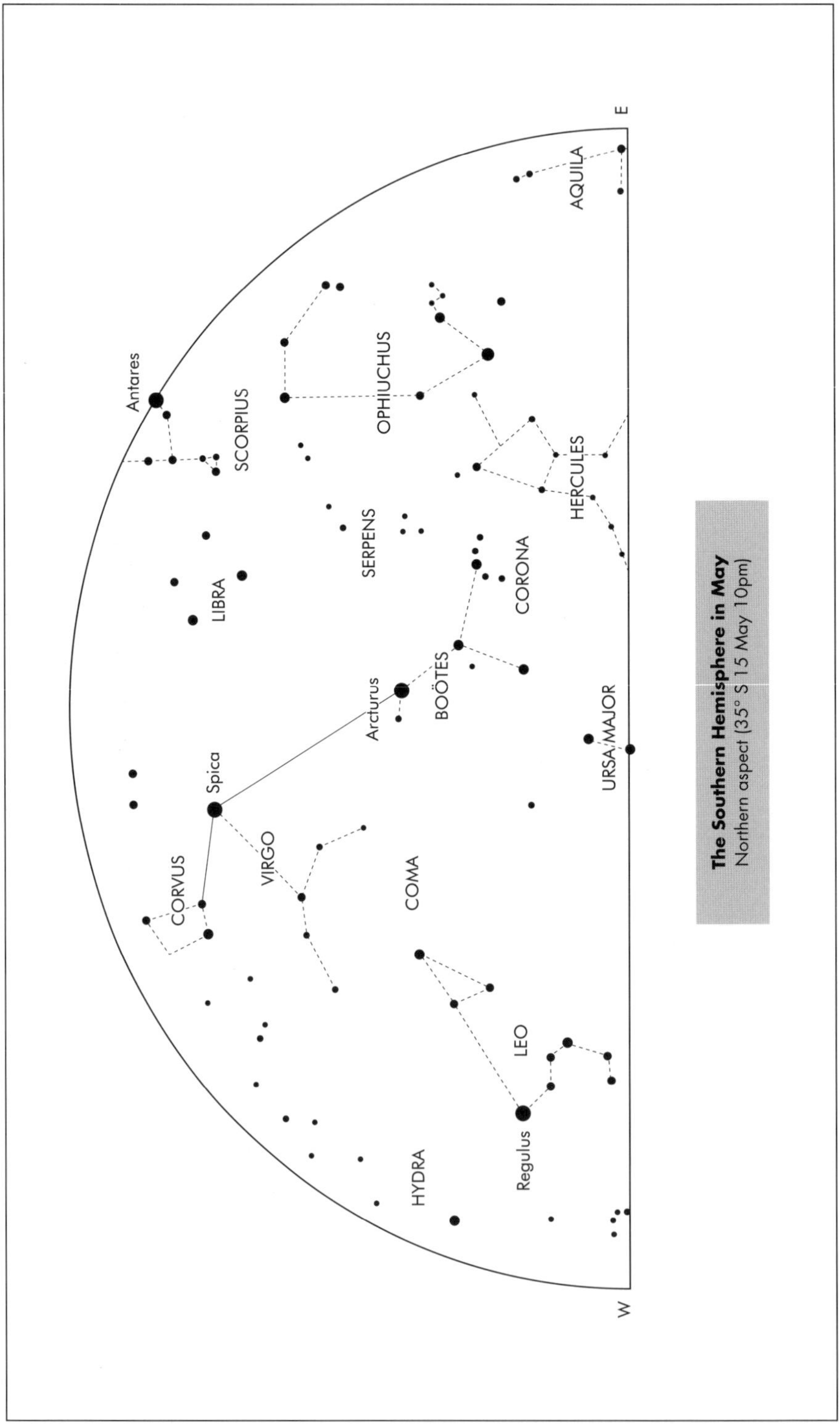

The Southern Hemisphere in May
Northern aspect (35° S 15 May 10pm)

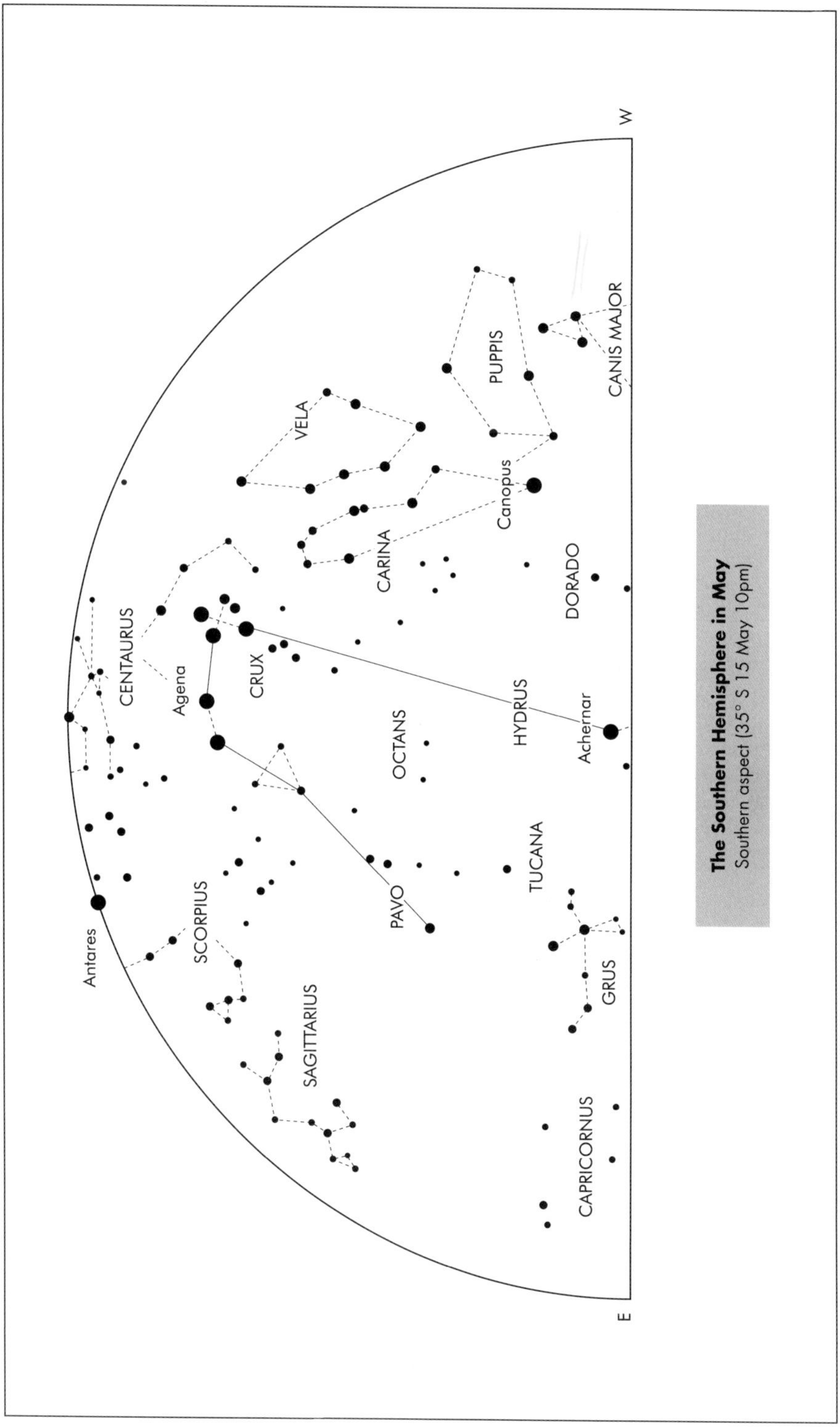

The Southern Hemisphere in May
Southern aspect (35° S 15 May 10pm)

May 1

The May Sky

In the northern hemisphere we are coming to the shorter nights; moreover we have lost Orion, and most of his retinue as well, though Procyon, the Twins and Capella can still be seen after sunset. (Capella never actually sets over Britain, though at its lowest it skirts the horizon.) Leo remains prominent, now to the west of south, and is followed round by Virgo, which is one of the largest of all the constellations even if by no means among the brightest. The Great Bear is almost overhead, and in the east Arcturus in Boötes, the Herdsman, is very prominent – it is actually the most brilliant star in the northern hemisphere of the sky, marginally outshining Vega and Capella. Vega is gaining altitude in the north-east, and Cygnus is conspicuous; Altair in Aquila comes into view later in the evening. The W of Cassiopeia is at its lowest in the north. Much of the eastern aspect is occupied by large, dim constellations: Hercules, Ophiuchus and Serpens.

Southern observers have also lost Orion, but they do have the glorious Scorpion, now high in the east; following it is Sagittarius, the Archer, with its lovely star-clouds. The Southern Cross is near the zenith, and Leo is disappearing in the west; Virgo is high, with Arcturus at a respectable height above the horizon. Canopus is dipping in the south-west, and the Magellanic Clouds are now below the pole. Very low in the north, observers in parts of South Africa and Australia (not New Zealand) should just be able to make out some of the stars of Ursa Major.

May 2

Tracing the Zodiac

The ecliptic may be defined as the projection of the Earth's orbit on to the celestial sphere – though an easier way of putting it is to say that it is marked by the apparent yearly path of the Sun among the stars. It passes through twelve accepted Zodiacal constellations: Aries (the Ram), Taurus (the Bull), Gemini (the Twins), Cancer (the Crab), Leo (the Lion), Virgo (the Virgin), Libra (the Balance or Scales), Scorpius (the Scorpion), Sagittarius (the Archer), Capricornus (the Sea-goat), Aquarius (the Water-bearer) and Pisces (the Fishes). A thirteenth constellation, Ophiuchus (the Serpent-bearer) intrudes into the Zodiac between Scorpius and Sagittarius. These constellations are very unequal in size and brilliance; some, such as Taurus and Scorpius, are magnificent in every way, while Cancer and Pisces are very obscure indeed. Astrologers are obsessed with the Zodiac – blissfully regardless of the fact that the constellation patterns are entirely arbitrary and mean nothing at all!

Now is a good time to trace the Zodiacal band. Begin in the west; the line passes south of Castor and Pollux. It then crosses Cancer, near Præsepe, and past Regulus in Leo; it lies south of the Leo triangle, and then past Spica in Virgo. It crosses the Balance, close to the star Zubenelgenubi or Alpha Libræ, and on into the Scorpion, just north of Antares; thence into Sagittarius, slightly south of the bright star Nunki or Sigma Sagittarii (17 July), and into Capricornus. This is as far as we can follow it tonight.

May 3

Areas of the Zodiacal Constellations

We have been tracing the constellations of the Zodiac; let us now pause to see how they compare with each other. Here is a list giving the areas of the constellations, together with 'star density', that is to say, the number of stars above the fifth magnitude per 100 square degrees.

The Zodiacal Constellations

Constellation	Area (square degrees)	Star density
Aries	441	2.5
Taurus	797	5.5
Gemini	514	4.5
Cancer	506	1.2
Leo	947	2.7
Virgo	1294	2.0
Libra	538	2.4
Scorpius	497	7.6
Sagittarius	867	3.8
Capricornus	414	3.9
Aquarius	980	3.2
Pisces	889	2.7
(Ophiuchus	948	3.8)

Therefore Scorpius is easily the richest of the Zodiacal groups, and Cancer the most barren. In the entire sky, the Southern Cross is the smallest constellation (68 square degrees) and has the highest star density (19.1). The most obscure of all the constellations is the southern Mensa (the Table), which has no star as bright as fifth magnitude, though it is redeemed by the fact that a small part of the Large Cloud of Magellan extends into it.

May 4

Elusive Mercury

Around this date in 1998, the planet Mercury will be well placed in the morning sky before dawn; in 2002 it will be visible in the evening sky after sunset. If there are no clouds or mist near the horizon, Mercury should be easily visible with the naked eye. This is not the case in other years; Mercury is quick-moving, which is why the ancients named it in honour of the messenger of the gods. The best times for observing it during the period covered in this book are:

Mornings

1998 6 Jan, 4 May, 31 Aug, 20 Dec
1999 16 Apr, 14 Aug, 3 Dec
2000 28 Mar, 27 July, 15 Nov
2001 11 Mar, 9 July, 29 Oct
2002 21 Feb, 21 June, 13 Oct
2003 4 Feb, 3 June, 27 Sept

Evenings

1998 20 Mar, 17 July, 11 Nov
1999 3 Mar, 28 June, 24 Oct
2000 15 Feb, 9 June, 6 Oct
2001 28 Jan, 22 May, 18 Sept
2002 11 Jan, 4 May, 1 Sept, 26 Dec
2003 16 Apr, 14 Aug, 9 Dec

Mercury is close to the Sun, moving at a mean distance of 36 000 000 miles; the orbital period is 88 days. It is small, with a diameter of only 3030 miles, as against 2160 miles for the Moon and 4219 miles for Mars. It has a low surface gravity, and practically no atmosphere. It shows phases from new to full in the same way as the Moon, and for much the same reason; at elongation – on the dates given above – it appears as a half-disk (dichotomy). In the sky it can never be as much as 30° of arc away from the Sun, so that with the naked eye it always has to be seen against a bright background. Before the first space-probe went there, we knew very little about its surface features.

When found, Mercury is brighter than might be expected; it outshines even Sirius, and can attain magnitude −1.9, though of course it can never be seen to advantage.

Future Points of Interest

1998: Mercury at western elongation.

2002: Mercury at eastern elongation.

Anniversary

1991: Fall of the Glatton Meteorite.

Future Points of Interest

Maximum of the Eta Aquarid meteor shower, which begins officially on 24 April and ends on 20 May. The maximum is rather broad; the ZHR can reach 35 to northern-hemisphere observers and 50 from southern countries, where the radiant is higher in the sky. The Eta Aquarids are associated with Halley's Comet.

May 5

Meteorites over Britain

It is quite wrong to associate meteors with meteorites. Meteors are cometary débris, and burn away when they are still at least 40 miles above the ground. Meteorites come in the main from the asteroid belt, and in fact there is no difference between a large meteorite and a small asteroid – except that the term 'meteorite' is used only for an object which has landed on the Earth.

The latest fall over Britain was that of 5 May 1991, at Glatton, Cambridgeshire. The fall was seen by a local man, A. Pettifor, who located the object in a shallow depression; it missed him by about sixty feet. It was the ninth 'British meteorite' of the twentieth century; the full list is given below.

The largest known meteorite is still lying where it fell, in prehistoric times, at Hoba West, near Grootfontein in Namibia (southern Africa). It weighs at least 60 tons.

British Meteorites

Date	Location	Weight (kg)
1902 September 13	Crumlin, Northern Ireland	4.1
1914 October 13	Appley Bridge, Lancashire	33
1917 December 3	Strathmore, Perthshire, Scotland	13 (4 stones)
1923 March 9	Ashdon, Essex	0.9
1931 April 14	Pontlyfni, Wales	120 g
1949 September 21	Beddgelert, Wales	723 g
1965 December 24	Barwell, Leicestershire	46
1969 April 25	Bovedy, Northern Ireland	? (Main mass fell in the sea)
1991 May 5	Glatton, Cambridgeshire	767 g

Future Points of Interest

2002: Penumbral eclipse of the Moon. The time of greatest obscuration is 12 h 0.5 m GMT, when 69% of the surface will be contained in the penumbra. The eclipse will not be seen from anywhere in Britain.

May 6

Mapping Mercury

For obvious reasons, Mercury is difficult to study from Earth. It is small; it never comes much within 50 000 000 miles of us, and when visible with the naked eye it is always low over the horizon, so that seeing conditions are bound to be poor. The best pre-Space-Age map was compiled by E.M. Antoniadi, using the powerful 33-inch refractor at the Meudon Observatory, near Paris; his map was published in 1934, but was not good enough for his nomenclature to be retained.

The first (and so far, the only) space-craft to encounter Mercury has been Mariner 10, which made three active passes of the planet on 29 March and 21 September 1974, and 16 March 1975. It had been launched on 3 November 1973, and had bypassed Venus on 5 February 1974 en route for its rendezvous with Mercury.

As expected, the Mercurian surface looked superficially rather like that of the Moon. There are craters, mountains and valleys, and one huge basin, the Caloris Basin, which is over 800 miles in diameter and is bounded by a ring of smooth mountain blocks rising to more than a mile above the surrounding landscape.

Unfortunately Mariner 10 could map less than half of the total surface; the same areas were in sunlight during each active pass. Contact with the probe was finally lost on 24 March 1975. No doubt it is still in solar orbit, and still making periodical passes of Mercury, but we have no hope of finding it again.

May 7

Transits of Mercury

Mercury and Venus, the two planets closer to the sun than we are, can sometimes pass in transit against the solar disk. This does not happen at every inferior conjunction, because their orbits are inclined to that of the Earth – by 7° in the case of Mercury, and 3° 23′ for Venus. Transits of Venus are very rare; the last occurred in 1874 and 1882, while the next will be in 2004 and 2012. Mercury, however, transits more often. Recent and future transits are on:

1970	May 9
1973	November 10
1986	November 13
1993	November 6
1999	November 15
2003	May 7
2006	November 8
2016	May 9
2019	November 11

Transits of Mercury can occur only in May and November. May transits occur with Mercury near aphelion; November transits of Mercury are near perihelion, and November transits are the more frequent in the ratio of 7 to 3. The longest transits (those of May) may last for almost 9 hours. The time of mid-transit on 7 May 2003 is 7 h 53 m GMT.

During transit, the planet will appear as a black disk against the Sun. When Venus crosses the Sun, its atmosphere

Future Points of Interest

2003: Transit of Mercury.

makes its edges appear blurred; not so with Mercury, where the atmosphere is so thin that it can be disregarded. Mercury is jet-black, and this shows up when it is compared with any sunspot which happens to be visible. If a sunspot could be seen shining on its own, its surface brightness would exceed that of an arc-lamp.

During transit Mercury cannot be seen with the naked eye, and so the best method of observing it is to project the image on to a screen held or fastened behind the eyepiece of the telescope.

Future Points of Interest

1999: Occultation of Uranus by the Moon (7 h GMT).

May 8

Virgo

To both northern and southern observers Virgo, the Virgin, is now prominent in the evening sky. It is the second largest of all the constellations; it covers 1294 square degrees, slightly more than Ursa Major (1280 square degrees). Only Hydra (1303 square degrees) is larger. In mythology Virgo represents Astræa, the goddess of justice, daughter of Jupiter and Themis.

There is only one first-magnitude star, Spica, but there are several more which are bright enough to be conspicuous, and the shape of the constellation makes it easy to identify. The leaders are listed below.

To locate Virgo, northern observers can take a line from the tail of the Great Bear past Arcturus; continued for some distance and curved somewhat, this will lead to Spica.

Virgo is Y-shaped, with Spica at the base of the Y; the 'bowl' is bounded on the far side by Denebola in Leo (2 April), and is crowded with faint galaxies. Virgo is, of course, a Zodiacal constellation, and is crossed by the celestial equator; both Zeta and Gamma lie within 2° of the equatorial line.

Virgo

Greek letter	Name	Magnitude	Luminosity (Sun=1)	Distance (light-years)	Spectrum
α Alpha	Spica	1.0	2100	260	B1
β Beta	Zavijava	3.6	3.4	33	F8
γ Gamma	Arich	2.7	7	36	F0 + F0
δ Delta	Minelauva	3.4	120	147	M3
ϵ Epsilon	Vindemiatrix	2.8	60	104	G9
ζ Zeta	Heze	3.4	17	75	A3
η Eta	Zaniah	3.9	26	104	A2

May 9

Spica

As Virgo is our 'constellation of the week' let us turn now to its leader, Spica, which is so close to the ecliptic that it can at times be occulted by the Moon. It has had several proper names in various countries; for instance 'Salkim' in Turkey and 'Shebbelta' in Syria, each of which can be loosely translated as 'the Ear of Wheat'. It lies in a rather sparse region, and was sometimes called 'Al Simak al Azel', or the Defenceless One, in contrast to the other Simak, Arcturus, which was provided with a lance. (See map, June 12).

Telescopically, Spica appears as a single white star, but in 1889 it was found to be a very close binary. The primary has a diameter of around 9 500 000 miles and a mass 11 times that of the Sun; the secondary is about 5 000 000 miles across and four times as massive as the Sun. The real separation is about 11 000 000 miles, so that no ordinary telescope will split them; Spica is an excellent example of a spectroscopic binary. The orbital period is just over four days. At each revolution a small part of the primary is blocked out by the fainter star, and this causes a slight drop in brilliancy; moreover the primary is itself slightly variable. However, the combined effects are so slight that no changes can be detected without sensitive measuring equipment.

Spica is not very far from the celestial equator; its declination is 11° S, so that it can be seen from every inhabited part of the world.

May 10

Arich

The star at the bottom of the 'bowl' of Virgo is Gamma Virginis, Arich (also known as Porrima or Postvarta). With the naked eye it looks unremarkable; it is in fact a famous binary, and is still a spectacular object even though not to the same extent that it was a few decades ago.

The two components are identical twins; each is of type F, and therefore yellowish-white, and each is considerably more luminous than the Sun. The orbital period is 171.4 years, and the orbit is decidedly eccentric; the real separation between the two ranges from 280 million miles to over 6500 million miles.

At present the apparent separation is still about 2 seconds of arc, but as time goes by we are seeing the pair at a less and less favourable angle, and by 2016 Gamma Virginis will appear single except in very large telescopes. After that it will start to open out once more. Early in the twentieth century it was one of the widest and easiest identical doubles in the sky. At 36 light-years, Arich is one of our nearer stellar neighbours.

Use a telescope to look at it tonight, and do the same in a few years' time. You will notice an obvious difference in both separation and in position angle.

May 11

Star Names

Last night something was said about Gamma Virginis, which has several alternative proper names. For example, the Romans knew it as Porrima or Postvarta – names of two ancient goddesses of prophecy. To the Arabs it was Zawait al Awwa, 'the Angle', and to the Chinese it was Shang Shang, the High Minister of State.

In fact, proper names of stars are hardly ever used nowadays except for those of above second magnitude, plus a few special cases such as Mizar (Zeta Ursæ Majoris) and Mira (Omicron Ceti). No new names are ever given. Unfortunately there are various unscrupulous bogus agencies which claim to be able to name stars – on payment of a sum of money, of course! These so-called names mean nothing at all, and all schemes of this sort should be avoided.

There are some stars which have, at various times, been given more than one proper name, and it may be diverting to list a few of these:

Alternative Star Names

Alpha Andromedæ	Alpheratz, Sirrah
Alpha Centauri	Toliman, Bundula, Rigel Kent
Beta Centauri	Agena, Hadar
Beta Ceti	Diphda, Deneb Kaitos
Gamma Cephei	Alrai, Errai
Alpha Coronæ Borealis	Alphekka, Gemma
Beta Draconis	Alwaid, Rastaban
Alpha Hydræ	Alphard, Cor Hydræ
Gamma Lyræ	Sulaphat, Jugum
Beta Ophiuchi	Cheleb, Celbalrai
Alpha Persei	Mirphak, Algenib
Tau Pegasi	Kerb, Salma
Alpha Piscium	Al Rischa, Kaïtain, Okda
Alpha Serpentis	Unukalhai, Cor Serpentis
Beta Scorpii	Graffias, Akrab
Gamma Velorum	Regor, Al Suhail al Muhlif
Gamma Virginis	Arich, Porrima, Postvarta
Gamma Ursæ Majoris	Phekda, Phad
Eta Ursæ Majoris	Alkaid, Benetnasch

There are many names which can be spelled in different ways (Betelgeux may be Betelgeuse or Betelgeuze, Vega may be Wega). Certainly some of the old names are attractive, even if somewhat cumbersome.

May 12

A Missing Link?

Still with Virgo, let us next turn our attention to a very insignificant-looking star: 70 Virginis. It lies near the edge of the constellation, close to the boundary with Coma, roughly between Epsilon Virginis and the brilliant orange Arcturus in Boötes (June 7). It is only of the fifth magnitude, and its spectral type is G4, so that it is not highly coloured; at first sight there is nothing to single it out. Its position is RA 13 h 28 m 26 s, declination +13° 46′ 43″.

70 Virginis is around 33 light-years away, which by cosmical standards is nearby. What makes it interesting is that delicate spectroscopic observations have shown that it is 'wobbling' slightly, and is being pulled upon by a companion which is too faint for us to see. However, this is not an

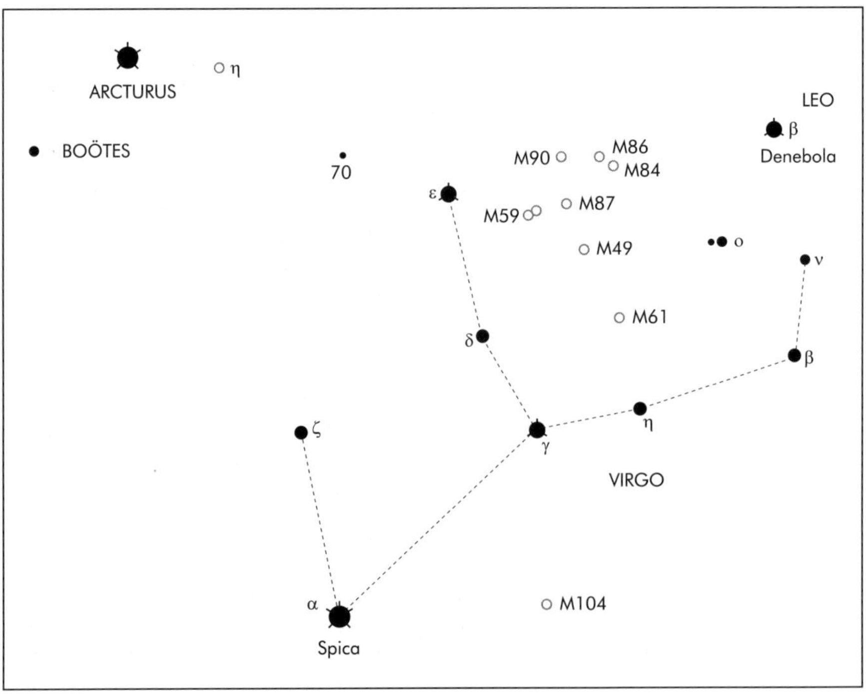

ordinary binary. The companion body seems to have a mass only 8 times that of Jupiter, the most massive planet in our Solar System. The orbital period is 116.7 days; the distance from the parent star is 40 000 000 miles (slightly greater than the distance between our Sun and Mercury), and the diameter of the companion is probably about the same as Jupiter's.

Just what is it? Probably not a true star; it is not massive enough. Its surface temperature is thought to be about 85 °C, which is not enough to make it radiate strongly in the optical range; it could be regarded as a heavyweight planet, but it may also be a kind of 'missing link' – not quite a star, not quite a planet. Bodies of this kind have been suspected for a long time, and are known, rather misleadingly, as brown dwarfs.

Until we can actually see it, uncertainties are bound to remain and it is possible that the observed effects are inherent in the star itself rather than due to an orbiting body. At any rate, it is worth seeking out 70 Virginis, which itself is not too unlike the Sun.

May 13

The Bowl of Virgo

The Bowl of Virgo is outlined by Epsilon, Delta, Gamma, Eta and Beta Virginis, and Denebola or Beta Leonis (6 April). This is an area which is crowded with galaxies. We are in fact looking at a genuine grouping – the famous Virgo Cluster of galaxies, a huge aggregation of systems at a mean distance of around 50 000 000 light-years. Our own Galaxy is a member of a cluster, the Local Group, which includes the Andromeda and Triangulum spirals and the Clouds of Magellan; but the Local Group is very insignificant compared with the giant, populous Virgo Cluster.

Telescopic observers enjoy themselves in the Bowl. There are a dozen Messier objects. There are several other galaxies with integrated magnitudes of above 10, but to find and identify them means using a detailed chart and, preferably, a telescope equipped with setting circles. Do not confuse the Virgo Cluster with the open star-cluster of Coma Berenices (19 May), which lies close to it in the sky but is a relatively near neighbour in our own Galaxy.

Also in Virgo, outside the Bowl, is M104 (NGC 4594), known as the Sombrero Hat Galaxy from its appearance; it is crossed by a dark band of obscuring material which really does recall a South American hat, though it has also been nicknamed the Saturn Nebula because of a slight resemblance to the form of Saturn's rings.

Galaxies in Virgo

M	NGC	RA h	RA m	Dec. °	Dec. ′	Magnitude	Type
61	4303	12	21.9	+04	28	9.7	Spiral
85	4382	12	22.9	+18	28	9.3	Spiral
84	4374	12	25.1	+12	53	9.3	Elliptical
86	4406	12	26.2	+12	57	9.2	Elliptical
49	4472	12	29.8	+08	00	8.4	Elliptical
87	4486	12	30.8	+12	24	8.6	Elliptical (Virgo A)
89	4552	12	35.7	+12	33	9.8	Elliptical
90	4569	12	36.8	+13	10	9.5	Spiral
58	4579	12	37.7	+11	49	9.8	Spiral
59	4621	12	42.0	+11	39	9.8	Elliptical
60	4649	12	43.7	+11	33	8.8	Elliptical

May 14

M87: a Giant Radio Galaxy

The most massive galaxy in the Virgo Cluster is M87 (NGC 4486). This is not a spiral; it is an elliptical (type E1), and from it issues a curious jet of material several light-years long. Visually it is none too easy to identify, because it lies in a crowded area, but of its exceptional importance there is no doubt at all. Its mass has been estimated as at least 300 thousand million times that of the Sun, so that it is a giant by any standards. It is surrounded by a halo of at least 800 globular clusters, many more than are associated with our Galaxy. The real diameter of M87 is of the order of 120 000 light-years, and it is about 50 000 000 light-years away from us.

Measurements of the motions of the stars near the centre of the system show that they are moving at tremendous speeds round the core. From this it is generally agreed that the core contains a gigantic black hole (20 July). Moreover, M87 is a powerful source of radio waves, so that radio astronomers refer to it as Virgo A.

Radio waves, like light, are electromagnetic vibrations; they are of long wavelength, and do not affect our eyes, so that they are collected by instruments known as radio telescopes. The name is somewhat misleading, because a radio telescope is in the nature of a large aerial, and it does not produce a visible picture; one certainly cannot look through it. Radio telescopes are of various forms; some are 'dishes', of which the most famous is probably the 250-foot Lovell Telescope at Jodrell Bank in Cheshire. It has now been in use for over 40 years. It is named in honour of Sir Bernard Lovell, who was responsible for building it.

Anniversary

1973: Launch of *Skylab*, the American space-station which was manned by three successive crews, and remained in orbit until 1979.

May 15

Minelauva and Stellar Evolution

Before we take our temporary leave of the Virgin, look at Delta Virginis or Minelauva, on the border of the Bowl. It is a lovely red star of type M, 147 light-years away and 120 times as luminous as the Sun. Some way from it, making a triangle with it and Spica, is Zeta Virginis or Heze, which is white, 75 light-years away, and only 17 times as powerful as the Sun. Look at them alternately, with binoculars or even with the naked eye, and you will see the difference in colour.

Minelauva, of course, has the lower surface temperature; it is a red giant. When the colours of the stars were first studied carefully, it was thought – correctly – that the colour of a star depends upon its age. Unfortunately, one very fundamental mistake was made.

It was assumed that when a star condensed out of the gas and dust in a nebula, it was large, cool and red. It shrank, under the influence of gravity, and heated up, becoming hot and white or bluish-white. It then began to cool down again, and became in turn orange, yellow and red before losing its light and heat. On this picture, a red giant such as Minelauva would be young.

But, as we have seen when discussing Betelgeux in Orion (15 January), this is wrong. A star does indeed condense out of nebular material and heat up, becoming hot and white (if it has sufficient initial mass), but it is using up its nuclear 'fuel' all the time, and when this starts to run low the star has to change its structure, blowing out and becoming large and cool. This is what has already happened to Minelauva, so that instead of being less advanced in its life-story than a white star, such as Heze, it is 'older'.

May 16

Future Points of Interest

2003: Total eclipse of the Moon. Mid-eclipse, 03.41 GMT. Totality lasts for 26 minutes, and the partial phase for 1 h 37 m.

Nu Virginis: Red Star in the Virgin

Let us pause to locate another red star in Virgo – Nu Virginis, which lies between Beta or Zavijava, at the end of the 'Bowl', and Denebola in Leo. Nu Virginis is of magnitude 4.0, so that it is far from conspicuous. It has an M-type spectrum, and like Minelauva or Delta Virginis (15 May) its colour is very evident in binoculars; with the naked eye the redness is less obvious, because of the star's comparative faintness.

In fact, it is just about the same luminosity as Minelauva – 120 times that of the Sun – but at 166 light-years it is further away. It is interesting to compare the two.

May 17

Close Planetary Conjunctions

Planetary conjunctions are not in the least important – they are mere line of sight effects – but when two brilliant planets are concerned they can be spectacular. Generally, the brightest planets are Venus and Jupiter, and at the conjunction of 17 May 2000 the separation between the two will be only 42 seconds of arc. Astronomical photographers will certainly be on the alert for some days to either side of closest approach.

On 3 January 1818 Venus actually occulted Jupiter. This will not happen again until 22 November 2065.

The only really close planetary conjunctions during the period covered by this book are:

2000 March 4	Venus–Uranus. Minimum separation 234 seconds of arc (00.12 GMT)
2000 May 17	Venus–Jupiter. Minimum separation 42 seconds of arc (10 h 35 m GMT)
2000 August 10	Mercury–Mars. Minimum separation 289 seconds of arc (12.59 GMT)
2003 March 28	Venus–Uranus. Minimum separation 157 seconds of arc (13.03 GMT)

Between 29 October and 4 November 2004 Mercury and Venus will be less than one degree apart, and on 6 December 2002 Venus will be less than two degrees from Mars.

Anniversary

1836: Birth of Sir (Joseph) Norman Lockyer. Lockyer is best remembered for his discovery, in 1868, that the solar prominences can be observed spectroscopically at any time, without waiting for a total eclipse. He founded the Norman Lockyer Observatory at Sidmouth in Devon, which is still in full operation, but perhaps his major contribution to science was his founding of the periodical *Nature*. He died in 1920.

Future Points of Interest

2000: Close conjunction of Venus and Jupiter.

May 18

The Legend of Berenice's Hair

In the north of the sky, in the region bounded by the Great Bear, Arcturus and the Bowl of Virgo, can be seen what looks like a large, faint star-cluster. It is very obvious as seen from northern latitudes, though from countries such as Australia it is always low down. This is Coma Berenices, or Berenice's Hair. It is not one of the original 48 constellations listed by Ptolemy, but there is a legend attached to it.

It is said that around 247 BC Berenice, the daughter of the King of Cyrene, married the Pharaoh Ptolemy III of Egypt, with whom she was very much in love. Two years later Ptolemy had to set off on a campaign against the Syrians. Berenice, naturally, was very anxious, and so she went to the Temple at Zephyrium and made a solemn vow that if her

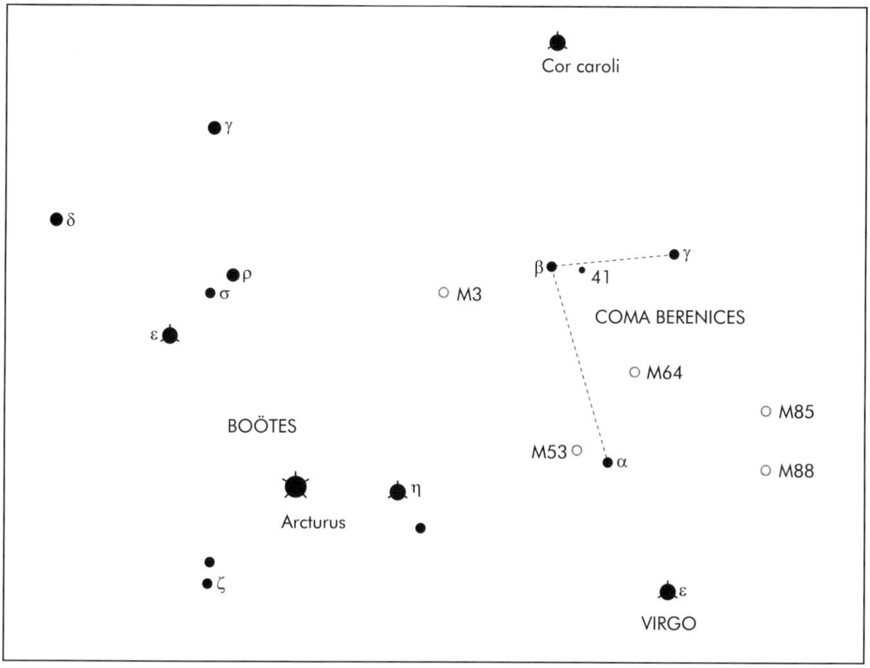

husband returned unharmed, she would cut off her lovely golden hair and place it on the altar of Venus. There are two versions of what happened next. One is that Ptolemy did return, and Venus transferred the hair to the sky. The other is that someone stole the hair from the temple, and the astronomer of the time, Conon, placated Berenice by explaining that the hair had been changed into stars.

Originally the constellation may have been part of Leo's bushy tail, or even part of the sheaf of grain carried by Astræa (Virgo). However, Ptolemy – Claudius Ptolemy, the astronomer, not the Pharaoh of Egypt – referred to the stars as 'hair'. The first full use of the name Coma Berenices was due to the Danish astronomer Tycho Brahe, who officially added the constellation to the sky in 1590.

May 19

The Coma Cluster

Coma looks like a star-cluster, and indeed it is. It consists of a very scattered group of naked eye stars, about midway between Alkaid in the Great Bear and Denebola in Leo; it occupies the space enclosed by the large triangle formed by Arcturus, Denebola and Cor Caroli. It is not listed either by

Messier or in the NGC, but some catalogues refer to it as Melotte 111. Altogether it covers an area about 5° in diameter. There are more than a dozen stars above the fifth magnitude, though a few of them are in the foreground and are simply superimposed against the cluster. The brightest star in the constellation is Alpha or Diadem: RA 13 h 10 m, declination +17° 32′. It is a very close binary; the components are almost equal, and the period is 26 years, but as the separation is never more than 0.3 of a second of arc this is a very difficult object. Each component is of type F. Alpha Comæ forms a triangle with Eta Boötis (7 June) and Epsilon Virginis (10 May), and is not hard to identify, as there are no other naked eye stars close to it.

Anniversary

1961: Probable pass of Venus by the Russian space-craft Venera 1. This was the first of all interplanetary probes, and was launched on 12 February 1961. Unfortunately contact with it was lost at a distance of less than 5 000 000 miles. It is likely that on the following 19 May the probe passed Venus at about 60 000 miles, but we will never really know.

May 20

Nebular Objects in Coma

Coma contains some interesting objects. In particular there is a bright globular cluster, M53, close to Alpha; its position is RA 13 h 13 m, declination +18° 10′, actually in the same binocular field with Alpha. It can be glimpsed with binoculars, and telescopes show it to be a beautiful, rather small system, with the stars so closely packed toward its core that they cannot be resolved. It is 69 000 light-years away, and of course far more remote than the Coma cluster.

Also in the background are many galaxies, including five in Messier's list and three in the Caldwell list.

The brightest of these galaxies is M64, the 'Black-Eye Galaxy', which is visible with binoculars and lies within a degree of the fifth-magnitude star 35 Comæ. It is about 44 000 000 light-years away. The nickname comes from the dark region north of its centre, but a telescope of fair aperture is needed to show the details well. M64 was discovered by J.E. Bode as long ago as 1779.

Galaxies in Coma

M	C	NGC	RA h	RA m	Dec. °	Dec. ′	Magnitude	Type
98		4192	12	13.8	+14	54	10.1	Spiral
99		4254	12	18.8	+14	25	9	Spiral
100		4321	12	22.9	+15	49	9.4	Spiral
88		4501	12	32.0	+14	25	9.5	Barred spiral
	36	4559	12	36.0	+27	58	9.8	Spiral
	38	4565	12	36.3	+25	59	9.6	Spiral
64		4826	12	56.7	+21	41	6.6	Spiral
	35	4889	13	00.1	+27	59	11.4	Elliptical

May 21

The Sun – from Beta Comæ

Beta Comæ, the second star in Berenice's Hair, is very undistinguished in apprarance; its magnitude is 4.3. Beside it is a fainter star, 41 Comæ, of magnitude 4.8. There is no real connection between the two. Beta, at 27 light-years, is one of our nearer neighbours; 41 lies in the background, at 360 light-years. They also differ in colour. Beta has an F5-type spectrum, and is not too unlike the Sun, while 41 is a K-type orange giant.

The absolute magnitude of a star is the apparent magnitude that it would have if it could be seen from a standard distance of 32.6 light-years. The absolute magnitude of the Sun is +4.8; of Beta Comæ, +4.7. This means that if we could observe the Sun from a planet in the Beta Comæ system, it would look marginally brighter than Beta Comæ does to us. But from this standard distance, 41 Comæ would shine at magnitude –0.3, and would outrank every star in the sky apart from Sirius and Canopus!

Look at Beta and 41 with binoculars, and it is not easy to appreciate how diverse they really are. As so often in astronomy, superficial appearance can be highly misleading.

Future Points of Interest

2001: Mercury at eastern elongation.

May 22

The Twilight Zone of Mercury

Around 22 May, Mercury is well placed as a morning object in 1997 and as an evening object in 2001. This may therefore be the moment to say something about its fascinating but, alas, non-existent 'twilight zone.'

The first serious attempt to map Mercury was made between 1881 and 1889 by the Italian observer G.V. Schiaparelli, using a fine nine-inch refractor from Milan. As we have seen (6 May) he worked during the daytime, when both Mercury and the Sun were high in the sky. He came to the interesting conclusion that the rotation period was the same as the orbital period: 88 Earth days. If so, the rotation would be synchronous, and Mercury would keep the same face turned toward the Sun – just as the Moon does with respect to the Earth. E.M. Antoniadi, using the great Meudon 33-inch refractor in the 1920s and 1930s, came to the same conclusion.

However, the situation would not be quite straightforward, because Mercury's orbit is not circular; it is decidedly elliptical.

This means that there would be a slight, slow 'rocking' motion, leading to an area of permanent day, an area of permanent night, and in between a 'twilight zone', from which the Sun would bob up and down over the horizon. Science fiction writers made great use of the twilight zone; it would be the only part of the planet which would be neither impossibly hot or unbearably cold.

Then, in the 1960s, infra-red observations showed that the night side is not nearly so cold as it would be if it never received any sunlight. The real rotation period is not 88 Earth days, but 58.6 Earth days, or two-thirds of a Mercurian year, and there is no twilight zone. Temperatures are extreme, ranging from a maximum of +350 °C down to a minimum of −180 °C. Obviously, life of any kind is out of the question.

May 23

Corvus

Our next constellation is Corvus, the Crow, which is surprisingly prominent even though it has no star above magnitude 2.5. It lies on the border of Hydra, not far from Spica in Virgo. During evenings in late May it is well above the southern horizon for observers in British latitudes, and to Australians and New Zealanders it is not very far from the zenith.

Corvus is one of the original 48 constellations listed by Ptolemy, and there are several mythological legends associated with it. According to one version, the crow was called in by the god Apollo, who had fallen in love with Coronis, the

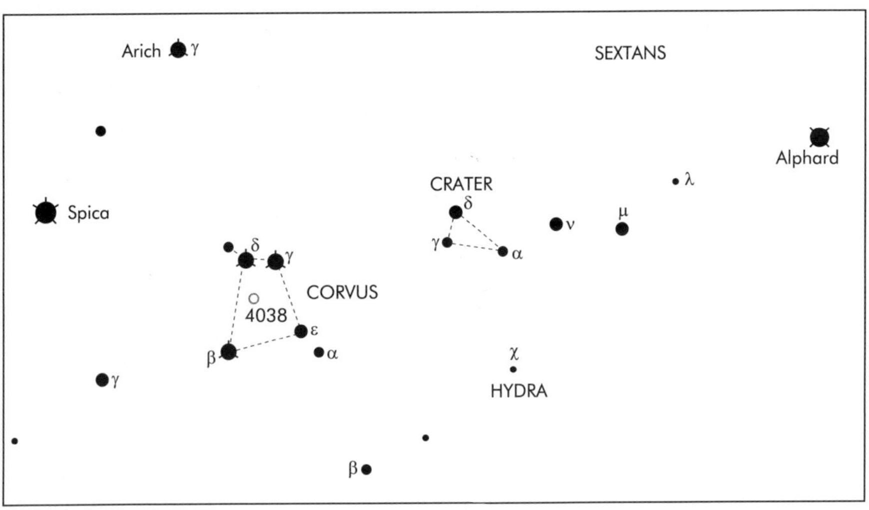

Corvus					
Greek letter	Name	Magnitude	Luminosity (Sun=1)	Distance (light-years)	Spectrum
β Beta	Kraz	2.6	600	290	G5
α Gamma	Minkar	2.6	250	185	B8
δ Delta	Algorel	2.9	60	117	B9
ϵ Epsilon	–	3.0	96	104	K2

mother of the great doctor Æsculapius. The crow's mission was to watch Coronis and report on her behaviour. To be candid, the bird's report was decidedly unfavourable, but Apollo rewarded it with a place in the sky.

The four main stars make up a well-formed quadrilateral. They are listed above.

The brightest star in a constellation is usually lettered Alpha, but Alpha Corvi (Alkhiba) is only of magnitude 4; it lies close to Epsilon, which is clearly orange in colour. Delta has a 9th-magnitude companion at a separation of 24 seconds of arc.

The main objects of interest in Corvus are the galaxies C60 and C61 (NGC 4038 and 4039), known as the Antennæ; they are however too faint to be seen to advantage with small telescopes (24 May).

May 24

The Antennæ

The Antennæ, less than 4° west-south-west from Gamma Corvi, are interacting spiral galaxies. Details are given below.

We seem certainly to be seeing a collision between two spiral galaxies. As they approached each other, with their disks rotating in opposite directions, they interacted, and new stars were produced; 'tails' made up of displaced stars were formed, giving the pair the unusual shape which has led to the nickname. (Originally it was thought that we were dealing with only one exceptional galaxy, which was sometimes called the Ring-tail.)

The Antennæ are perhaps 100 million light-years away. They are well worth finding, though their relative dimness makes them elusive.

The Antennæ							
M	NGC	RA		Dec.		Magnitude	Type
		h	m	°	′		
60	4038	12	01.9	−18	52	11.3	Spiral
61	4039	12	01.9	−18	53	13	Spiral

May 25

Crater, the Cup

Adjoining Corvus, and also bordering on Hydra, is the obscure constellation of Crater (the Cup), which, rather surprisingly, is one of Ptolemy's 48 originals. One legend associates it with the wine goblet of Bacchus. Another story tells how Apollo gave a cup to a white bird (Corvus) with instructions to fetch some drinking-water; the bird dallied for so long that at last the god lost patience, changed Corvus from white to black, and threw both it and the cup into the sky.

The brightest star is Delta (RA 11 h 19 m, declination –14° 47′), magnitude 3.6. It makes up a small triangle with Alpha or Alkes (4.1) and Gamma (also 4.1). The best way to identify Crater is by means of the nearby third-magnitude star Nu Hydræ (30 March), but there is little of interest in the constellation.

Anniversary

1834: Birth of John Tebbutt, one of Australia's greatest amateur observers, who lived at Windsor, NSW. He made many important contributions; he discovered the Great Comet of 1861, and another bright comet in 1881. He died in 1916.

May 26

The Sun – and Richard Carrington

Richard Carrington was an amateur astronomer who made major contributions to studies of the Sun. He set up his observatory at Redhill, in Surrey, and equipped it with first-class instruments. In particular, he is remembered for two main discoveries.

He was the first to make a visual observation of a solar flare. Flares are violent, short-lived outbreaks usually associated with active sunspot groups. They consist of hydrogen, and are generally detected only with spectroscopic equipment; visual sightings are very rare indeed, but Carrington made such an observation on 1 September 1859. Today many amateurs as well as professionals make regular records of flares, mainly by using filters which cut out all the incoming light except that of hydrogen.

Carrington's other discovery concerned the distribution of sunspots. This discovery was made independently by F.G.W. Spörer, and is – perhaps rather unfairly – known solely as Spörer's Law (23 October).

Remember never to look directly at the Sun through any telescope or binoculars, even with the addition of a dark filter. Irreversible damage to the eye is certain to result. The only sensible way to look at sunspots is by the method of projection (22 February).

Anniversary

1826: Birth of Richard Carrington, pioneer observer of the Sun.

May 27

R Hydræ

Hydra, the largest constellation in the sky, contains only one bright star, Alphard (8 March), but there are some interesting objects. One of these is the very red variable R Hydræ.

R Hydræ is in the southern part of the constellation, so that from British latitudes it is always low down; its position is RA 13 h 30 m, declination −23° 17′. It is of the Mira type, and at its best it can rise to magnitude 4, so that it can be seen with the naked eye; at minimum it never falls below magnitude 10, so that it is always an easy telescopic object, and for much of the time can be followed with binoculars. It is a huge red giant; its colour makes it easy to identify. According to the Cambridge catalogue, its distance from us is 100 light-years. Its variability was discovered by the Italian astronomer Maraldo in 1704; previously only three variables had been identified – Mira itself, Chi Cygni (22 August) and the eclipsing binary Algol (12 November).

In the early eighteenth century it seems that the period was almost 500 days, but it has steadily shortened, and is now around 390 days, though – as is usual with Mira stars – no two cycles are exactly alike. At any rate, it may well be that we are seeing a real change in the star's evolution, and it is worth monitoring. The 1997 maximum fell on 20 April.

R Hydræ has a 12th-magnitude companion at a separation of 21″.2. The two seem to share a common motion through space, and there is probably a true physical connection, but they are a very long way apart.

Future Points of Interest

1998: Opposition of Pluto.

May 28

Lockyer and Stellar Evolution

Earlier this month we noted the anniversary of the birth of a great Victorian astronomer, Sir Norman Lockyer; we also noted a possible observable change in the evolution of a star, R Hydræ (27 May). There is a connection between the two, though on this score we have to admit that Lockyer was completely wrong.

Until well into our own century, the source of stellar energy was very much of a mystery. It was generally supposed that a star shone because it was contracting under the influence of gravity, releasing energy in the process.

Lockyer had other ideas, and in 1890 he outlined his own theory. He believed that the Sun and other stars shone because they were being constantly bombarded by particles from space; each time a particle hit, it would produce a spark of energy. We now know that this 'meteoritic hypothesis' is untenable; it could not possibly provide enough heat to power the Sun, even if there were sufficient numbers of meteorites. The Sun shines by nuclear reactions going on deep inside it. Still, Lockyer had at least provided a novel theory.

May 29

Anniversary

1794: Birth of the German astronomer Johann von Mädler

Der Mond

It is time to take another look at the Moon, and this is a suitable moment, because it is the anniversary of the birth of one of the greatest of all pioneer lunar observers, Johann Heinrich von Mädler.

Both Mädler's parents died when he was young, and it was not until he was in his twenties that he managed to join Berlin University. He then became a teacher, and it was fortunate that he met a wealthy banker, Wilhelm Beer (brother of Meyerbeer, the composer) who was interested in astronomy and had set up a private observatory near Berlin, equipping it with a fine 3.75-inch refracting telescope.

Beer took lessons from Mädler, and the two then joined forces to carry out a systematic programme of mapping the Moon and planets. Their book on the Moon, *Der Mond*, was published in 1838, and was a masterpiece of careful, accurate observation; it was accompanied by a description of every named formation. It remained the best lunar map for many years. Mädler left Berlin in 1840 to become Director of the observatory at Dorpat, in Estonia; he made many other contributions to astronomy, but he will always be remembered for *Der Mond*. He died in 1874; Beer had died long before, in 1850.

Both men have lunar craters named after them. Crater Mädler, 20 miles across, is irregular but prominent, on the Mare Nectaris near Theophilus; Beer, 8 miles in diameter, lies between Archimedes and Timocharis on the Mare Imbrium, forming a neat pair with its 'twin', Feuillée. A small telescope will show them both well.

Anniversary

1966: Surveyor 1 launched to the Moon. It landed in the Mare Nubium on 2 June of the same year, and returned over 11 000 images. Contact was finally lost in January 1967.

1971: Mariner 9 launched to Mars. It entered Martian orbit on 13 November, and returned over 7000 images. Contact was maintained until 27 October 1972.

May 30

The Lunar Atmosphere

Early lunar observers believed that the Moon might have an appreciable atmosphere. Beer and Mädler believed otherwise. The Moon's low escape velocity means that it could not retain a substantial atmosphere, and it is now known that the residual atmosphere is incredibly rarefied; it is in the form of what is called a 'collisionless gas'. The total weight of the lunar atmosphere is of the order of 30 tons. At normal room temperature and pressure, it would just about fill a 210-foot cube. Therefore, the simple statement that 'the Moon is an airless world' is to all intents and purposes true. This is why the surface features appear sharp and clear-cut.

Future Points of Interest

1999: Pluto at opposition.

2003: Annular eclipse of the Sun.

May 31

Icelandic Eclipse

The 2003 eclipse will be well worth seeing, so make your plans well in advance! Clouds permitting, the best view should be obtained from Iceland, but the track also covers North Scotland.

From Reykjavík, the Sun rises at 3.26 GMT; the eclipse starts two minutes later. Mid-eclipse is at 4.04, and the end of the eclipse is at 5.01. Annularity lasts for 3 minutes 36 seconds, and the Sun is 1°.4 above the horizon. Annularity can also be seen from the Faroes, Orkneys and Shetland. On the British mainland, Aberdeen just misses annularity, but it can be seen from Fort William and Inverness.

This is the last British annularity for some time, so make the most of it!

June

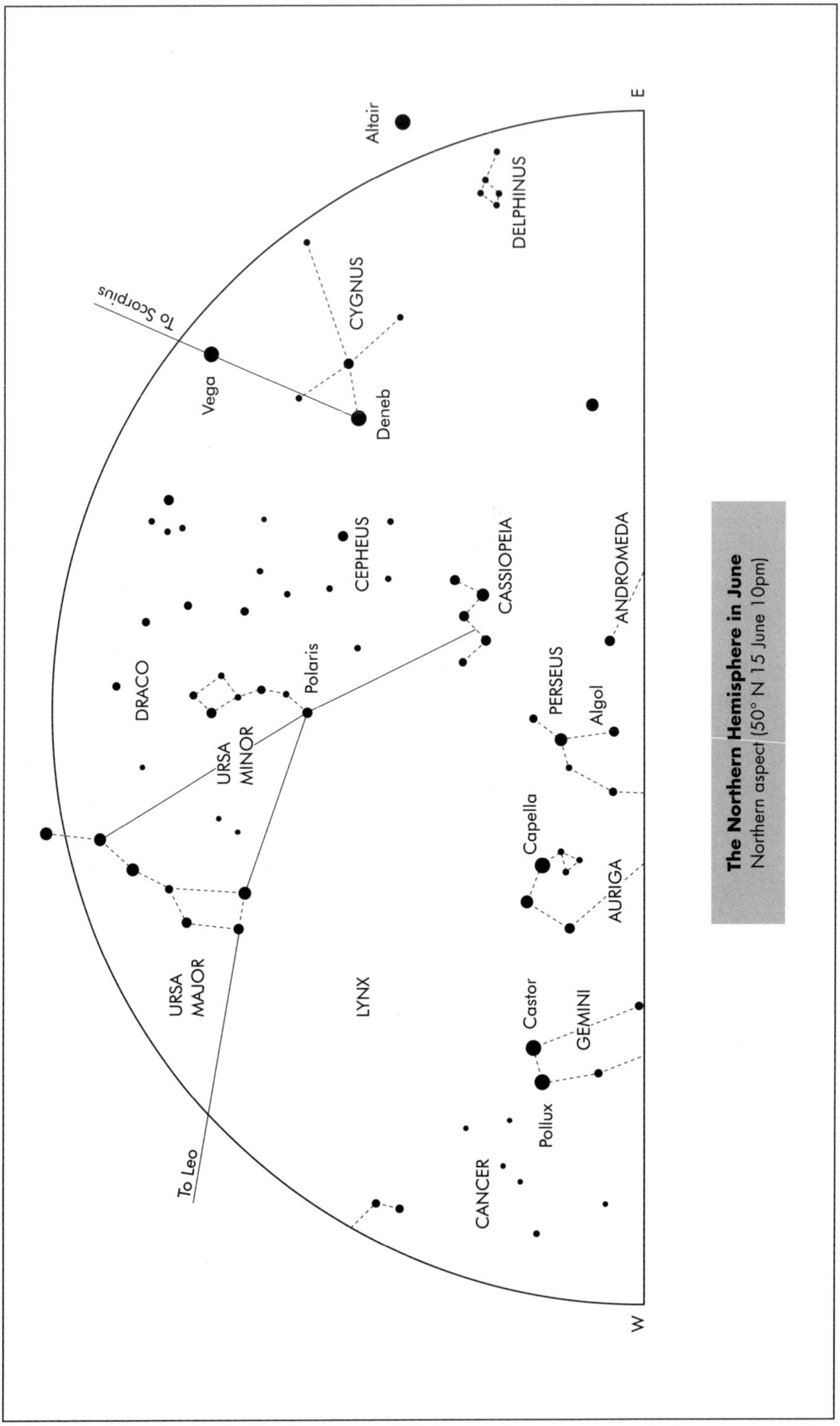

The Northern Hemisphere in June
Northern aspect (50° N 15 June 10pm)

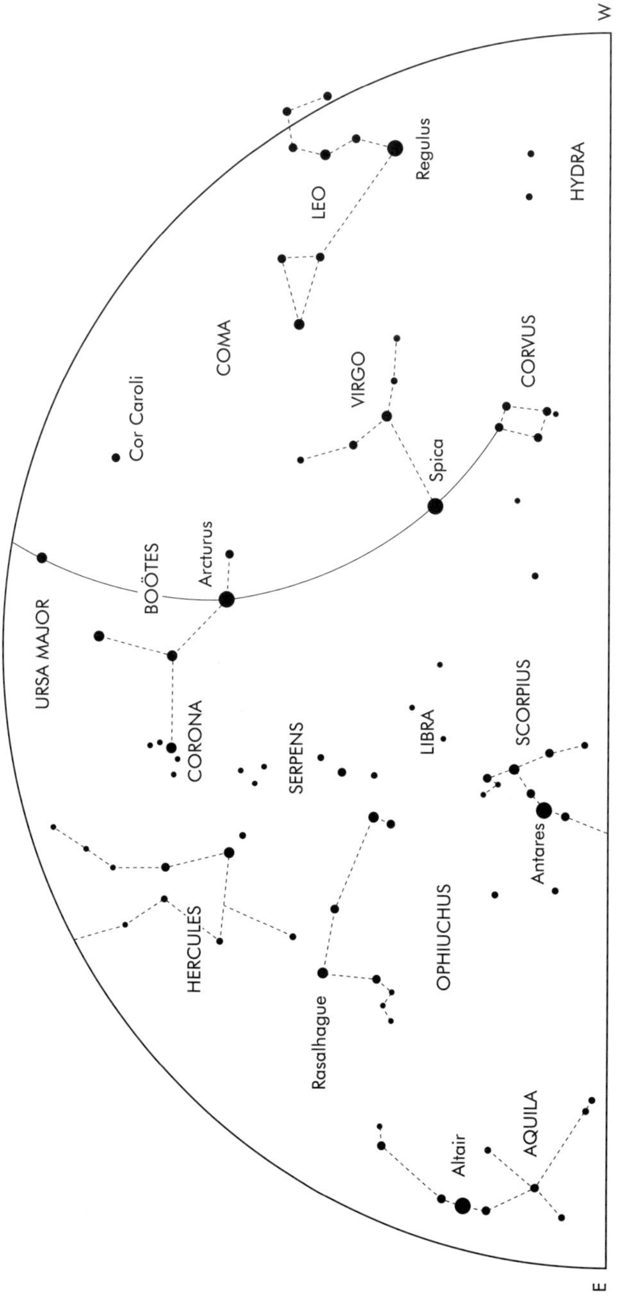

The Northern Hemisphere in June
Southern aspect (50° N 15 June 10pm)

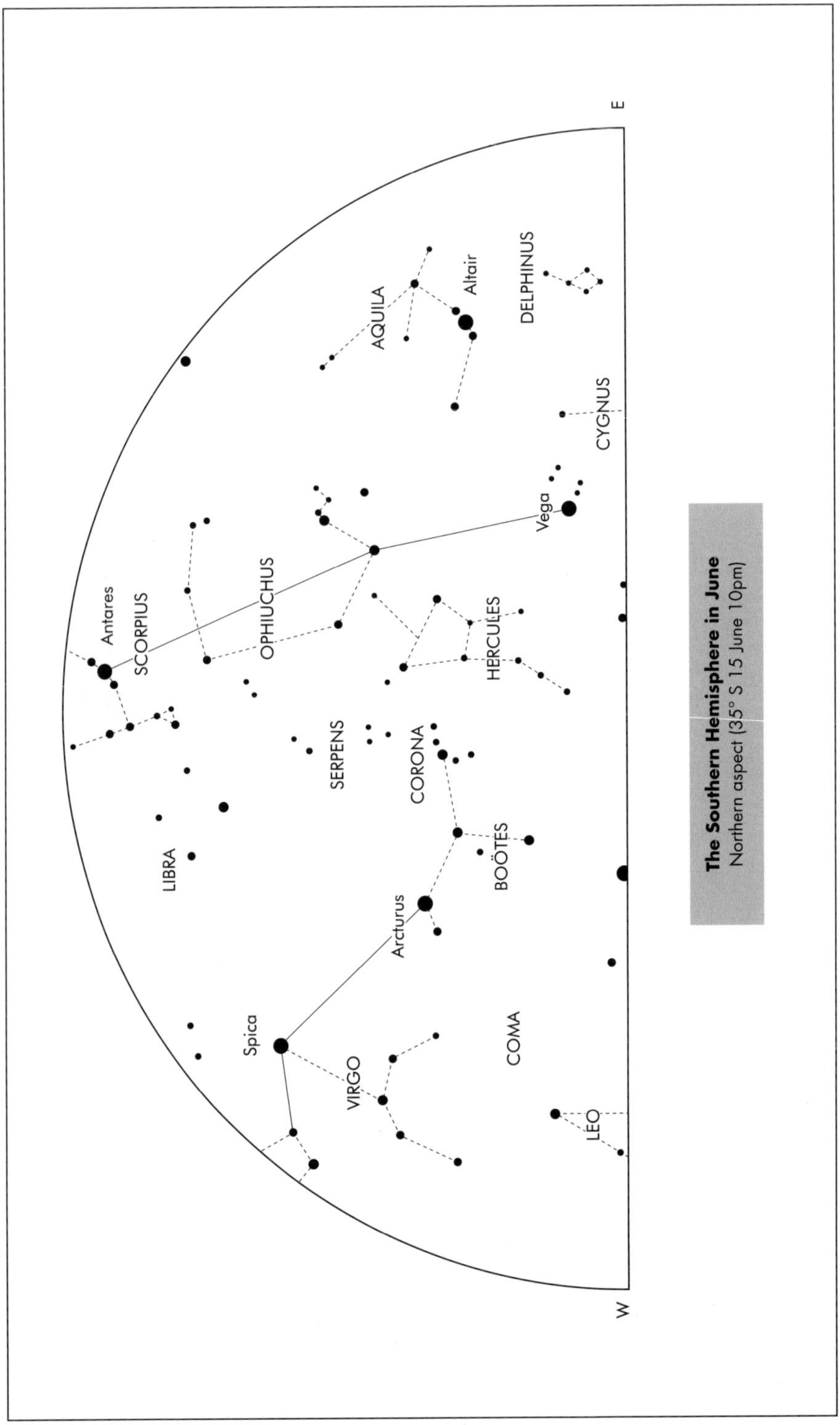

The Southern Hemisphere in June
Northern aspect (35° S 15 June 10pm)

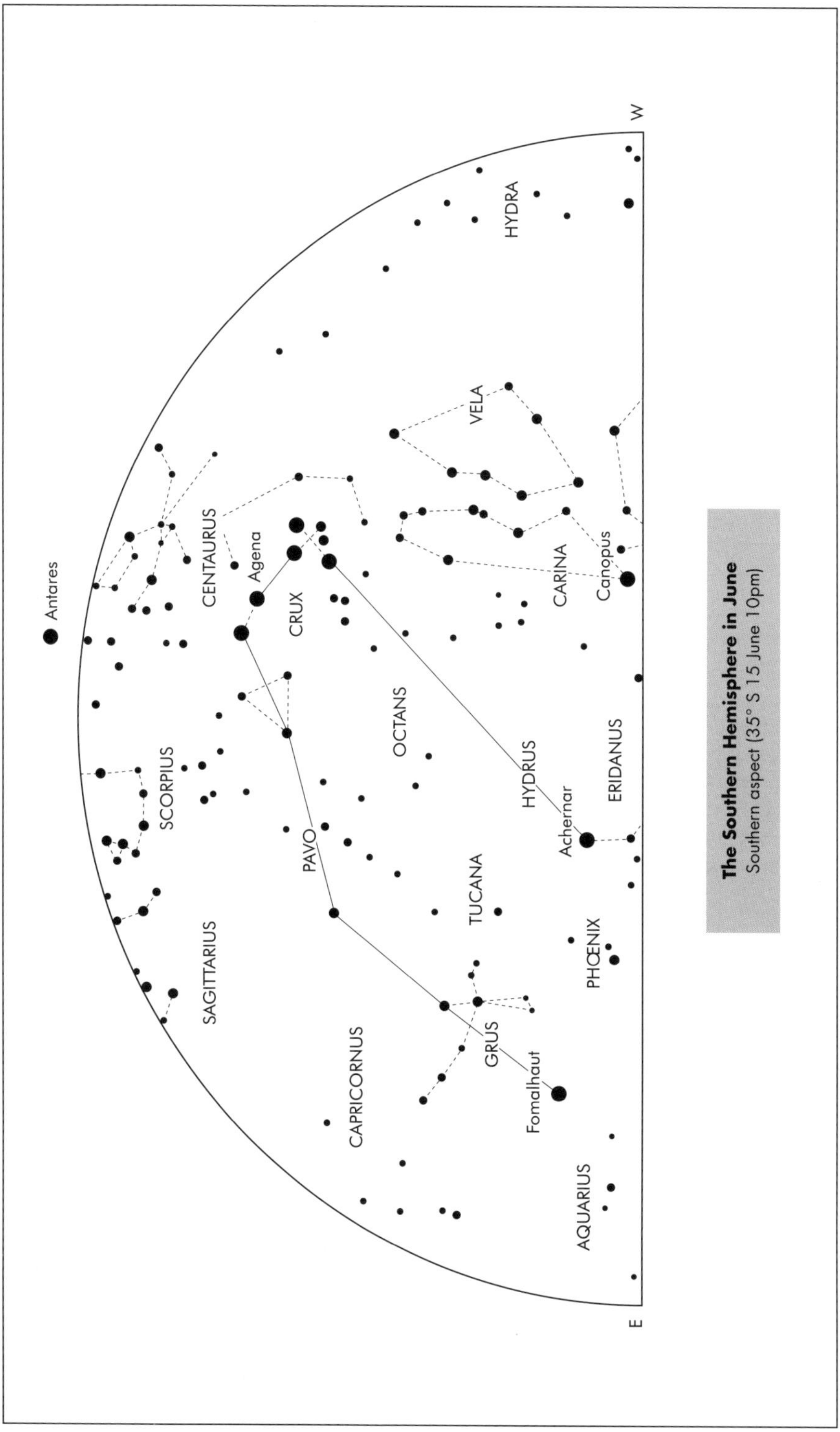

The Southern Hemisphere in June
Southern aspect (35° S 15 June 10pm)

June 1

The June Sky

To northern-hemisphere observers, the evening sky is now dominated by the so-called Summer Triangle, consisting of Vega in Lyra, Deneb in Cygnus and Altair in Aquila. This name is completely unofficial; it arose from a casual remark of mine, made in a *Sky at Night* television programme more than 30 years ago, and in any case it is now winter in the southern hemisphere – but it has come into general use. Ursa Major is high in the north-west, Cassiopeia rather low in the north-east. The lovely orange Arcturus, in Boötes, is high up, and Virgo remains prominent, but Leo is starting to merge into the evening twilight. Much of the south-eastern aspect is occupied by the large but rather dim constellations of Hercules, Ophiuchus and Serpens. Antares, the red supergiant in the Scorpion, is visible in the south, though from British latitudes it is never seen to advantage.

To southern-hemisphere observers, Scorpius is near the zenith, and is followed round by Sagittarius, with its lovely star-clouds. The Southern Cross is west of the zenith, while Canopus is so low in the south that it may not be seen. Arcturus attains a respectable altitude in the north, but Vega and Altair are very low, and Deneb does not rise until after midnight. This is a splendid time for seeing the Milky Way, which runs right across the sky from one horizon to the other.

Future Points of Interest

2000: Pluto at opposition.

June 2

Dark Adaptation

It is now summer in the north, and around the time of full moon there is no real darkness at all. Lunar phases are given on page ix; for convenience, the June full moons for the period covered here are:

1998, June 10 1999, June 28 (and May 30) 2000, June 16
2001, June 6 2002, June 24 2003, June 14.

If you go straight outdoors from a brightly lit room, well after dark, you will be able to see very few stars. If you remain outside for some while, the stars will become brighter and brighter. One's eyes need time to adapt; the period necessary varies from person to person, but it must always be borne in mind.

There are bright stars on view tonight – Arcturus and the Summer Triangle, for example but even these seem pale to a non-adapted eye. Also, recording observations means using

Anniversary

1858: Discovery of Donati's Comet. It was then a feeble glow near the star Lambda Leonis, but brightened up, and by October was brilliant; it is said to be the most beautiful comet ever seen, with its straight gas-tail and scimitar-like dust-tail. Telescopically it was followed until 4 March 1859. The period is probably about 2000 years.

1966: Landing of Surveyor 1 on the Moon. It returned 150 images.

some sort of light, and can affect the adaptation. A red light is better than a white one; a good hint is to buy or make a board equipped with a light which will cast a soft, preferably red glow over the area being used for recording.

I suggest an experiment, if the sky is clear. Wait inside a lit room. Then go out and look at the sky. Come indoors again. Then go out once more and do not look upward for twenty minutes or so. When you do, you will be amazed at the numbers of stars which have flashed into view.

Anniversary

1948: Official opening of the Palomar 200-inch reflector.

Future Points of Interest

2003: Mercury at western elongation.

June 3

The Hale Reflector

This is the anniversary of the opening of what was for decades the largest and most powerful telescope in the world: the 200-inch Hale reflector at Palomar, in California.

The man after whom the telescope is named was George Ellery Hale, an American who became an expert in studies of the Sun. He knew that large telescopes were needed for research, and he had an amazing knack of persuading friendly millionaires to finance them. He set up the Yerkes Observatory, with its 40-inch refractor, and then the Mount Wilson Observatory, near Los Angeles, with 60-inch and 100-inch reflectors. The Hooker 100-inch was completed in 1917, and was in a class of its own; using it, Edwin Hubble was able to prove that the 'spiral nebulæ' are not nebulæ at all, but independent galaxies.

Still Hale was not satisfied, and he master-minded a 200-inch reflector. Sadly. he did not live to see it completed – he died in 1938 – but the telescope certainly achieved all that he could have hoped.

It is no longer the world's largest telescope, but it is even more effective than it used to be, because it is now used with electronic equipment rather than photographic plates; it remains in the forefront of astronomical research.

Future Points of Interest

2001: Pluto at opposition.

June 4

Peculiar Pluto

Pluto, usually regarded as the outermost member of the Sun's family of planets, is a slow mover. It takes almost 248 years to complete one journey round the Sun, and its synodic period – that is to say, the mean interval between successive oppositions – is 366.7 days. Thus in the period covered here, Pluto comes to opposition on May 25 (1997), May 28 (1998), May 31 (1999), June 1 (2000), June 4 (2001), June 7 (2002) and June 9 (2003). The opposition magnitude is about 14, so that Pluto is well beyond the range of binoculars. During this period it lies in the region of Ophiuchus.

In fact, Pluto's orbit is much more eccentric than those of the other planets, and when near perihelion its distance from the Sun is less than that of Neptune – though the relatively high inclination of Pluto's orbit (17°) means that there is no fear of a collision with Neptune.

Pluto is smaller than the Moon, and there are serious doubts as to whether it should be regarded as a true planet; in size, it is more like a satellite. Data are as follows:

Data for Pluto	
Distance from the Sun (miles)	
maximum	4583 000 000
mean	3666 000 000
minimum	2 766 000 000
Orbital period	247.7 years
Rotation period	6 days 9 hours
Orbital eccentricity	0.248
Orbital inclination	17.1°
Diameter (miles)	1444
Mass (Earth)	0.002
Surface gravity (Earth = 1)	0.06

Pluto has one attendant, Charon, which has a diameter more than half that of Pluto itself, so that the two make up a unique pair. Truly Pluto has presented astronomers with plenty of problems! It was discovered in 1930 by Clyde Tombaugh, from the Lowell Observatory in Arizona.

June 5

Pluto and Charon

Pluto's discovery was not accidental. In 1846 the planet Neptune had been tracked down by the irregularities it produced in the movements of Uranus (24 August); in the same way, tiny irregularities in the movements of the outer giants led Percival Lowell to work out a position for yet another planet, and Pluto was found not far from the position which Lowell had given. Yet when the size of Pluto was measured, it became evident that the situation was anything but clear-cut. If Pluto were really small and lightweight, it could not possibly pull a large world such as Neptune measurably out of position.

It became important to find out how large – or how small – Pluto really is. Careful studies were made, and Pluto was photographed as often as possible. Then, in 1977, it was found that Pluto is not a solitary traveller in space; it is accompanied by a second body, now named Charon, which is over 750 miles across. Yet when when the masses of Pluto and Charon are combined, they are insignificant by planetary standards. Either Lowell's reasonably correct forecast was sheer luck (which is hard to believe), or still another planet awaits discovery.

When observed through an ordinary telescope, Pluto looks exactly like a faint star, and it can be identified only by its very slow movement. The best way to locate it is by means of photography. If photographs are taken over a period of a few weeks, Pluto's motion will betray it.

Charon moves round Pluto in a period of 6 days 9 hours – the same as Pluto's rotation period, so that the two are 'locked' in a unique manner; Charon would appear fixed in the Plutonian sky. The distance between the two bodies, centre to centre, is only 12 000 miles.

No space-craft have been anywhere near Pluto, but surface details have been recorded by the Hubble Space Telescope. There are brighter and dark areas; the darkest province lies directly under Charon. The surface seems to be coated with frozen nitrogen, methane and carbon monoxide. There is an extensive but very thin atmosphere, with a ground pressure of about 100 000 times less than that of our own air at sea-level; the likely constituent is nitrogen. However, the atmosphere may not be permanent. The present surface temperature of Pluto is of the order of –230 °C, but Pluto passed perihelion in 1989 and is now moving outward; before it reaches its next aphelion, in 2113, it will have become so cold that the atmosphere will have frozen out on to the surface. Charon is thought to be coated with ordinary ice, and there is no evidence of any atmosphere.

Numbers of smaller bodies, no more than around 200 miles across, have now been found moving near and beyond Pluto's orbit; these make up what has become known as the Kuiper Belt (after the late G.P. Kuiper, who first suggested its existence). It may well be that Pluto and Charon are nothing more than Kuiper Belt objects.

Pluto must be a gloomy place, though sunlight there is still 1500 times as strong as full moonlight on Earth. To a Plutonian, Charon would have a magnitude of –9, and would be one-fifth the size of the Moon seen from Earth.

June 6

Boötes, the Herdsman

Our 'constellation of the week' is Boötes, the celestial Herdsman. It is an original constellation, and is marked by

Boötes					
Greek letter	Name	Magnitude	Luminosity (Sun=1)	Distance (light-years)	Spectrum
α Alpha	Arcturus	−0.04	115	36	K2
β Beta	Nekkar	3.5	58	137	G8
γ Gamma	Seginus	3.0	53	104	A7
δ Delta	–	3.5	58	140	G8
ε Epsilon	Izar	2.4	200	150	K0
ζ Zeta	–	3.8	105	205	A2
ρ Rho	–	3.6	105	183	K3

Arcturus, the brightest star in the northern hemisphere of the sky (its only superiors, Sirius, Canopus and Alpha Centauri, are all well south of the equator). Yet there seem to be few mythological legends attached to it. It is merely recorded that Boötes was a herdsman who invented the plough drawn by two oxen, for which contribution to human welfare he was placed in the sky. Note, however, that he seems to have taken hold of the two hunting dogs, Asterion and Chara (23 March) – possibly to stop them chasing the Bears across the cosmos?

The leading stars of Boötes are listed above.

There is an easy way to find Arcturus; simply follow round the tail of the Great Bear. This will lead straight to Arcturus. There can be no confusion with Vega, the only star of comparable brilliance in the same general area of the sky, because Vega is decidedly blue, while the light orange hue of Arcturus is evident at a glance. In binoculars, Arcturus is one of the most beautiful of all stars. Its declination is 19° 11′ N, so that it is close enough to the equator to be seen from every inhabited country; it can even be seen from Invercargill, at the southernmost tip of New Zealand.

June 7

Arcturus

Arcturus is one of the most famous of all stars. There are many mentions of it in ancient writings ('Canst thou guide Arcturus with his sons?') and in 460 BC Hippocrates even claimed that it influenced human health; a dry season after Arcturus' first appearance in the dawn sky 'agrees best with those who are naturally phlegmatic, with those who are of a humid temperament, and with women, but it is inimical to the bilious'. Seamen regarded it as unlucky; to Pliny it was 'horridum sidus'.

It is a giant, with a diameter of around 20 000 000 miles, but it is very insignificant compared with stars such as Betelgeux; it looks brighter only because it is so much closer

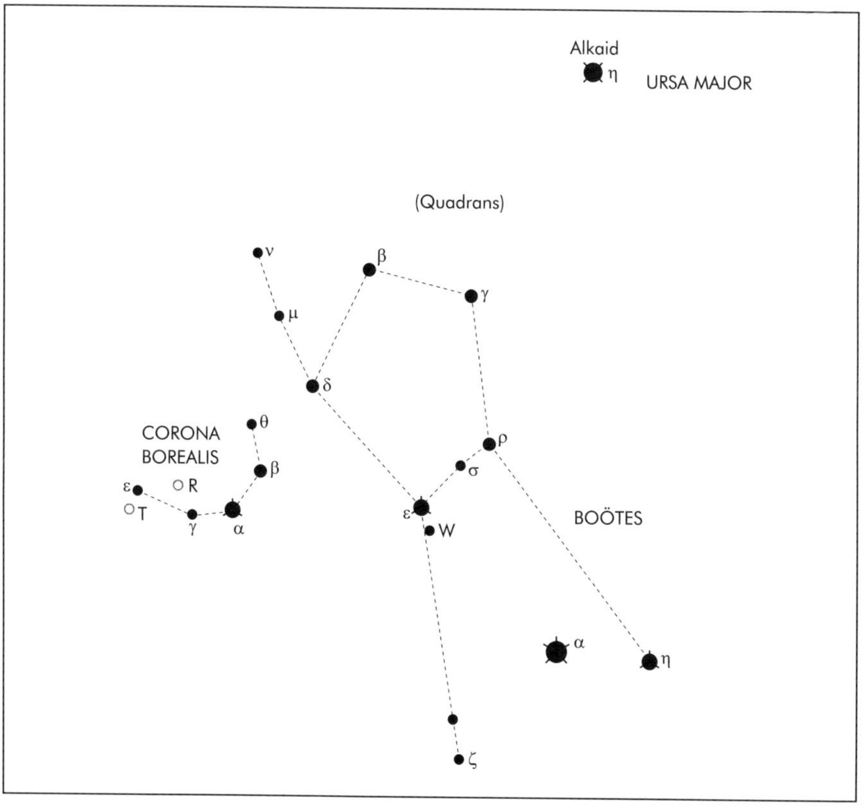

to us. It is at present approaching at 3 miles per second, but this will not continue indefinitely, and in the future it will start to draw away. In about half a million years' time it will have fallen below naked-eye visibility.

Interestingly, modern instruments can measure the heat sent to us by Arcturus. It works out at about equal to the heat received from an ordinary candle five miles away.

June 8

Transits of Venus

We have already discussed transits of Mercury across the disk of the Sun (7 May). Thirteen transits took place during our own century. Transits of Venus are much less common. They occur in pairs, separated by eight years, after which there are no more for over a century. Thus there were transits in 1761, 1769, 1874 and 1882; we now have transits in 2004 and 2012, after which we must wait until 2117 and 2125.

During transit, Venus (unlike Mercury) can be seen with the naked eye – but remember never to look direct through any telescope; the only sensible way of seeing the transit is to project the Sun's image on to a screen held or fastened behind the eyepiece of the telescope.

Transits of Venus were once regarded as important, because they provided a means of measuring the distance of the Earth from the Sun. As this method is now obsolete, future transits will be regarded as of no more than academic interest, but they are worth watching. The time of mid-transit in 2004 is 08.24 hours GMT, so that – clouds permitting – it will be well seen from Britain and Europe.

Future Points of Interest

2001: Venus at western elongation.

2004: Transit of Venus.

Anniversary

1625: Birth of G.D. Cassini, an Italian astronomer who was called to France to become the first Director of the Paris Observatory. He discovered four of Saturn's satellites (Iapetus, Rhea, Dione and Tethys) as well as the main division in the rings; he made pioneer observations of Mars, and was the first to give a reasonably good measurement of the distance of the Earth from the Sun (86 000 000 miles). He died in 1712.

June 9

The Ophiuchids

Meteor watching is a fascinating but exhausting process. One never knows where or when a meteor will appear; though there are many annual showers, such as the January Quadrantids (4 January) and the August Perseids (11 August) there are also many non-shower or sporadic meteors.

The June Ophiuchids are not usually rich, and the maximum is not well defined; officially the shower begins on May 19 and lasts until well into July. The radiant lies in Ophiuchus, the Serpent-bearer (8 July), which is now in the eastern part of the sky. It is from here that the meteors seem to issue – perhaps five or six per hour – and so a good idea is to set your camera for a time exposure and point it toward Ophiuchus. If you are lucky, you may snare a meteor or two, and certainly the Ophiuchids can sometimes produce brilliant fireballs.

Of course, the meteors are not confined to Ophiuchus. They may reach any part of the sky. But if you see one and plot its path 'backwards', so to speak, you will find that if it is a shower member, it will indeed have come from the region of the Serpent-bearer.

It is possible – though by no means certain – that the Ophiuchids are associated with Lexell's Comet, which passed within 1.5 million miles of the Earth in 1770 and reached second magnitude. The period was given as 5.6 years. Unfortunately the comet subsequently made a close approach

Future Points of Interest

Maximum activity of the Ophiuchid meteor shower.

2000: Mercury at eastern elongation.

2003: Pluto at opposition.

to Jupiter, and its orbit was altered so drastically that it was never seen again.

Future Points of Interest

2002: Annular eclipse of the Sun. The time of mid-eclipse is 24 h GMT, so that the best views will be from the Pacific area; of course the eclipse will not be seen from Britain or Europe. Annularity will last for 1 minute 13 seconds.

June 10

Red Variable: W Boötis

Having flirted with meteors, let us return to our 'constellation of the week', and look at the red variable W Boötis.

Many red stars are variable in light, because they have used up their main nuclear 'fuel' and have become unstable; they pulsate, swelling and shrinking, sometimes regularly and sometimes erratically. Those stars with small ranges and somewhat unpredictable behaviour are known as semiregular variables, and W Boötis, close to Epsilon Boötis (Izar) is an excellent example on view at the moment.

Its spectrum is of type M, so that its colour and temperature are much the same as for Betelgeux in Orion. Its distance is not certainly known, but must be well over 100 light-years. The position is RA 14 h 43 m, declination 26° 32′.

The magnitude range is only from 4.7 to 5.4, so that the star is always visible with the naked eye under good conditions and is easy in binoculars; simply put Izar in the binocular field, and W Boötis will show up, distinguished by its ruddy hue. On average it takes 450 days to pass from one maximum to the next, but, as with all stars of its kind, both period and amplitude are markedly variable.

Future Points of Interest

1999: Venus at eastern elongation.

June 11

The Vanishing Star

Still with Boötes, let us now turn to the strange case of a star which was seen only for a brief period and then disappeared. Just what it was, we still do not know. It was very close to Arcturus, at a separation of only 25 minutes of arc, so that the two were in the same low-power telescopic field.

It was discovered in April 1860 by a well-known astronomer, Joseph Baxendell. On 9 April its magnitude was 9.7, and it remained the same on the following nights. By 22 April it had faded to magnitude 12.8. On 23 April it could not be found – and it has never been seen again. It has vanished as completely as the hunter of the Snark.

It may have been an unusual sort of variable star, but in this case it must surely have reappeared. It is improbable that an observer as skilled as Baxendell could have made such a gross mistake, and it may well be that the star was a nova; it may even have been a recurrent nova, such as T Coronæ (16 June), flaring up only at long intervals. In this case it may be seen again in future, and it is always worth checking on the field to see whether it is there. In variable star catalogues it is officially listed as T Boötis.

June 12

Double Stars in Boötes

Boötes is not striking apart from Arcturus. It is marked by a sort of Y-pattern of stars, with Arcturus at the base, but one of these stars, Alphekka, belongs to the adjacent constellation of Corona Borealis (14 June). The others are Epsilon Boötis or Izar, and Gamma Boötis or Seginus.

Izar, of the second magnitude, is an excellent double star, though none too easy to split with a telescope below about three inches in aperture. The primary star is orange; there is little colour in the secondary, but it has often been regarded as bluish or greenish by contrast with its companion. The separation is 2.8 seconds of arc; the secondary is of magnitude 4.9. Izar itself is 200 times as luminous as the Sun, and even the companion could match 50 Suns. The orbital period must be very long.

Zeta Boötis has a dim companion; the primary is itself a very close binary with identical-twin components and a period of 123 years, but the separation never exceeds one second of arc. Xi, in the same region, is a neat binary; the magnitudes are 4.7 and 6.8, and the present separation is 6.6 seconds of arc. The period is 150 years. Mu, in the northern part of Boötes, is a very wide, easy pair; magnitudes 4.3 and 7.0, separation 171 seconds of arc. You should have no trouble in splitting this pair with almost any telescope.

Anniversary

1843: Birth of Sir David Gill, a Scottish astronomer who had a long tenure as Director of the Cape Observatory in South Africa. It was his photograph of the bright comet of 1882 which showed him the value of mapping the sky by means of photography, since his plate showed many faint stars as well as the comet. Gill made many important contributions to astronomy, and raised the Cape Observatory to its position of eminence. He died in 1914.

June 13

Quadrans: the Forgotten Constellation

There was a time when almost every celestial cartographer felt bound to add new constellations to the sky. Some of these have endured (and, of course, there was no choice but to add new constellations to the far southern stars, because Ptolemy

Future Points of Interest

2001: Mars at opposition. The magnitude is −2.1, and the apparent diameter 20.8 seconds of arc. Minimum distance from Earth is just over 41 000 000 miles, on 21 June. Mars is in Sagittarius, and therefore inconveniently low for British and European observers.

spent all his life in Europe or Egypt and could never have seen down beyond the Southern Cross). Others have not, and had cumbersome names; typical examples are Sceptrum Brandenburgicum (the Sceptre of Brandenburg), Machina Electrica (the Electrical Machine), Officina Typographica (the Printing Press) and Honores Frederici (the Honours of Frederik). All these were proposed by J.E. Bode in his maps issued in 1775. None has survived. Also rejected was Quadrans Muralis, the Mural Quadrant, which has at least left us a legacy.

Quadrans lay in what is now the northern part of Boötes. It adjoined Nekkar or Beta Boötis, and included the three fainter stars between Nekkar and Alkaid in the Great Bear. During the reorganisation of the constellations by the International Astronomical Union, in our own century, Quadrans vanished; but the meteor shower of early January radiates from the position where it used to be (4 January), and this is still called the Quadrantid shower. There are some people, mainly meteor enthusiasts, who regret that Quadrans is no longer to be found on our maps.

June 14

The Northern Crown

As we have noted, the Boötes pattern is rather Y-shaped, but one of the stars in it belongs to Corona Borealis, the Northern Crown. Its leader, Alpha Coronæ or Alphekka, is of magnitude 2.2, and therefore brighter than any of the stars of Boötes apart from Arcturus.

Corona really does give the impression of a crown, as its five chief stars are arranged in a semicircle, with Alphekka in the central position. There are several variants of the mythological association. In one, Corona was the crown given by the wine-god Bacchus to Ariadne, daughter of King Minos of Crete; in another, the crown was given to Ariadne by the hero Theseus, who had disposed of a particularly unpleasant monster known as the Minotaur, and was on his way home.

The five stars of the 'Crown' are listed below.

Corona Borealis

Greek letter	Name	Magnitude	Luminosity (Sun=1)	Distance (light-years)	Spectrum
α Alpha	Alphekka	2.2	130	78	A0
β Beta	Nusakan	3.7	28	59	F0
γ Gamma	–	3.8	110	210	A0
ϵ Epsilon	–	4.1	100	240	K3
θ Theta	–	4.1	250	360	B5

Alphekka – sometimes known as Gemma – is very slightly variable, but the changes are too slight to be detected without sensitive measuring equipment. Beta is a magnetic variable of the same type as Cor Caroli (23 March), but here too the changes in magnitude are very slight.

June 15

R Coronæ: the Sooty Star

Look inside the bowl of Corona, and you may be able to see a faint star. Use binoculars, and you will certainly be able to see one star, possibly two. One of these is an ordinary star of magnitude 6.6, rather below naked-eye visibility; the other, R Coronæ, is one of the most remarkable variables in the sky.

R Coronæ is usually of around magnitude 5.8, but at unpredictable intervals it starts to fade, becoming so dim that a powerful telescope is needed to show it. The magnitude may fall below 15. After a while the star slowly recovers, returning to its normal brightness.

Spectroscopes tell us that R Coronæ has less hydrogen in its outer layers than most stars, but more carbon. Periodically, clouds of soot accumulate in the star's atmosphere, and dim the light; only when the clouds disperse does the star brighten up again. It has also been found that R Coronæ is surrounded by a shell of dust 33 light-years across; its origin is uncertain.

If binoculars show only one visible star in the bowl, you may be sure that R Coronæ is going through one of its minima – so keep looking, night after night, until it reappears, though admittedly this may not be for many weeks or even months. There are other stars of the same type, but they are very rare, and R Coronæ is much the brightest of them.

Anniversary

1985: Vega 2 dropped a balloon into the atmosphere of Venus. Vega 2 (no connection with the star Vega) was the Russian probe sent to encounter Halley's Comet. Its predecessor, Vega 1 had bypassed Venus on June 10 and dropped a balloon into the atmosphere; it was tracked for 46 hours, and sent back useful data. The Vega 2 balloon was also successful. Both probes subsequently went on to Halley's Comet.

June 16

T Coronæ: the Blaze Star

Though Corona Borealis is a small constellation, it contains a number of interesting objects. One of these is T Coronæ, nicknamed the Blaze Star. Usually it is far below naked-eye visibility, and since the mean magnitude is between 10 and 11 even ordinary binoculars will not show it. However, in 1866 it suddenly flared up to second magnitude, so that for a few nights it equalled Alphekka. It then sank back to its normal obscurity, but flared up again in 1946, this time reaching magnitude 3.5 before declining once more.

T Coronæ is one of those unusual stars known as recurrent novæ. An ordinary nova is a binary system in which one component is a white dwarf (21 February); the dwarf pulls material away from its less dense companion, and eventually the situation becomes unstable, so that an outburst takes place and there is a temporary flare-up. With most novæ there is only one observed outburst, but T Coronæ has already shown two, and there may be another at any time. Eighty years elapsed between 1866 and 1946, so that variable star observers will be on the alert around 2026. T Coronæ is in the same telescopic field with Epsilon Coronæ, one of the members of the main semicircle. Only one other recurrent nova (RS Ophiuchi) ever becomes visible with the naked eye; usually it is below magnitude 12, but on several occasions (1901, 1933, 1958, 1967) it has reached magnitude 5.3.

Anniversary

1800: Birth of the third Earl of Rosse.

June 17

The 'Leviathan of Parsonstown'

This evening let us return to Ursa Major, and take another look at M51, the Whirlpool Galaxy. Photographs show its beautiful spiral form – and it is linked with a remarkable man, the third Earl of Rosse, who was born on 17 June 1800.

Lord Rosse was an Irish landowner, and was deeply involved in the affairs of his country, but he was also enthusiastic about astronomy, and he decided to build a large telescope. He completed a 36-inch reflector, and set it up in the grounds of his home, Birr Castle in central Ireland. It worked well, and he then began constructing a 72-inch reflector, far larger than any previously made (William Herschel's largest telescope had a 49-inch mirror). Lord Rosse had no helpers apart from workmen whom he trained on his estate; he had to do everything on his own. The mirror was made of metal, and he even had to build a furnace to cast the blank.

The mirror was completed; but what about mounting it? Lord Rosse knew that he could not make a telescope manœuvrable enough to cover the whole of the sky, and so he slung the tube between two massive stone walls. This meant that he could see for only a short distance to either side of the meridian. Yet the telescope was a triumph, and with it Lord Rosse discovered the spiral forms of the objects we now know to be galaxies. His drawings of them are amazingly accurate, and his representation of the Whirlpool is very like a modern photograph.

Lord Rosse died in 1867. His son, the fourth Earl, continued the astronomical work, but was more interested in measuring the tiny quantity of heat sent to us from the Moon. The 72-inch 'Leviathan' was overtaken by modern-type instruments, and was dismantled in 1909. However, it has now been restored, and was brought back to use in 1997.

The story of Birr Castle astronomy is unique in scientific history. Alone, one man built what was by far the world's largest telescope, and used it to make fundamental discoveries. Nothing like it had ever happened before; nothing like it can ever happen again.

June 18

Scorpius

Our constellation this week is Scorpius, the Scorpion (often, incorrectly, called Scorpio). It is one of the most brilliant of the Zodiacal groups, and indeed one of the most magnificant constellations in the entire sky, rivalling Orion. It is also one of the few groups to give at least a vague impression of the object it is meant to represent; it is not difficult to picture a scorpion in the long line of bright stars, with the red Antares in the position of the 'heart', the head marked by a well-formed pattern, and the very prominent 'sting' which includes Lambda Scorpii or Shaula, only just below the first magnitude. The leaders are listed in the table below.

It is unfortunate for northern observers that Scorpius lies so far south of the equator; from Britain it is always low down, and part of it never rises at all. During June evenings, look for it above the southern horizon; it is easy to recognize,

Anniversary

1799: Birth of William Lassell, an English amateur astronomer (by profession, a brewer) who became an expert observer. One of his telescopes was a 24-inch reflector, with which he discovered several hundred nebulæ. He also discovered Triton, the largest satellite of Neptune; Ariel and Umbriel, two of the satellites of Uranus; and (independently of Bond, in America) Hyperion, the seventh satellite of

(Continued opposite)

Scorpius

Greek letter	Name	Magnitude	Luminosity (Sun=1)	Distance (light-years)	Spectrum
α Alpha	Antares	1.0	7500	330	M1
β Beta	Graffias	2.6	2600	815	B0+B2
δ Delta	Dschubba	2.3	3800	550	B0
ϵ Epsilon	Wei	2.3	96	65	K2
ζ^2 Zeta^2	–	3.6	110	160	K5
η Eta	–	3.3	50	68	F2
θ Theta	Sargas	1.9	14 000	900	F0
ι^1 Iota^1	–	3.0	200 000	5500	F2
κ Kappa	Girtab	2.4	1300	390	B1
λ Lambda	Shaula	1.6	1300	275	B2
μ^1 Mu^1	–	3.0	1300	520	B1
π Pi	–	2.9	2100	620	B1
σ Sigma	Alniyat	2.9	5000	590	B1
τ Tau	–	2.8	3800	780	B0
υ Upsilon	Lesath	2.7	16 000	1560	B3
'G	–	3.2	96	150	K2

Saturn. He died in 1880. His first observatory was in Liverpool; in 1996 his 24-inch telescope was reconstructed, and is now in use at Liverpool Museum.

because Antares is marked out not only by its brightness and redness but also by the fact that it is flanked to either side by a fainter star. (So is Altair in Aquila – 7 August – but Altair is white, and much higher up). To southern-hemisphere observers, Scorpius is not far from the overhead point. It dominates the night sky, and will continue to do so well into southern spring.

In mythology, Scorpius has been identified with the insect which stung Orion in the heel and caused his untimely demise (12 January). When transferred to the sky, it was placed as far away from Orion as possible, so that there could be no further unpleasantness!

June 19

Antares: the Scorpion's Heart

It is usually said that Antares is the reddest of all the brilliant stars; its very name means 'the Rival of Mars' – Ares being the Greek equivalent of the war-god Mars. Certainly it is a superb sight in binoculars or a telescope. Its magnitude is 1.0, though very slightly variable. Though not so powerful or so vast as Betelgeux in Orion, it still ranks as a supergiant; its diameter is about 200 000 000 miles. Yet its outer layers, at

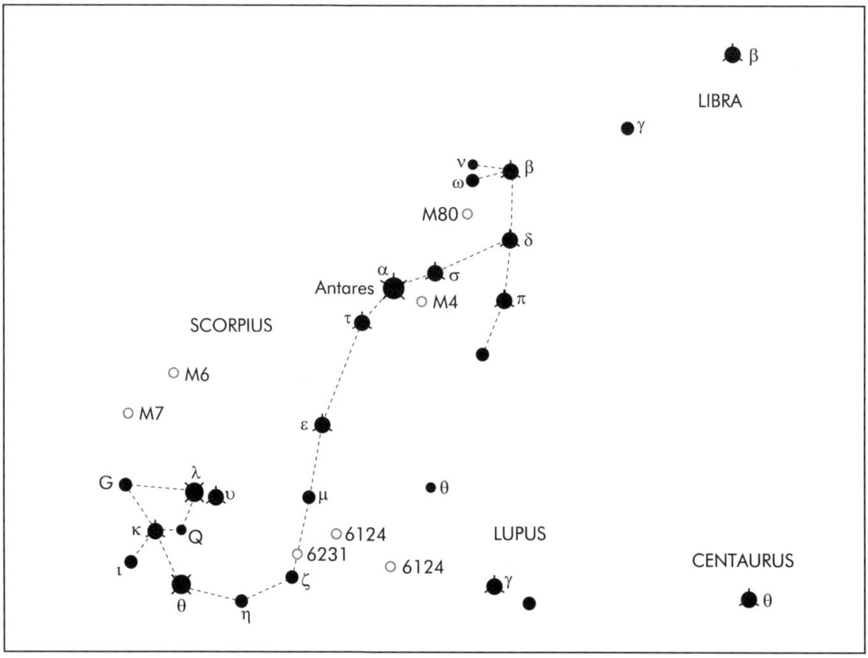

least, are very rarefied, and it is not likely to be more than ten times as massive as the Sun.

Antares has a companion of magnitude 5.4, at a separation of 2.6 seconds of arc. The companion would be an easy object were it not so overpowered; in fact it is rather elusive, and if you can see it with a telescope, of below four inches aperture you are doing well. It looks green, but this may be chiefly because of contrast with the red primary. It is also a source of radio waves. It is a true binary companion of Antares, but the period is very long, and has been given as 878 years.

The declination of Antares is (in round numbers) 27° S. This means that from latitudes on Earth north of 63° N, it can never be seen.

June 20

Neighbours of Antares

To make sure that there has been no mistake about identifying Antares, note that it has a fainter star to either side of it – a characteristic shared only by Altair in Aquila (7 August).

The stars flanking Antares are Tau, to the south, and Sigma or Alniyat, to the north. (Tau Scorpii never seems to have been given a proper name, and that of Sigma is seldom used.) Both are hot, bluish-white stars of type B, and both are over 500 light-years away. Sigma is rather the closer of the two, and is very slightly variable.

Certainly they appear dim when compared with Antares, but this is only because they are more remote. Earlier (21 May) we discussed absolute magnitude, which is the apparent magnitude which a star would have if it could be viewed from a standard distance of 32.6 light-years. From that range, Antares would shine at magnitude −4.7, Sigma −4.4 and Tau −4.1. From Earth, the maximum apparent magnitude of the planet Venus is −4.4, so that all three stars would be comparable with it, and in the case of Antares surpass it. They would indeed make an imposing trio, particularly as the redness of Antares would contrast with the bluish-white of its neighbours.

June 21

Globular Clusters in Scorpius

Continuing with Scorpius, our targets tonight are the two globular clusters M4 and M80. Of the two M4 is the more

Anniversary

1863: Birth of Max Wolf, a German astronomer who made notable contributions; he was the first to hunt for asteroids by means of photography, he observed comets (discovering the periodical comet which bears his name) and also discovered many dark nebulæ. He spent most of his life in Heidelberg. He died in 1932.

conspicuous, and it is very easy to find, because it lies only 1.5° west of Antares and is in the same binocular field with it. It is on the fringe of naked-eye visibility, and it is less condensed than most globulars, so that it is easier to resolve into stars.

M80, between Antares and Beta Scorpii, is rather elusive with binoculars, though very easy in a small telescope. It is small, round and condensed, with a bright nucleus; it is not too easy to resolve. In 1860 a bright nova appeared in it, and temporarily outshone it; the nova (T Scorpii) has never been seen again.

M4 is one of the closest of all globular clusters, and seems to be only 7500 light-years away; the distance of M80 is about 36 000 light-years.

Future Points of Interest

June solstice: The Sun reaches its northernmost point in the sky.

2001: Total eclipse of the Sun. Mid-eclipse falls at 12 h GMT. This is a long eclipse – totality lasts for 4 m 56 s – and should be well seen from the Atlantic and parts of Southern Africa.

2002: Mercury at western elongation.

June 22

The Scorpion's Head

The northernmost member of the long line of stars marking the Scorpion's body is Delta or Dschubba, magnitude 2.3. North of it is the Head, formed by three stars; Beta (magnitude 2.6), Nu (4.0) and Omega (also 4.0). All are double or multiple.

Beta, known either as Graffias or as Akrab, is a wide, easy pair; the magnitudes are 2.6 and 4.9, and the separation is over 13 seconds of arc, so that almost any telescope will split it. The brighter component is a very close binary.

Nu Scorpii is quadruple. The two main components are easily split; their magnitudes are 4.3 and 6.8, and the separation is over 41 seconds of arc. Each component is again double, the fainter component being the easier to resolve. A good six-inch telescope should show all four stars.

Finally there is Omega Scorpii, made up of two stars which can be seen separately with the naked eye; the magnitudes are 4.0 and 4.3. Here we have an optical pair, not a binary system; the fainter member is 170 light-years away, while the brighter lies at over 800 light-years. Omega Scorpii has an old proper name: Jabhat al Akrab.

June 23

The Scorpion's Sting

It is only right and proper for a scorpion to have a sting. In the sky it is marked by a group of stars of which one, Lambda or Shaula, is not far below the official first magnitude. The stars are: Lambda (magnitude 1.6), Upsilon (2.7), Kappa (2.4), G (3.2) and Q (4.3).

Shaula and Upsilon (Lesath) are so close together that they give the impression of being a wide double, but once again appearances are deceptive; Shaula is 275 light-years away from us, Lesath over 1500. Moreover Lesath is a real searchlight, equalling 16 000 Suns and therefore matching Betelgeux in Orion. Yet even Lesath pales before one of the apparently fainter stars close to the Sting, Iota[1]. It shines modestly at magnitude 3, but it is perhaps 200 000 times more luminous than the Sun – as powerful as Canopus.

The Sting is now very high as seen from Australia or New Zealand, but from Britain it is very low and hard to see. The declination of Shaula is –37°, so that it does not rise from anywhere north of latitude 53° N. From London it grazes the horizon; from the southernmost tip of England it rises to a few degrees, but it is invisible from the Midlands or anywhere further north. To see it at all from England you need a very clear sky right down to the horizon.

June 24

Open Clusters in Scorpius

This evening we will stay with Scorpius; it has so much to offer. North of the Sting, still in the southern part of the constellation but easily accessible from Britain, are two glorious open clusters, M6 and M7. Of course they are seen to best advantage from the southern hemisphere, and are then bright naked-eye objects.

M6 is known as the Butterfly Cluster. Its apparent diameter is rather more than half that of the full moon, and it contains over 50 stars; it has been said that the lines and chains of stars give the impression of the shape of a butterfly, but considerable imagination is needed. M7 is larger and brighter; it is indeed so big that it is best observed with binoculars, because it will not fit into the more restricted field of a telescope. It contains almost a hundred stars, and is one of the most impressive of all open clusters.

Another bright open cluster is C76 (NGC 6231), about 0.5° north of Zeta[2] Scorpii (which makes up a pair with its appar-

Future Points of Interest

2002: Penumbral eclipse of the Moon. Mid-eclipse occurs at 21 h 29 m GMT. However, as only 21% of the surface will enter the penumbra, it will not be easy to detect that anything unusual is happening.

ently fainter neighbour Zeta[1], which is however much more remote and luminous). C75 (NGC 6124) makes up a triangle with Zeta and the two stars Mu[1] and Mu[2]; it too is an easy object. Moreover the Milky Way flows right through Scorpius, and the whole area is very rich.

June 25

Scorpius X-1

Before taking temporary leave of Scorpius, let us look at a region a few degrees north of Antares. It is not marked by any bright star, but it contains Scorpius X-1, the first known celestial X-ray source.

X-rays from space are blocked by the Earth's atmosphere, so that they cannot be studied from ground level. In June 1962 American researchers launched a rocket carrying X-ray detectors, and found that they could locate what seemed to be a strong source unmarked by any visible object; it was named Scorpius X-1. Later in 1962 X-rays were detected from the Crab Nebula in Taurus (16 December), and in March 1966 Scorpius X-1 was identified with what looked like a faint bluish star. It is now known to be a binary system, in which one component is a neutron star. Many thousands of X-ray sources are known, but Scorpius X-1 remains the brightest of them.

June 26

Anniversary

1730: Birth of Charles Messier, a French observer who is best remembered for his catalogue of star-clusters and nebulæ, published in 1781. Ironically, Messier was not interested in nebulæ; he was a comet-hunter, and merely listed the nebular objects so that he could avoid wasting time by confusing them with comets. Few people now remember his thirteen comets, but everyone uses the M or Messier numbers. Messier died in 1817.

Libra

Libra, the Balance or Scales, is one of the least distinguished of the Zodiacal constellations, lying between Virgo and Scorpius; from Britain it is rather low in the south-west during evenings in late June. It is one of Ptolemy's original 48 constellations, but no mythology seems to be associated with it, apart from a very vague connection with Mochis, the inventor of weights and measures. It was once included in Scorpius as Chelæ Scorpii, the Scorpion's Claws, and the name survives, in modified form, in the two brightest stars. Beta Libræ is Zubenelchemale, the Northern Claw, while Alpha is Zubenelgenubi, the Southern Claw. Sigma Libræ (Zubenalgubi) was formerly included in Scorpius, as Gamma Scorpii.

Alpha, Beta and Gamma Libræ form a triangle. The leaders of the constellation are listed overleaf.

Alpha is a very wide double. The components are of magnitudes 2.8 and 5.2, and are separated by 231 seconds of arc,

Libra					
Greek letter	Name	Magnitude	Luminosity (Sun=1)	Distance (light-years)	Spectrum
α^2 Alpha[2]	Zubenelgenubi	2.7	28	72	A3
β Beta	Zubenelchemale	2.6	105	121	B8
γ Gamma	Zubenelhakrabi	3.9	16	75	G8
σ Sigma	Zubenalgubi	3.3	120	166	M4

so that this is a naked-eye pair. The two components share a common motion through space, but are a long way apart. The primary is white, the companion slightly yellowish.

Delta Libræ, near Beta, is an eclipsing binary of the Algol type. The range is from magnitude 4.9 to 5.9, in a period of 2.33 days, so that the star can always be followed with binoculars.

June 27

Beta Libræ: a Green Star?

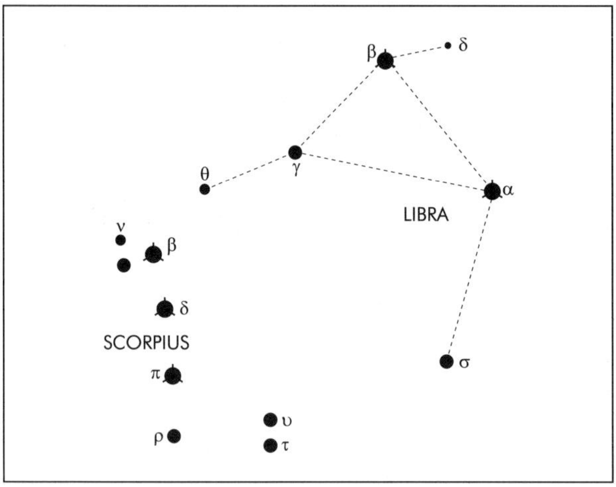

Beta Libræ, the 'Northern Claw' of the Scorpion, is actually the brightest star in Libra. Outwardly there is nothing exceptional about it, but there are two points which are worth mentioning. First, there is a suggestion that it may have faded since ancient times. Secondly, it is said to be the only single star which is green in colour.

Some of the early observers ranked it as being of the full second magnitude, but it is now well below this. However, the evidence in favour of change is very slender indeed.

The alleged greenness is more interesting. The Rev. T.W. Webb, author of a classic nineteenth-century book on astronomy (*Celestial Objects for Common Telescopes*) referred to its 'beautiful pale green hue', while T.W. Olcott, another very experienced observer, called it 'distinctly green'. Yet to most people it appears white, and I admit that I have never been able to see any colour in it. Look at it for yourself, and make up your mind!

Future Points of Interest

1999: Mercury at eastern elongation.

June 28

The First Point of Libra

We have seen that the Sun crosses the celestial equator twice per year, once when moving from south to north, in March and the other when moving from north to south (22 January). Originally these equinoctial points lay in Aries and Libra, respectively, and we still use these names, but the First Point of Aries has now been moved into Pisces, because of the effects of precession. Similarly, the First Point of Libra is no longer in Libra. It has moved well away, and now lies between Beta and Eta Virginis, in the Bowl of the Virgin (13 May).

Sweep through this area with binoculars, and you will see absolutely nothing of note; telescopically there are a few very faint galaxies in the general area, but that is all.

Anniversary

1818: Birth of Angelo Secchi, an Italian Jesuit. He was one of the great pioneers of stellar spectroscopy, and also made major contributions to solar research. He died in 1878.

1868: Birth of George Ellery Hale, the American astronomer who master-minded the construction of great telescopes (June 3). He was also a pioneer solar researcher. He died in 1938.

June 29

Secchi's Spectral Classification

Between 1864 and 1868 Angelo Secchi made the first really good spectroscopic survey of the brighter stars, and divided them into four types. It is interesting to look back at these and to check on some of the examples which he gave.

Type I. White stars, showing prominent hydrogen lines in their spectra; weak metallic lines. Examples: Sirius, Vega, Spica, Altair, Alkaid.

Type II. Yellow stars, hydrogen lines less proninent, metallic lines more so. Examples: Capella, Pollux, Arcturus, Aldebaran, Deneb.

Type III. Orange stars, with complicated banded spectra. Examples; Antares, Betelgeux, Alpha Herculis, Eta Geminorum, Mu Geminorum, Mira.

Type IV. Red stars, with strong carbon lines. No bright examples; mainly variable. Examples: R Hydræ, R Lyræ, R Leonis, R Andromedæ.

Generally Type I includes stars of types B and A in the modern system; Type II corresponds to types F and G; Type III, to K and M; and Type IV, to R, N and S. However, there are some curious anomalies – both Arcturus and Aldebaran were included in Secchi's Type II, but both are strongly orange, and belong to the modern type K. It is interesting to range round with binoculars and look at some of these examples: many are on view during evenings at this time of the year.

June 30

Meteorite Craters

Numbers of meteorite craters are now known. The most famous is the Meteor Crater in Arizona, which is a well-known tourist attraction; it is well over 4000 feet in diameter and over 500 feet deep. It was certainly produced by a meteorite which hit the desert long ago; the age may be of the order of 50 000 years. Craters which are undoubtedly of impact origin include:

Meteorite Craters

	Diameter	Discovered
Meteor Crater, Arizona	4150	1891
Wolf Creek, Australia	2790	1947
Henbury, Australia	660×360	1931 (13 craters)
Boxhole, Australia	575	1937
Odessa, Texas	560	1921
Tswainga, South Africa	430	1991
Waqar, Arabia	330	1932
Oesel, Estonia	330	1927
Campo del Ciela, Argentina	250	1933
Dalgaranga, Australia	230	1928

In 1947 there was a major fall in the Vladivostok area of Siberia; over 100 craters were produced, up to nearly 100 feet in diameter.

It is not always easy to decide upon the nature of a crater; for instance, the Vredefort Ring, near Pretoria, is often listed as an impact structure, but geologists who have made close studies of it are almost unanimous in saying that it is of volcanic origin. It has been claimed that a very ancient crater in the Chixulub region of Yucatán represents the impact of a meteorite which destroyed the dinosaurs; whether or not this is true, it is at least an intriguing theory.

Anniversary

1908: Fall of the Siberian impactor. It came down in the (fortunately) uninhabited Tunguska region, and blew pine-trees flat over a wide area. It was seen during descent, and outshone the Sun. The first expedition did not reach the site until 1927. No crater was found, and it may well be that the object was either a cometary fragment or else the nucleus of a small comet.

July

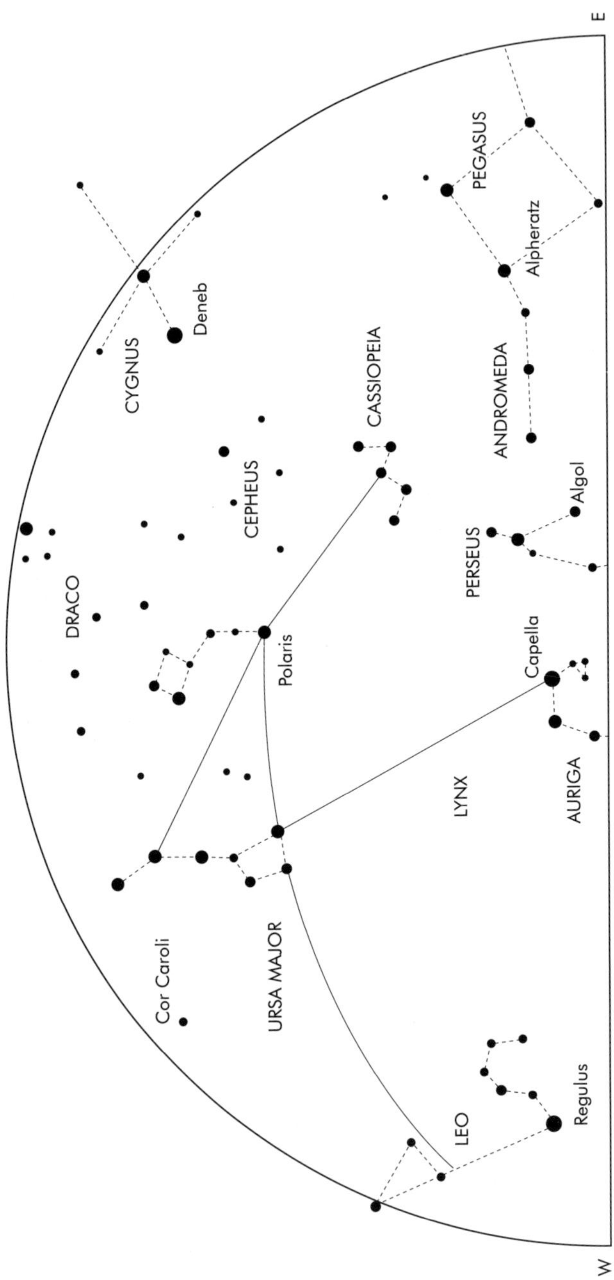

The Northern Hemisphere in July
Northern aspect (50° N 15 July 10pm)

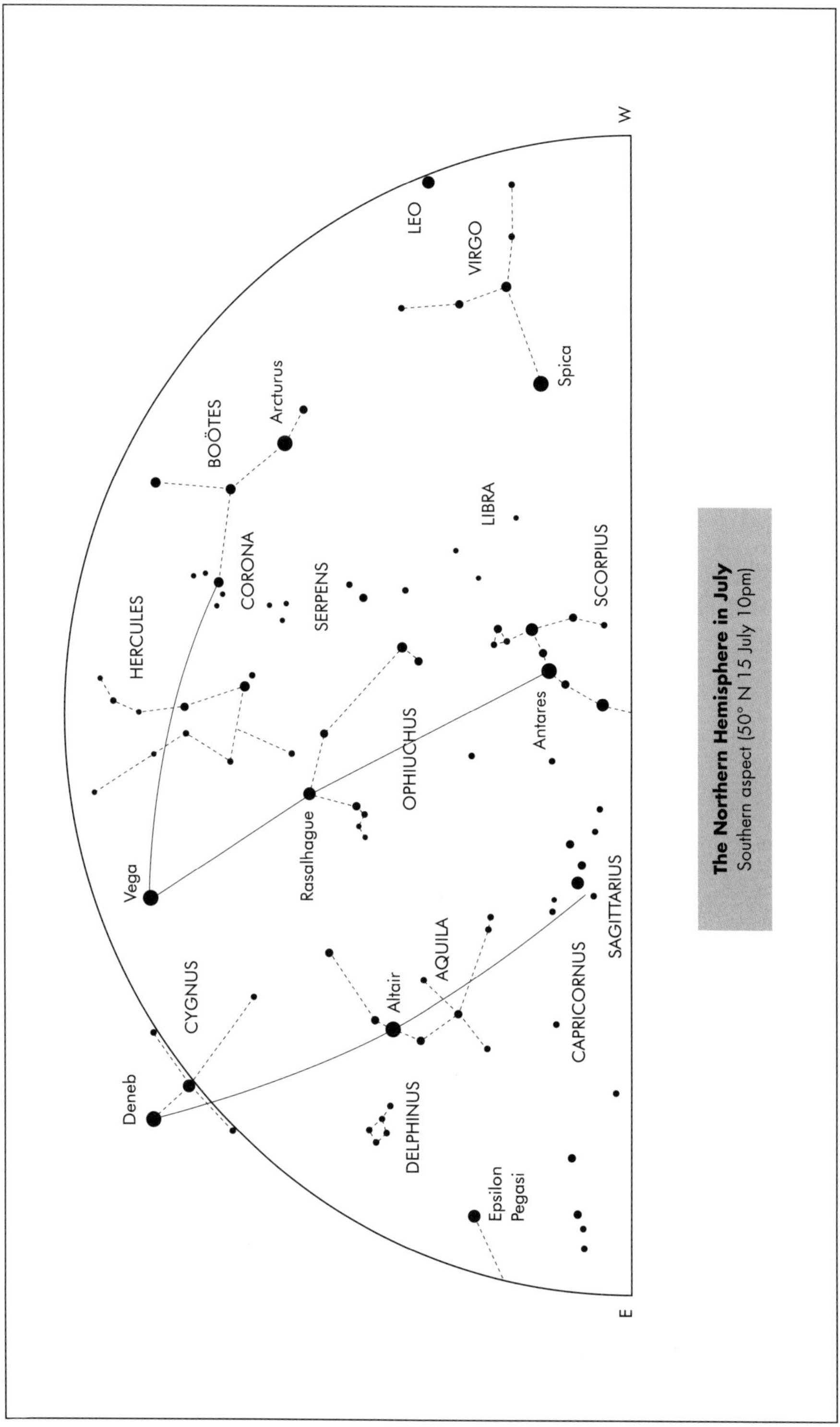

The Northern Hemisphere in July
Southern aspect (50° N 15 July 10pm)

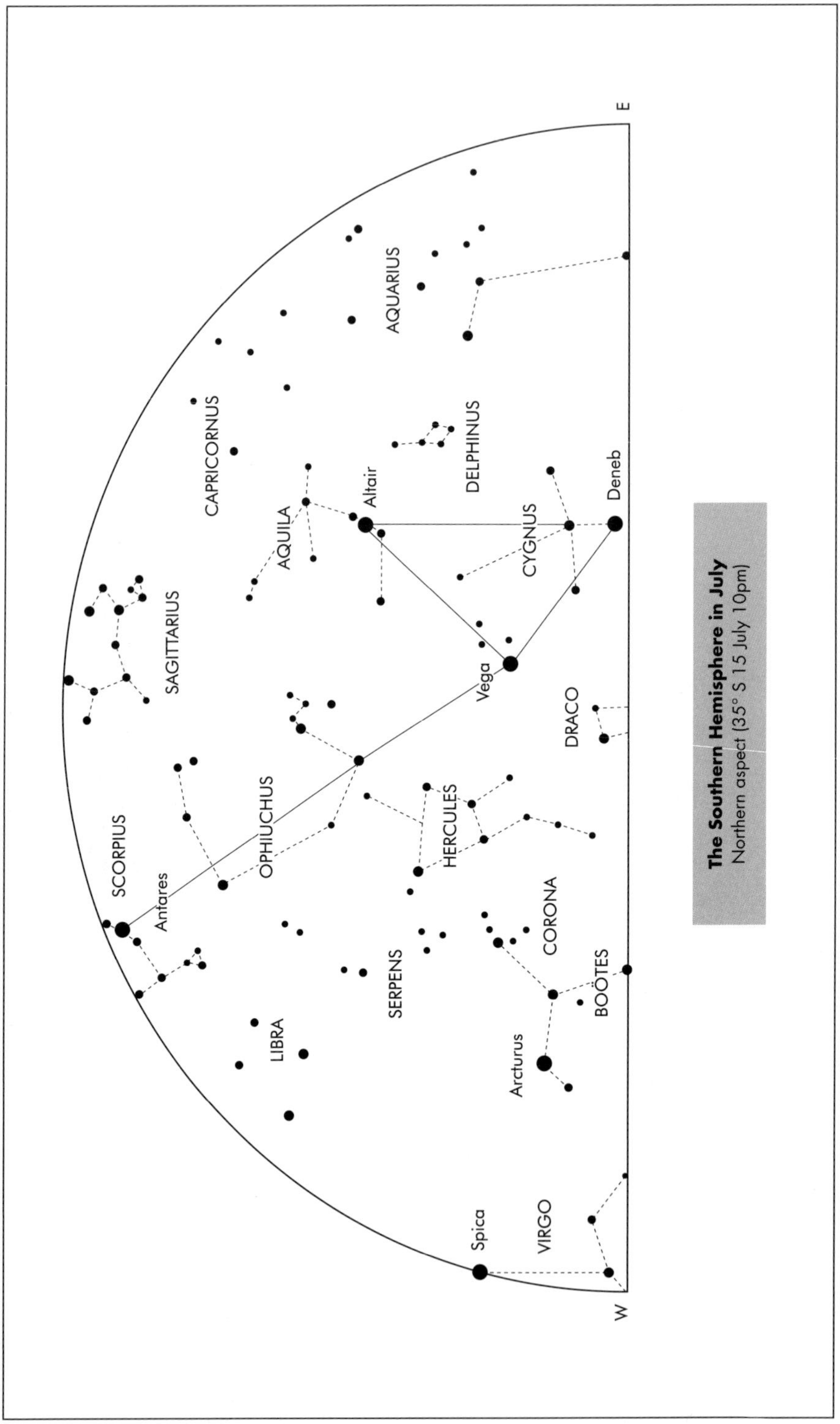

The Southern Hemisphere in July
Northern aspect (35° S 15 July 10pm)

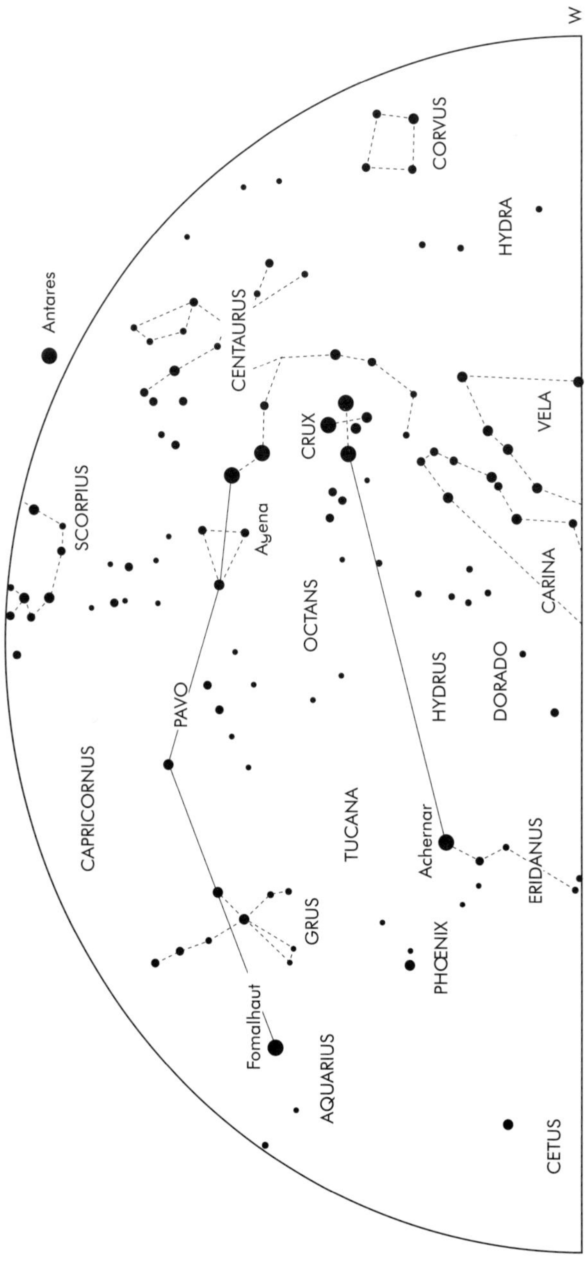

The Southern Hemisphere in July
Southern aspect (35° S 15 July 10pm)

July 1

The July Sky

July sees a marked change in the evening sky. The spring groups, Leo and Virgo, have faded into the twilight; Antares is very low in the south-west after dark. The Great Bear is in the north-west, with the W of Cassiopeia gaining altitude in the north-east. The brilliant blue Vega is almost overhead, which means that Capella, on the opposite side of the pole, is so low that it will probably not be seen, though from Britain it never actually sets. The 'Summer Triangle' of Vega, Deneb and Altair is now dominant; the star-clouds of Sagittarius can be seen over the southern horizon, but much of the southern aspect is filled with the large but relatively dim constellations of Hercules, Ophiuchus and Serpens. The Square of Pegasus is starting to come into view in the east; Arcturus shines in the west.

To southern-hemisphere observers, the scene is still dominated by Scorpius, while Sagittarius is not far from the zenith and the Milky Way is at its very best. The Southern Cross, with Centaurus, is now west of the overhead point; Fomalhaut in the Southern Fish is gaining altitude in the east. The Vega–Deneb–Altair triangle can be seen, but only Altair attains a respectable altitude; Spica has almost disappeared, and Arcturus is nearing the north-west horizon. Canopus is at its lowest in the south, and from Australia and South Africa it dips below the horizon, though from Wellington and the southernmost part of New Zealand it is circumpolar.

July 2

Hercules

Before concentrating on the Summer Triangle, it may be as well to dispose of some of the larger constellations now in the evening sky: Hercules, Ophiuchus and Serpens. Frankly, this is a rather confusing area, as there are few really well-defined patterns, and there is a marked paucity of bright stars. In the whole region there is only one (Alpha Ophiuchi) as bright as second magnitude.

Everyone must have heard of the legend of the Labours of Hercules, whose tasks ranged from cleansing the Augæan stables to killing the multi-headed Hydra, and dragging Cerberus, Pluto's dog, away from the Underworld. In mythology Hercules was a great hero, but in the sky we have to admit that he is not very prominent even though he occupies

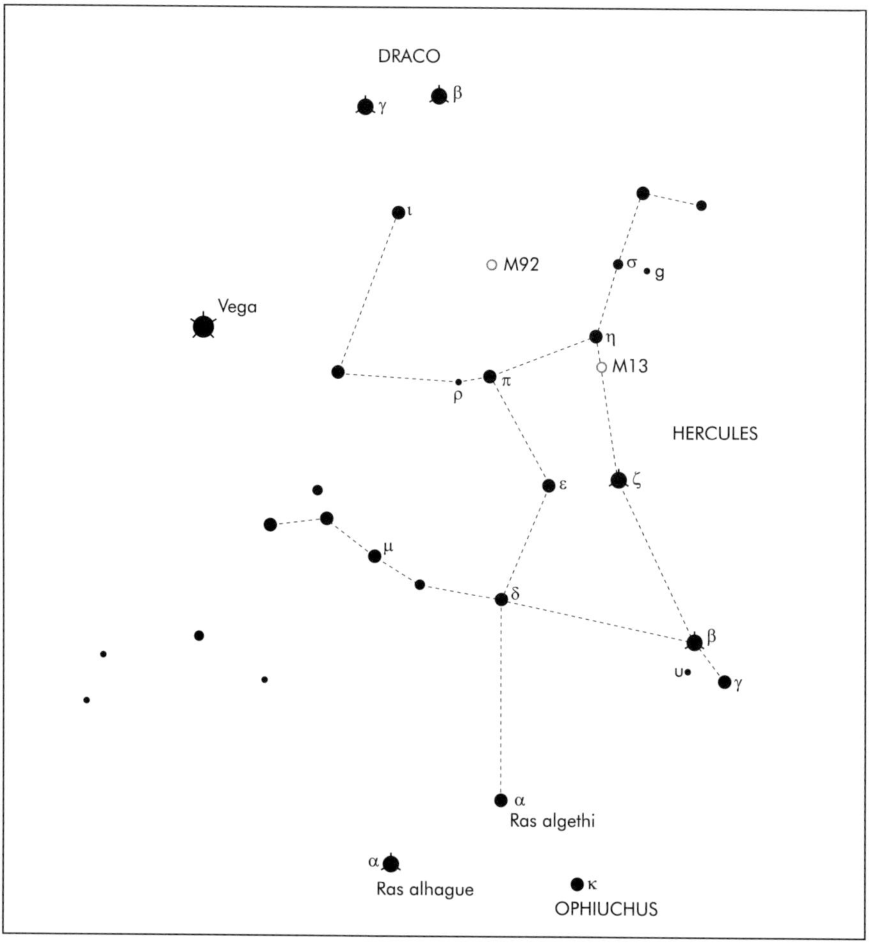

more than 1200 square degrees. The leading stars are listed below.

Hercules extends from near Vega and the head of Draco down as far as Ophiuchus; Alpha Herculis (Rasalgethi) is not far from Alpha Ophiuchi (Rasalhague) and is rather separated from the main constellation. Though the stars of

Hercules

Greek letter	Name	Magnitude	Luminosity (Sun=1)	Distance (light-years)	Spectrum
α Alpha	Rasalgethi	3 to 4	700	220	M5
β Beta	Kornephoros	2.8	58	100	G8
δ Delta	Sarin	3.1	37	91	A3
ζ Zeta	Rutilicus	2.8	5.2	31	G0
η Eta	–	3.5	16	68	G8
μ Mu	–	3.4	2.5	26	G5
π Pi	–	3.2	700	390	K3

Hercules are not bright, they make up a pattern which is not really hard to identify.

In 1934 a bright nova flared up in Hercules, near Iota (magnitude 3.8); it reached first magnitude, and was slow to fade. It was discovered by an amateur, J.P.M. Prentice of Suffolk, who had decided to go for a late-evening stroll after a spell of meteor watching, and suddenly noticed that, in his own words, 'there was something wrong with the head of the Dragon'. The nova, known officially as DQ Herculis, is now a faint object, and is an eclipsing binary system. Since then there have been only three comparable novæ: CP Lacertæ in 1936 (maximum magnitude 1.9), CP Puppis in 1942 (0.4) and V 1500 Cygni in 1975 (1.8), but all these faded much more quickly than Prentice's star.

July 3

Zeta Herculis

Having identified Hercules, it is worth taking a look at one of the two brightest stars in the constellation, Zeta – still sometimes known by its old proper name of Rutilicus. It is a fine binary, only just over 30 light-years away. The components are of magnitudes 2.9 and 5.5, respectively, and since the orbital period is a mere 34.5 years both the separation and the position angle alter quite quickly.

At present the separation is about 1.6 seconds of arc. The primary star is over five times as luminous as the Sun, while the secondary has about half the Sun's luminosity. If you use a telescope to observe Zeta Herculis now, and look again in a few years' time, you will notice a very obvious difference in appearance.

July 4

Globular Clusters in Hercules

Let us not be too disparaging about Hercules. True, the great hero is not brilliant, but it does contain two major globular clusters, M13 and M92. Of these M13 is the more spectacular, and is in fact the finest globular in the sky apart from the far-southern Omega Centauri (14 March, 25 April) and 47 Tucanæ (29 November).

M13 is just visible with the naked eye, but it is very dim, and it is not surprising that it was overlooked until Edmond Halley discovered it in 1714 – by chance, as he was the first to

Anniversary

1997: Landing of the Pathfinder probe on Mars. Pathfinder had been launched on 2 December 1996, and came down in the Ares Vallis region of Mars, which was an old flood-plain. It had been expected that the rocks carried down by a torrent of water in the ancient river would be of many different types, and so it proved. Instead of making a gentle landing, as the Vikings of 1976 had done, Pathfinder was encased in air-bags, and

(Continued opposite)

on impact bounced several times before coming to rest. Once upright, it released a small Rover, *Sojourner*, which crawled down the ramp on to the Martian surface and began to image and analyze the rocks. *Sojourner* was only about the size of a television set, but it was highly sophisticated, and sent back detailed information, while the main station (named in honour of the late Carl Sagan) provided a complete panoramic view of the whole area. The mission was a complete success, and paved the way for future probes designed to search for any trace of past or present life on Mars.

admit; at the time he was not cluster-hunting. He said that it was 'but a little patch but shows itself to the naked eye when the sky is serene and the Moon absent'. Telescopically it is magnificent; the edges are easy to resolve, but the centre is not and looks like a blaze of stars too close together to be seen individually. It is about 22 500 light-years away. To find it, look between Zeta and Eta, rather closer to Eta. It is unmistakable with binoculars.

M92 is comparable with M13, though just below naked-eye visibility; it is 37 000 light-years away, and lies between Eta and Iota Herculis. The nearest fairly bright star is the orange Pi, of magnitude 3.2.

Future Points of Interest

2001: Partial eclipse of the Moon. Mid-eclipse occurs at 14 h 57 m GMT, and 49% of the surface enters the main cone of shadow.

July 5

The Changing Moon

As there is a lunar eclipse on this date in 2001, it may be a good moment to look back at the Moon, which, after all, dominates the night sky for part of every month (admittedly to the intense annoyance of those who are anxious to observe objects such as faint nebulæ). During the period covered in this book, the Moon on 5 July appears as follows:

1998 Three days after First Quarter, and therefore gibbous.
1999 Two days before Last Quarter, and therefore gibbous in the morning sky.
2000 Three days after new, and therefore a slender crescent in the evening sky.
2001 Full (obviously).
2002 Three days after Last Quarter, and therefore a crescent in the morning sky.
2003 Two days before First Quarter, and well seen after sunset.

(Remember that First Quarter indicates the half-moon after new, and Last Quarter the half-moon after full.)

Lunar craters show marked changes in appearance according to the angle at which the sunlight strikes them. When a crater is near the terminator, it will be wholly or partly shadow-filled, and may look striking; near full, when there are almost no shadows, even a large crater may become hard to identify at all. The best way to learn one's way around is to select a number of craters, and draw them as often as possible. It does not take long to obtain a good working knowledge of the main surface features.

July 6

Rasalgethi

Alpha Herculis, often known by its old proper name of Rasalgethi, is variable; its fluctuations were discovered by Sir William Herschel as long ago as 1795. The range is between magnitudes 3 and 4, so that the star never becomes quite as bright as Beta or Zeta Herculis. It is a red supergiant, and is classed as a semi-regular variable with a period of between 90 and 100 days. Yet this period is very rough indeed, and for most of the time the magnitude seems to hover between 3.1 and 3.5. Naked-eye estimates are not difficult to make; good comparison stars are Delta Herculis (3.1), Gamma Herculis (3.8) and Kappa Ophiuchi (3.2).

Certainly the star is easy to locate, not far from the brighter Rasalhague or Alpha Ophiuchi. One sometimes feels that Rasalgethi itself ought to belong to Ophiuchus rather than Hercules.

According to the Cambridge catalogue it is 220 light-years away and 700 times as luminous as the Sun, though there are suggestions that these values may be underestimates. The redness is always very obvious, and the diameter cannot be far short of 200 000 000 miles.

There is a binary companion of magnitude 5.4, at a separation of 4.7 seconds of arc; the orbital period is of the order of 3600 years. This is a very easy pair, and is spectacular because the companion appears green, mainly no doubt by contrast with the ruddy hue of the primary. The companion is actually a very close binary.

One interesting feature of the system is that the primary star seems to be surrounded by a shell of incredibly tenuous expanding gas, which stretches out far enough to engulf the companion.

July 7

Telescopic Objects in Hercules

While we are with Hercules, let us look around for a few more objects within range of modest telescopes. There is, for example, a typical optical double: Delta Herculis (Sarin). The primary is of magnitude 3.1, the companion 8.2. There is absolutely no connection between them; the bright star is moving almost due south, the companion due west. When the first reliable measurement was made, in 1830, the separation was almost 26 seconds of arc. By 1960 it had decreased to less than 9 seconds of arc, and is now slowly widening again.

Rho Herculis, near the brighter Pi, is an easy double; the components are of magnitude 4.6 and 5.6, and the separation is over 4 seconds of arc. Mu Herculis is an even wider pair (separation 34 seconds of arc); the tenth-magnitude companion is itself a very close and difficult binary.

Close to Beta and Gamma is a red Mira-type variable, U Herculis, which can reach magnitude 6.5 at maximum; the period is around 406 days. The 1997 maximum fell on September 1. At minimum the star falls to well below the thirteenth magnitude, so that a powerful telescope is needed to show it. There is nothing special about it, but it is worth locating. The position is RA 16 h 25 m, declination +18° 54′. Finally you may care to seek out g (30) Herculis, near the fourth-magnitude Sigma; here we have a red semiregular star, with a magnitude range of 5.7 to 7.2 and a rough period of 70 days.

July 8

Ophiuchus

Ophiuchus, the Serpent-bearer – once known as Serpentarius – covers almost 950 square degrees, and is of course one of Ptolemy's original 48 constellations. In mythology it is said to represent Æsculapius, son of Apollo, who became so skilled in medicine that he could even restore the dead to life. To avoid depopulation of the Underworld, Jupiter reluctantly disposed of him with a thunderbolt, but in compensation elevated him to celestial rank and placed him in the sky. The leading stars are listed below.

Ophiuchus is crossed by the equator; Rasalhague lies at declination 12° 34′ N, Theta at 25° S. Rasalhague, not far from Rasalgethi in Hercules (July 6) is the only really bright

Ophiuchus

Greek letter	Name	Magnitude	Luminosity (Sun=1)	Distance (light-years)	Spectrum
α Alpha	Rasalhague	2.1	58	62	A5
β Beta	Cheleb	2.8	96	121	K2
δ Delta	Yed Prior	2.7	120	140	M1
ϵ Epsilon	Yed Post	3.2	58	104	G8
ζ Zeta	Han	2.6	5000	550	O9.5
η Eta	Sabik	2.4	26	59	A2
θ Theta	–	3.3	1320	590	B2
κ Kappa	–	3.2	96	117	K2
ν Nu	–	3.3	60	137	K0

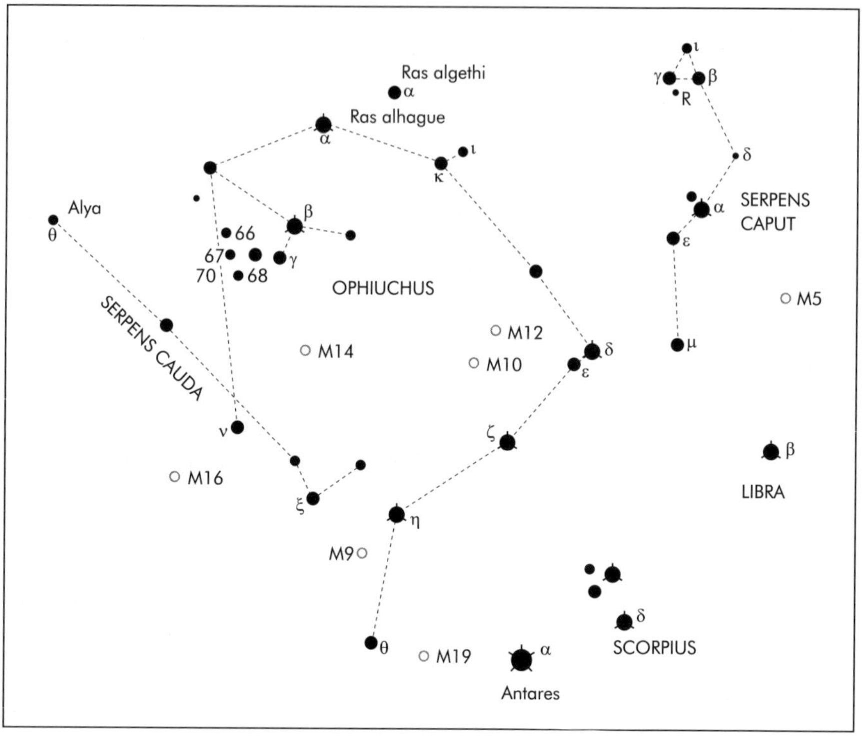

star, but the two Yeds, Delta and Epsilon, form a notable pair with strongly contrasting colours; Epsilon is only slightly yellowish, while Delta is orange-red. It is worth looking at this pair with binoculars.

July 9

> ### *Anniversary*
> 1979: The American probe Voyager 2 passed Jupiter, at a range of 443 750 miles. It sent back excellent data, and then went on to encounter Saturn (1981), Uranus (1986) and Neptune (1989). It is now on its way out of the Solar System.

Poniatowski's Bull

Some distance south of Rasalhague it is easy to identify Beta Ophiuchi or Cheleb (magnitude 2.8) and Gamma (3.7). Adjoining Gamma is a little group of stars, none of which have Greek letters; they are 66 Ophiuchi (magnitude 4.6), 67 (4.0), 68 (4.4) and 70 (4.0). These were once formed into a separate constellation, Taurus Poniatowski or Poniatowski's Bull. To quote Admiral Smyth, in his famous book *Cycle of Celestial Objects*:

> Taurus Poniatowski is a small asterism placed in the heavens, in 1777, by the Abbé Poczobut, of Wilna, in honour of Stanislaus Poniatowski, King of Poland; a formal permission to that effect having been obtained from the French Academy.

Future Points of Interest

2001: Mercury at western elongation

> It is between the shoulder of Ophiuchus and the Eagle, where some stars form the letter V, and from a fancied resemblance to the zodiac-bull and the Hyades, became another Taurus. Poczobut was content with 7 component stars, but Bode has scraped together no fewer than 80.

Alas, Poniatowski's Bull has not survived; but one of its stars, now known as 70 Ophiuchi, is a notable binary. The components are of magnitudes 4.2 and 6.0, and the separation is almost 2 seconds of arc; the orbital period is 88 years. Both components are dwarfs, and together amount to about half the Sun's luminosity. The system is relatively nearby, at a mere 16.5 light-years. It is worth identifying, even if only in memory of the forgotten Bull.

Anniversary

1992: The Giotto spacecraft, having encountered Halley's Comet (13–14 March 1986) passed within 130 miles of the nucleus of a much fainter comet, Grigg-Skjellerup, which has an orbital period of 5.1 years. Giotto's camera was no longer working, but it obtained valuable data from Grigg-Skjellerup, which is a much older comet than Halley's; the density of gas near the nucleus was greater than expected, and there was a considerable quantity of fine 'dust'. Giotto is still in solar orbit, though it does not retain enough gas to send it on to another comet rendezvous, as had originally been hoped.

July 10

Ophiuchus in the Zodiac

In addition to being crossed by the celestial equator, Ophiuchus is also crossed by the ecliptic, which runs slightly north of Theta Ophiuchi. This means that the Serpent-bearer contains part of the Zodiac, between Scorpius and Sagittarius, and planets may pass through it as well as the Sun and Moon.

Of course this has always been known, but it suddenly became 'headline news' in January 1995, following a misleading programme on British television. It was solemnly claimed that 'astronomers had found a new Zodiacal sign', which would affect astrological predictions. In fact the astrological 'signs' are wrong in any case, because of the effects of precession (18 October), and moreover the constellation patterns are wholly arbitrary and meaningless; it so happens that we use the Greek system rather than, say, the Chinese or the Egyptian. It is fair to say that astrology proves only one scientific fact: 'There's one born every minute'!

July 11

Barnard's Star

There are some rich regions in Ophiuchus, notably the lovely star-clouds near Rho; but before we take our leave of the Serpent-bearer, let us pause to seek out a much fainter object – Barnard's Star, otherwise known by its catalogue number of Munich 15040. Its position is RA 14 h 55 m, declination +04° 33′, in the region of the rejected Taurus Poniatowski (9 July), but the magnitude is only 9.5, so that a telescope is essential.

Barnard's Star is so named because it was first noted in 1916 by E.E. Barnard, from America. It is a dim red dwarf with only 1/2500 the luminosity of the Sun; the diameter is probably of the order of 140 000 miles, about twice that of Saturn. By stellar standards it is cool, with a surface at less than 3000 °C. Very slight irregularities in its motion indicate that it may be attended by a body which is too lightweight to be a normal star, and could well be a planet.

At only 6 light-years, Barnard's Star is our nearest stellar neighbour apart from the Alpha Centauri trio. Its rapid proper motion – 10.29 seconds of arc per year – means that in only 180 years it crawls across the sky by an apparent distance equal to the diameter of the full moon; no other star has a proper motion of as much as 5 seconds of arc annually.

If you have the necessary equipment, it is interesting to locate Barnard's Star, but I warn you that it is not too easy to identify.

Finally, Ophiuchus is rich in globular clusters, and there are no less than seven in Messier's catalogue: Nos. 9, 10, 12, 14, 19, 62 and 107. All can be found with a small telescope.

July 12

Serpens: the Broken Constellation

Serpens is one of Ptolemy's original constellations. There seem to be no definite mythological associations, but certainly there is a struggle going on with Ophiuchus, and the luckless reptile seems to be having very much the worst of the encounter, because it has been pulled in half. The head (Caput) is quite separate from the body (Cauda), and Ophiuchus lies in between. Caput begins near Corona Borealis (15 June), while Cauda adjoins Aquila (5 August).

Caput is much the more prominent of the two sections, and has one reasonably bright star, Alpha or Unukalhai. The leaders are listed below.

The actual head is made up of Beta, Gamma, and the reddish Kappa (magnitude 4.1). Directly between Beta and Gamma, and therefore easy to find when near maximum, is the red Mira variable R Serpentis. The range is from magnitude 5.1

Serpens Caput

Greek letter	Name	Magnitude	Luminosity (Sun=1)	Distance (light-years)	Spectrum
α Alpha	Unukalhai	2.6	96	85	K2
β Beta	–	3.7	50	124	A2
γ Gamma	–	3.8	3.2	39	F6
δ Delta	Tsin	3.8	17	88	F0
ϵ Epsilon	–	3.7	19	107	A2
μ Mu	–	3.5	50	144	A0

to below 14, and the period is 356 days, so that the star reaches maximum nine days earlier each year and there are spells when maximum occurs when the star is too near the Sun to be seen at all (it is interesting to consider a star with a period of exactly one year!). The 1997 maximum fell on February 15, and from this it is easy to work out the relevant dates for the period covered in this book, bearing in mind that, as with all Mira stars, the period is not absolutely constant.

Delta, south of the Head, is an easy double; the components are of magnitudes 4.1 and 5.2, and the separation is 4.4 seconds of arc. This is a binary, with an orbital period of 3168 years.

Anniversary

1995: First light on the VATT (Vatican Advanced Technology Telescope) on Mount Graham, in Arizona. This is a 1.8-metre reflector of the most modern type. When it was planned, local conservationists objected strongly, because the area is a habitat for some rare red squirrels, and for a time arguments raged, so that astronomers felt that they were being driven nuts. Eventually a compromise was reached; the observatory was erected, and the squirrels and the astronomers exist happily side by side.

July 13

The Serpens Globular

The Serpent's head contains M5, one of the finest globular clusters in the sky. It is not far from the reddish Unukalhai, and is only just below naked-eye visibility, so that in binoculars it is unmistakable. It is difficult to better the description of it given long ago by Mary Proctor, a well-known populariser of astronomy. After using a powerful telescope to look at it, she wrote: 'Myriads of glittering points shimmering over a soft background of starry mist, illuminated as though by moonlight, forming a striking contrast to the darkness of the night sky. For a few blissful moments, as the watcher gazed upon this scene, it suggested a veritable glimpse of the heavens beyond.'

A small telescope will resolve the outer parts of M5. It is 27 000 light-years away, and seems to be very old, so that it contains a large number of highly evolved, variable red stars. It is also rich in short-period variables. All in all, M5 is the most magnificent globular visible from Britain, apart from M13 in Hercules (4 July). It lies only 2° north of the celestial equator, and is therefore equally well seen from the northern and the southern hemispheres of the Earth.

Anniversary

1965: Mariner 4 passed Mars at a range of 6084 miles returning 21 images – the first close-range views of Mars, showing craters. Contact with Mariner 4 was lost on 20 December 1967.

July 14

The Serpent's Body

Serpens Cauda is entwined with Ophiuchus. Its leaders are tabulated overleaf.

The most notable object in Serpens from the point of view of the owner of a small telescope is Theta, or Alya. It is easy to find, as it lies more or less in line with the three stars in Aquila south of Altair: Theta, Eta and Delta Aquilæ (5

Serpens Cauda					
Greek letter	Name	Magnitude	Luminosity (Sun=1)	Distance (light-years)	Spectrum
η Eta	Alava	3.3	17	52	K0
θ Theta	Alya	3.4	12+12	102	A5+A5
ξ Xi	–	3.5	17	75	F0

August). Alya is one of the finest doubles in the sky. The components are equal at magnitude 4.4, and the separation is over 22 seconds of arc, so that the pair can be split even with good binoculars. No doubt the two Thetas are associated, but the orbital period must be immensely long.

Future Points of Interest

1998: Occultation of Jupiter by the Moon. This occurs at 19 h GMT, and will be observable from New Zealand, the South Pacific, and Antarctica.

July 15

The Eagle Nebula

The Hubble Space Telescope has taken many spectacular pictures, but none more striking than that of M16, the Eagle Nebula in Serpens (once also referred to as the Star Queen Nebula). This is a region where new stars are being formed; it is 7000 light-years away, and consists of a bright open cluster immersed in a vast cloud of glowing nebulosity. It lies at the very edge of Serpens, close to the border with the little constellation of Scutum, the Shield (14 August); the most convenient guide to it is Gamma Scuti, of magnitude 4.7. M16 lies one degree north and 2.5° west of Gamma.

The newly forming stars are contained in masses of gas known as ECGs or Evaporating Gaseous Globules, which are themselves contained in vast columns of gas and dust which look uncannily like elephants' trunks. Gradually the trunks are eroded by short-wave radiation from nearby hot stars, and eventually even the ECGs are eroded, so that the embryo stars can emerge. M16 has been known for a long time – it was first noted in 1746 by de Chéseaux – but only the Hubble Space Telescope has been able to show it in its full glory.

July 16

The Central Bay

Sinus Medii, at the centre of the Earth-turned hemisphere of the Moon, can be seen whenever the phase is greater than half; to check whether it is visible tonight, refer to the table on page ix. This was the target of two Surveyors; No. 4, which failed, and No. 6, which landed on 9 November 1967 and returned 29 000 images.

Anniversary

1967: The American probe Surveyor 4 crashed on to the Moon, in the region of the Sinus Medii (Central Bay). No data were returned.

Future Points of Interest

2000: Total eclipse of the Moon. Mid-eclipse occurs at 17 h 57 m GMT. Totality lasts for exactly one hour.

This is a relatively small bay, 217 miles across and covering 20 000 square miles. It leads off the Mare Nubium or Sea of Clouds, north of the great walled plain Ptolemæus (9 November), It is easy enough to identify; the floor is comparatively smooth, with two small but well-formed craters, Bruce (4 miles in diameter) and Blagg 3.3 miles). There is no high mountain border, but, as with all the lunar maria, the floor is darkish.

Anniversary

1894: Birth of Georges Lemaître, a Belgian priest. He was a leading mathematician, and it was his paper, published in 1927, which led to what is now called the Big Bang theory of the origin of the universe. During the war of 1914–18 he won the Croix du Guerre. Later he became a professor at Louvain University. He died in 1966.

July 17

Sagittarius

Let us now turn to Sagittarius, the Archer, the southernmost constellation of the Zodiac. It follows Scorpius round. From British latitudes it is now very low in the south – part of it never rises at all – though from Australia or New Zealand it is near the zenith. The sequence of Greek letters is wildly out of order here; the brightest stars are Epsilon, Sigma and Zeta, with Alpha and Beta very much 'also rans'! The leaders of Sagittarius are listed in the table below.

Of the brightest stars only Nunki (Sigma) reaches a respectable height above the British horizon. One way to locate Sagittarius is to use Deneb and Altair as pointers; a line through them, continued to the horizon, will reach Sagittarius. From southern latitudes, of course, it cannot be missed. There is no really distinctive shape, though the pattern is often nicknamed 'the Teapot'. Note also the little curved line of stars marking Corona Australis (the Southern Crown).

Future Points of Interest

1998: Mercury at eastern elongation.

Sagittarius

Greek letter	Name	Magnitude	Luminosity (Sun=1)	Distance (light-years)	Spectrum
γ Gamma	Alnasr	3.0	60	117	K0
δ Delta	Kaus Meridionalis	2.7	96	81	K2
ϵ Epsilon	Kaus Australis	1.8	110	85	B9
ζ Zeta	Ascella	2.6	50	78	A2
η Eta	–	3.1	800	420	M3
λ Lambda	Kaus Borealis	2.8	96	98	K2
μ Mu	Polis	3.9	60 000	3900	B8
ξ^2 Xi^2	–	3.5	82	144	K1
π Pi	Albaldah	2.9	525	310	F2
τ Tau	–	3.3	82	130	K1
ϕ Phi	–	3.2	250	244	B8
σ Sigma	Nunki	2.0	525	209	B3

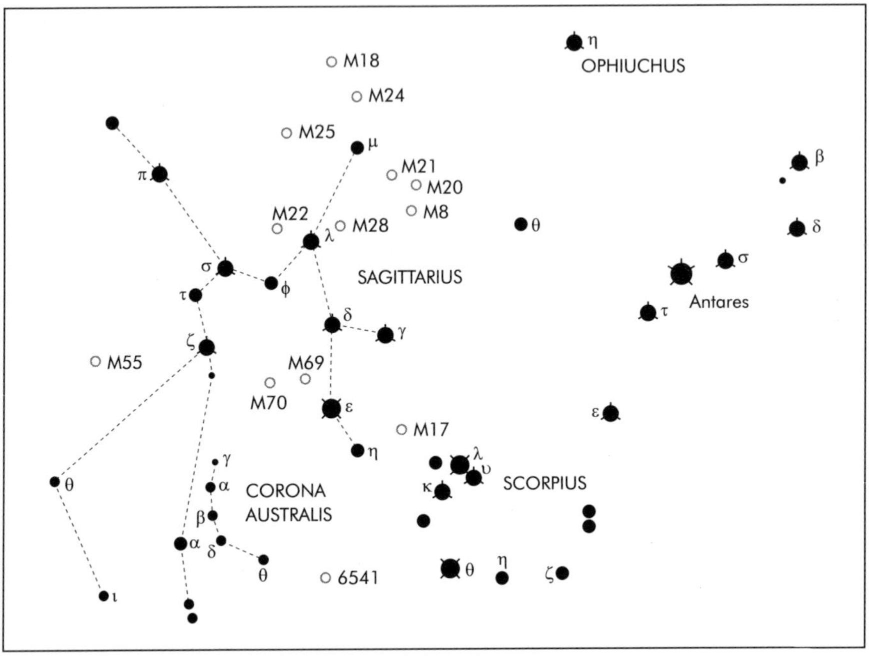

July 18

Southernmost Sagittarius

British observers always regret that Sagittarius is so far south; it is a truly magnificent constellation, and contains the richest part of the Milky Way.

To recapitulate for a moment: declination is the celestial equivalent of latitude, so that the north pole is at +90°, the equator at 0° and the south pole at –90°. To see what stars are available subtract your latitude on Earth from 90°. If your latitude is +51°, as in parts of southern England, the key figure is 90–51 = 39. Any star north of declination +39° will be circumpolar, and no star south of declination –39° will rise.

In Sagittarius the brightest star, Epsilon (Kaus Australis) is at declination –34°, so in theory it will rise to an altitude of 5°, though this is complicated by the effects of refraction. Rukbat or Alpha Sagittarii, magnitude 4.0, is at declination –44°, and will be inaccessible. Declinations of the main stars of Sagittarius are as follows:

Pi	–21°	(in round numbers, of course)
Sigma	–26°	
Delta	–30°	
Zeta	–30°	

Epsilon	–34°
Eta	–37°
Alpha	–44°

From these figures, it is easy to work out which stars will rise above your horizon and which will not.

Anniversary

1846: Birth of Edward Charles Pickering, a leading American astronomer who became Director of the Harvard College Observatory and held office for 43 years. He specialized in studies of stellar spectra, and made outstanding contributions. He was the elder brother of W.H. Pickering, the lunar and planetary observer. E.C. Pickering died in 1919.

1969: Neil Armstrong and Buzz Aldrin, in *Eagle* (the lunar module of Apollo 11) landed on the Moon; they came down in the Mare Tranquillitatis (latitude 00° 67′ N, longitude 23° 49′ E). They actually set foot on the surface in the early hours of 20 July.

July 19

The Star-Clouds of Sagittarius

The Milky Way is at its very best in Sagittarius, and the 'star-clouds' are unrivalled. We are in fact looking toward the heart of the Galaxy, and we cannot see the actual centre only because there is too much obscuring material in the way.

Looking at the star-clouds, it is not easy to realise that the stars there are not so closely packed that they almost touch. We are dealing with a simple line-of-sight effect. The Galaxy is a flattened system, and the centre itself lies beyond the star-clouds, at a distance of rather less than 30 000 light-years.

The Galaxy is rotating, and the Sun is in orbit round the centre, carrying the Earth and the other members of the Solar System with it. The orbital period is about 225 000 000 years, sometimes termed the 'cosmic year'. One cosmic year ago, the most advanced life-forms on Earth were amphibians; even the dinosaurs had yet to make their entry. It is interesting to speculate as to what conditions may be like one cosmic year hence.

Anniversary

1976: The lander of the Viking 1 probe landed on Mars. The site was Chryse; latitude 22°.4 N, 470°.5 W. The experiment was a complete success, and provided our first views of Mars from the actual surface. No definite signs of Martian life was found.

July 20

The Centre of the Galaxy

For many years it was thought that the Sun must lie close to the centre of the Galaxy. This was disproved by Harlow Shapley, in 1917. He studied the distribution of the globular clusters, which form a sort of 'outer surround', and realized that their distribution was not symmetrical; most of them lie in the southern hemisphere of the sky, with a marked concentration in Sagittarius. It followed that we are having a lop-sided view, and that the Sun is a long way from the galactic centre. The distance is rather less than 30 000 light-years.

But how can we investigate the centre, if we cannot see through the star-clouds? The answer is to use 'invisible astronomy'; infra-red and radio waves are not blocked by the interstellar matter. There is for example the VLA or Very Large Array in New Mexico, which involves 27 separate antennæ. Using this and other installations, it has been possible to locate a small, compact source at the galactic centre which is called Sagittarius A-* (pronounced Sagittarius A-star). This may well mark the actual centre. Its precise nature is still uncertain, but there is every reason to suppose that there is a massive black hole in the central region.

Much remains to be learned, but at least we are sure that when we view in the direction of the Sagittarius star-clouds we are looking in the right direction. They are always worth sweeping with binoculars or a wide-field telescope.

July 21

Nebulæ in Sagittarius

Sagittarius contains more Messier objects than any other constellation. Bearing in mind that the whole region is so rich, it is not too easy to sort them out, but three gaseous nebulæ are particularly notable: M8 (the Lagoon Nebula), M17 (the Omega Nebula) and M20 (the Trifid Nebula). All three are within range of binoculars, though of course telescopes are needed to show them really well, and to bring out the fine details means using photography.

M8, the Lagoon, lies not far from Lambda Sagittarii, at RA 18 h 04 m, declination –24° 23′. It is an emission nebula 6500 light-years away, associated with an open cluster; it contains a number of small dark masses known as Bok globules in honour of Bart J. Bok, the Dutch astronomer who first drew attention to them. They are believed to be embryo stars. Near it is M20, the Trifid Nebula (RA 18 h 03 m, declination –23° 02′), another emission nebula – a large mass of glowing gas excited by very hot stars in or near it, so that it sends out some light on its own account. Its diameter seems to be about 30 light-years. M17, the Omega Nebula, is also known as the Swan or the Horseshoe; its position is RA 18 h 21 m, declination –16° 11′, so that it is more readily accessible from British latitudes. It lies on the border of Sagittarius and Scutum, and a useful guide star is the fifth-magnitude Gamma Scuti. It shows up as a bright 'bar' across the centre of an open cluster. Camille Flammarion, the great French planetary observer and populariser of astronomy, described it as being 'like a smoke-drift, fantastically wreathed by the wind'.

Future Points of Interest

1997: Neptune at opposition. The synodic period of Neptune is only 367.5 days, so that all the opposition dates in the period covered by this book are in July or August.

Anniversary

1784: Birth of Friedrich Wilhelm Bessel, the German astronomer who was the first man to measure the distance of a star (61 Cygni, in 1838, see 21 August). In 1810 he became Director of the Königsberg Observatory, retaining the position until his death in 1846. He determined the positions of 75 000 stars, and predicted the positions of the then-unknown companions of Sirius and Procyon.

1962: Attempted launch of the Mariner 1 probe toward Venus. Unfortunately it fell into the sea immediately after take-off.

July 22

Messier 22

Still with Sagittarius, let us look next at the globular cluster M22, which lies 1.5° north and 2° east of Lambda. Its position is RA 18 h 36 m, declination –23° 54′.

M22 was the first globular cluster to be discovered (though no doubt Omega Centauri and 47 Tucanæ had been seen earlier by southern-hemisphere observers who did not know what they were). M22 was first reported in 1665 by a totally obscure astronomer named Abraham Ihle, about whom virtually nothing is known except that he was German. Edmond Halley described M22 as 'small and luminous'. It is in fact one of the most spectacular globular clusters in the sky, and binoculars show it well; it is easy to resolve, even near its centre – indeeed, this is probably the easiest of all globulars to resolve into stars, because it is unusually close. Its distance from us is no more than 9600 light-years, which is much closer than most globulars. The only problem in identifying it is that it lies in a very rich area.

Sagittarius abounds in globular and open clusters. Many are within range of modest telescopes; all in all, this is one of the most rewarding areas of the sky from the viewpoint of the stellar enthusiast.

Future Points of Interest

Beginning of the annual Perseid meteor shower.

1998: Opposition of Neptune.

July 23

The Missing Messier Objects

Look on your star-maps for M24, and you will find it, rather to the north of Mu Sagittarii; RA 18 h 15 m, declination –18° 26′. But it is not a nebular object at all; it is merely a star-cloud in the Milky Way, so that logically it has no place in Messier's list. It is not too easy to locate, and it has no proper claim to separate identity.

Messier's catalogue contained 104 objects; six more have been added to it by later observers, though these additions are not always recognized. In addition there are a few numbers which do not correspond to either nebulæ or clusters. They are:

M24 (NGC 6603). Star cloud in Sagittarius.

M40 A pair of stars in Ursa Major; RA 12 h 21 m, dec. +58° 20′. No NGC number.

M73 (NGC 6994). RA 20 h 56 m, dec. –12° 50′. A group of four faint stars in Aquarius.

M91 Missing; possibly a comet which Messier failed to recognize for what it was. Efforts have been made to identify M91 with NGC 4571, a faint galaxy in Virgo, but the identification is at best dubious.

M102 Missing. Messier described it as 'a nebula between Omicron Boötis and Iota Draconis', but not identifiable; it may have been a case of confusion with M101 in Ursa Major.

There are also slight doubts about the identities of M47 with NGC 2422, an open cluster in Puppis, and M48 with NGC 2548, an open cluster in Hydra (March 9). But in view of the fact that Messier used a tiny telescope, we have to admit that he acquitted himself well – even if he listed the nebulæ as 'objects to avoid' during his searches for comets.

July 24

Corona Australis: the Southern Crown

The Southern Crown, Corona Australis (sometimes known as Corona Austrinus), is not so conspicuous as its northern namesake, and it is moreover too far south to be seen from British latitudes. Its brightest star, Alpha (Meridiana), lies at declination –38°, but is only of fourth magnitude. It can be seen from southern Europe.

Corona Australis adjoins Alpha and Beta Sagittarii (Rukbat and Arkab), which, as we have seen, are surprisingly faint in view of the fact that they have been allotted the first two Greek letters in Sagittarius. Arkab is a wide optical double, easily separable with the naked eye; Rukbat is a single star of type A. Both are of the fourth magnitude.

The Southern Crown consists of a gently curved line of stars; Gamma, Alpha, Beta, Delta, Zeta and Theta, all of which are between magnitudes 4.1 and 4.7. Despite their dimness, their arrangement makes them easy to identify when they are sufficiently high above the horizon. On the border of the constellation, between Theta Coronæ Australis and the bright Theta Scorpii (19 June) lies a bright globular cluster, C78 (NGC 6541), which has an integrated magnitude of 6.6 and is therefore an easy binocular object. Its position is RA 18 h 08 m, declination –43° 42′.

July 25

The Summer Triangle

It is now time to discuss the 'Summer Triangle' of Vega, Deneb and Altair, which will occupy us for quite a number of nights. Perhaps we ought to have come to it earlier, because to northern-hemisphere observers it dominates the scene for a major part of the summer, and even from southern countries it is very much in evidence – except from the southern part of New Zealand, where it is almost lost.

As we have seen (25 July) the term is quite unofficial, and does not apply to the southern hemisphere, where the 'triangle' is at its best in winter; I must plead guilty to introducing the term, but nowadays it is widely used. Let us begin by giving some relevant details of the three brilliant stars.

In our skies Vega is much the brightest of the three, but both it and Altair owe their eminence to the fact that they are relatively close neighbours, while Deneb is very powerful and

The Southern Triangle

Star	Name	Apparent magnitude	Absolute magnitude	Distance (light-years)	Declination
Alpha Lyræ	Vega	0.0	0.6	25	+39°
Alpha Cygni	Deneb	1.2	–7.5	1800	+45°
Alpha Aquilæ	Altair	0.8	2.2	17	+9°

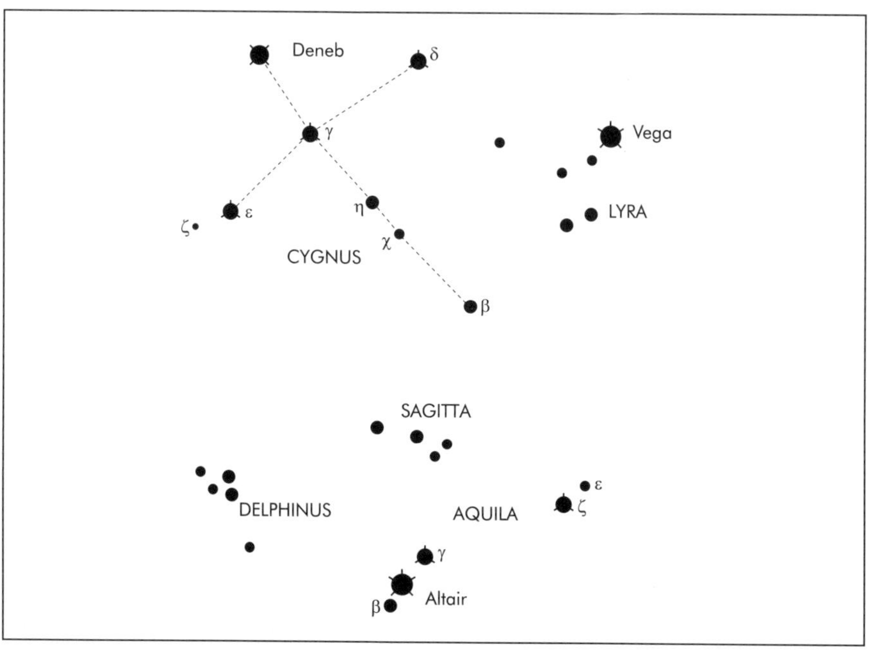

remote. Absolute magnitude is defined as the apparent magnitude that a star would have if it could be seen from a standard distance of 32.6 light-years (21 May). From this standard distance, Vega would be half a magnitude fainter than it actually looks, Altair would be about as bright as Mizar in the Great Bear appears to us, and Deneb would cast strong shadows. Look at Deneb this evening, and you are seeing it not as it is now, but as it used to be in the far-off days when Britain was occupied by the Romans.

July 26

Lyra

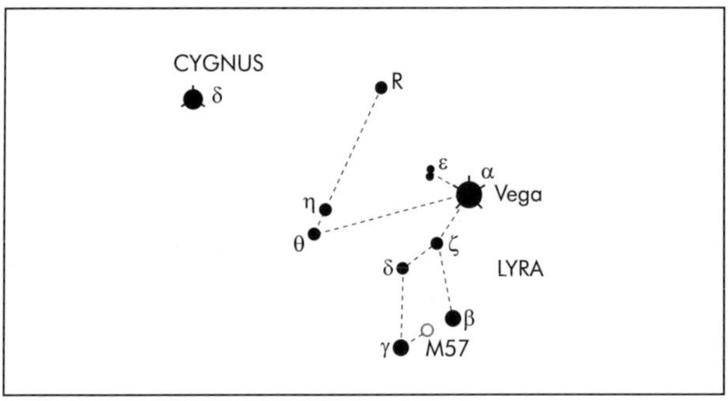

This is a small constellation, covering less than 300 square degrees, but it contains a surprising number of interesting objects. It is of course dominated by the brilliance of Vega, but there are many double stars and variables, a prototype eclipsing binary, and the most famous of all planetary nebulæ – M57, the Ring.

The leading stars are listed below.

According to mythological lore, the lyre was invented by Mercury, who made the first instrument of its kind by stretching the intestines of a cow over the shell of a turtle. The celestial lyre may well represent the harp which either Mercury or Apollo gave to the great musician Orpheus.

There is also an ancient Chinese legend which involves Vega and Altair. Altair was the 'herd-boy', while Vega was

Future Points of Interest

1999: Neptune at opposition

Lyra

Greek letter	Name	Magnitude	Luminosity (Sun=1)	Distance (light-years)	Spectrum
α Alpha	Vega	0.0	52	25	A0
β Beta	Sheliak	3.3–4.3	130	300	B7
γ Gamma	Sulaphat	3.4	180	192	B9

the 'weaving-girl'. The two fell in love and, not unnaturally, neglected their heavenly duties – so that at length they were placed firmly on opposite sides of the Milky Way, so that they could not meet except on the seventh night of the seventh moon, when a bridge of birds spans the 'River of Stars' and allows the lovers to spend a brief but blissful time together.

Anniversary

1801: Birth of Sir George Biddell Airy, a great astronomer who served as Astronomer Royal at Greenwich between 1835 and 1881. He modernised and re-equipped the Observatory, and raised it to a position of eminence. It is unfortunate that he is probably best remembered for his failure to instigate a search for Neptune (see 4 August). He died in 1892.

July 27

Vega

Vega, the brightest star in the northern hemisphere of the sky with the exception of Arcturus, is one of our nearer neighbours; indeed, among the first-magnitude stars only Sirius, Alpha Centauri, Procyon, Altair and Fomalhaut are closer. It seems to be a normal A-type star, but in 1983 a very interesting discovery was made about it.

This was the year in which IRAS, the Infra-Red Astronomical Satellite, was orbiting the Earth and sending back data. Its equipment was designed to pick up long-wavelength radiation from cool bodies, and proved to be very successful; though it operated for less than a full year, IRAS provided a complete infra-red map of the sky, and discovered thousands of new sources. While calibrating the instruments, the investigators found that Vega was associated with 'a huge

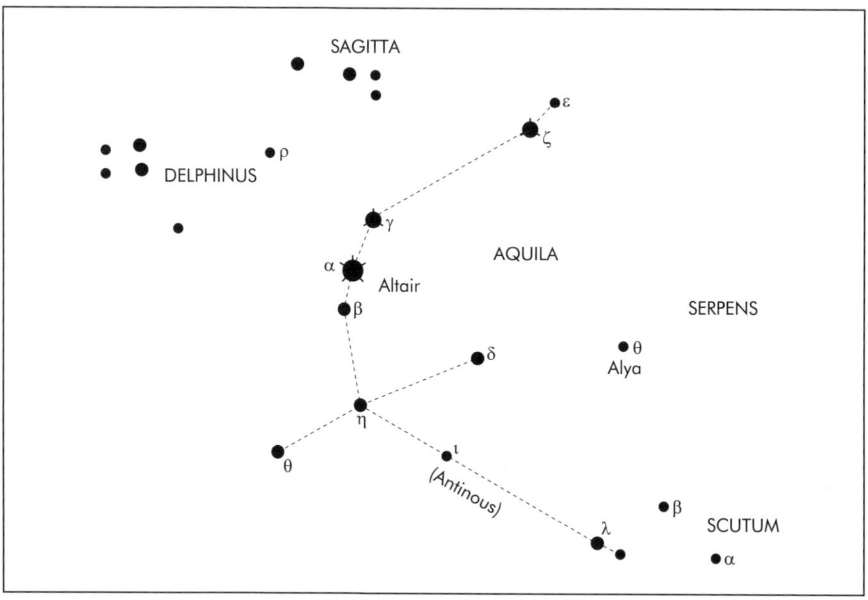

infra-red excess', showing that it was surrounded by cool, possibly planet-forming material. Subsequently other stars, notably Fomalhaut (2 October) and the southern Beta Pictoris (28 December) were found to have similar infra-red excesses.

Does this mean that there are planets orbiting Vega? This is not impossible, but we have no proof. At least the discovery makes this lovely star even more fascinating.

Future Points of Interest

2000: Neptune at opposition

2000: Mercury at western elongation.

July 28

The Changing Size of the Moon

Since we have two 'lunar days' during the month covered here, this may be the moment to say a little about the changes in apparent size. The Moon's orbit is not circular; it is markedly elliptical, and the centre-to-centre distance ranges between 221 500 miles at closest approach (perigee) and 252 700 miles at furthest recession (apogee). Therefore the apparent diameter changes, and this is not always appreciated. The maximum perigee diameter is 33′ 31″: the minimum apogee diameter, 29′ 22″. This, of course, is why some solar eclipses are total while others are annular (16 February).

Perigee and apogee dates in July during our period are:

Lunar Perigee and Apogee

Year	Perigee	Apogee
1997	July 23	July 9
1998	16	30
1999	11	23
2000	1	15
2001	21	9
2002	14	30
2003	10	22

Future Points of Interest

1999: Total eclipse of the Moon. Mid-eclipse occurs at 11 h 34 m GMT, when 40% of the disk is in shadow. The partial phase extends in all over 2 hours 22 minutes

July 29

Beta Lyræ

Beta Lyræ, close to Vega, is an eclipsing binary, but it is not of the same type as Algol in Perseus (12 November). It is the prototype of a different class of eclipsing binary.

The maximum magnitude is 3.3, so that the star is then practically equal to its neighbour Gamma (3.2). However, Beta is always in variation. The full period is 12.9 days, but there are alternate deep and shallow minima; at deep minimum the magnitude drops to 4.2, at secondary minimum only to 3.8.

The reason is that Beta Lyræ is made up of two components, rather unequal, and so close together that they almost touch. The orbit is almost circular, and the centre-to-centre distance between the two is no more than about 22 000 000 miles; each must be gravitationally pulled out into the shape of an egg. Primary minimum occurs when the brighter component is hidden, and shallow minimum when it is the fainter component which is covered up. Moreover, it seems that both components are contained in swirling gas-clouds, so that from close range the view would indeed be spectacular.

The fluctuations are easy to follow with the naked eye, Gamma makes an ideal comparison star, but during minima you can also use Kappa Lyræ (4.3) and Zeta (also 4.3).

Beta Lyræ type variables are rather rare, and Beta Lyræ itself is much the brightest of them. It has a companion of magnitude 8.6, at a separation of 46 seconds of arc. This is a true member of the Beta Lyræ system, but the orbital period must be immensely long.

July 30

Future Points of Interest

2001: Opposition of Neptune.

The Ring Nebula

Directly between Beta and Gamma Lyræ lies M57, the Ring Nebula, one of the most famous objects in the sky. It was discovered in 1777 by Antoine Darquier, a French astronomer, using a 2.5-inch refracting telescope. I admit that I have never been able to see it with binoculars, but others claim that it is not difficult, and certainly it is a very easy object telescopically.

It is a planetary nebula, but the term is misleading, because M57 has absolutely nothing to do with a planet, and is not truly a nebula. A planetary nebula marks a late stage in the evolution of a star. When the hydrogen 'fuel' is used up, different reactions begin, and in the end the star becomes a huge, bloated red giant. The outer layers are then thrown off, and move away into space; this is the planetary nebula stage. When they have dispersed, all that is left of the old star is its core, made up of the shattered pieces of atoms which are packed together. The star has become a very dense white dwarf – a fate which will eventually overtake our Sun, and has already overtaken the companion of Sirius (21 February).

M57 is in the nebular stage. Telescopically it looks like a tiny, luminous cycle tyre; the central star is of magnitude 14.8. (A second star, apparently in the nebula, is not in fact connected with it, and merely lies in the foreground.) The distance is around 1400 light-years, and the ring is expanding at about one second of arc per century.

Why do we see a ring? The reason is that the old star is surrounded by a shell of gas; we see it from a great distance, and so we observe more glowing material at the periphery than through the centre. Some other planetary nebulæ, such as the Dumbbell in Vulpecula (9 September) are much less symmetrical than M57.

July 31

Epsilon Lyræ, the Double-Double Star

Close beside Vega we find a superb example of a quadruple star: Epsilon Lyræ. People with keen sight can see that it is made up of two, with no optical aid at all; the magnitudes are 4.7 and 5.1, and the separation is 208 seconds of arc. Look through a telescope, and you will see that each component is again double. The brighter member of the pair is the easier to split; the separation is 2.6 seconds of arc as against only 2.3 seconds for the fainter component, but a three-inch telescope should split them both without much difficulty.

There is no doubt that all four stars of Epsilon Lyræ are associated, and have a common origin, but the two main pairs are a long way apart – probably as much as a fifth of a light-year. Each of the close pairs has a separation around 150 times the distance between the Earth and the Sun. The orbital periods are naturally very long; perhaps 590 years for the brighter pair, 1200 for the fainter. The distance from Earth is 175 light-years.

Other quadruple stars are known, but none can be regarded as quite so neat as Epsilon Lyræ. The only thing lacking is colour contrast; all four components are white.

Anniversary

1964: Ranger 7 crash-landed on the Moon. It had been launched on 28 July, and came down in the Mare Nubium (lat. 10°.7 S, long. 20°.7 W), returning 4306 high-quality images before impact. This was the first really successful American lunar mission.

1969: Mariner 6 flew past Mars at 2108 miles, returning 76 good images. It is now in solar orbit.

Future Points of Interest

2000: Partial eclipse of the Sun. Mid-eclipse is at 02 h GMT. 60% of the solar disk is obscured, but to see it you will have to go to the Arctic.

August

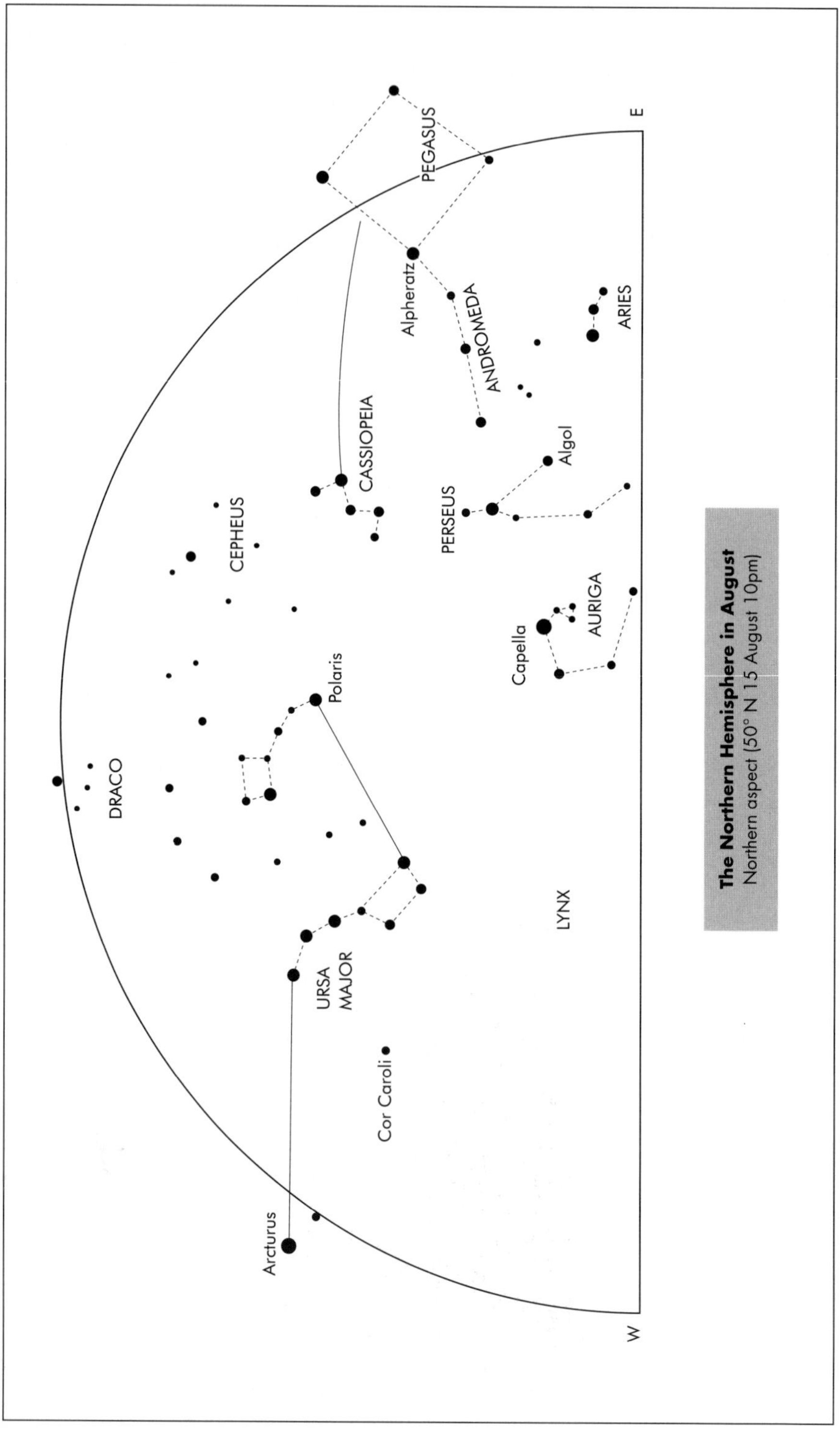

The Northern Hemisphere in August
Northern aspect (50° N 15 August 10pm)

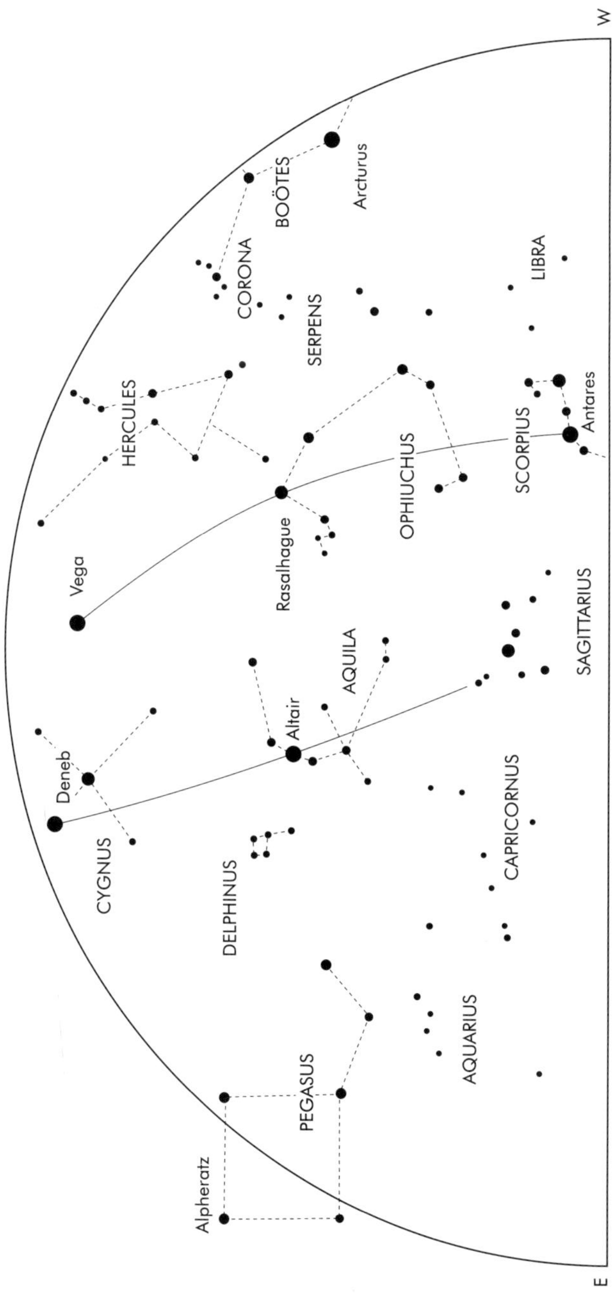

The Northern Hemisphere in August
Southern aspect (50° N 15 August 10pm)

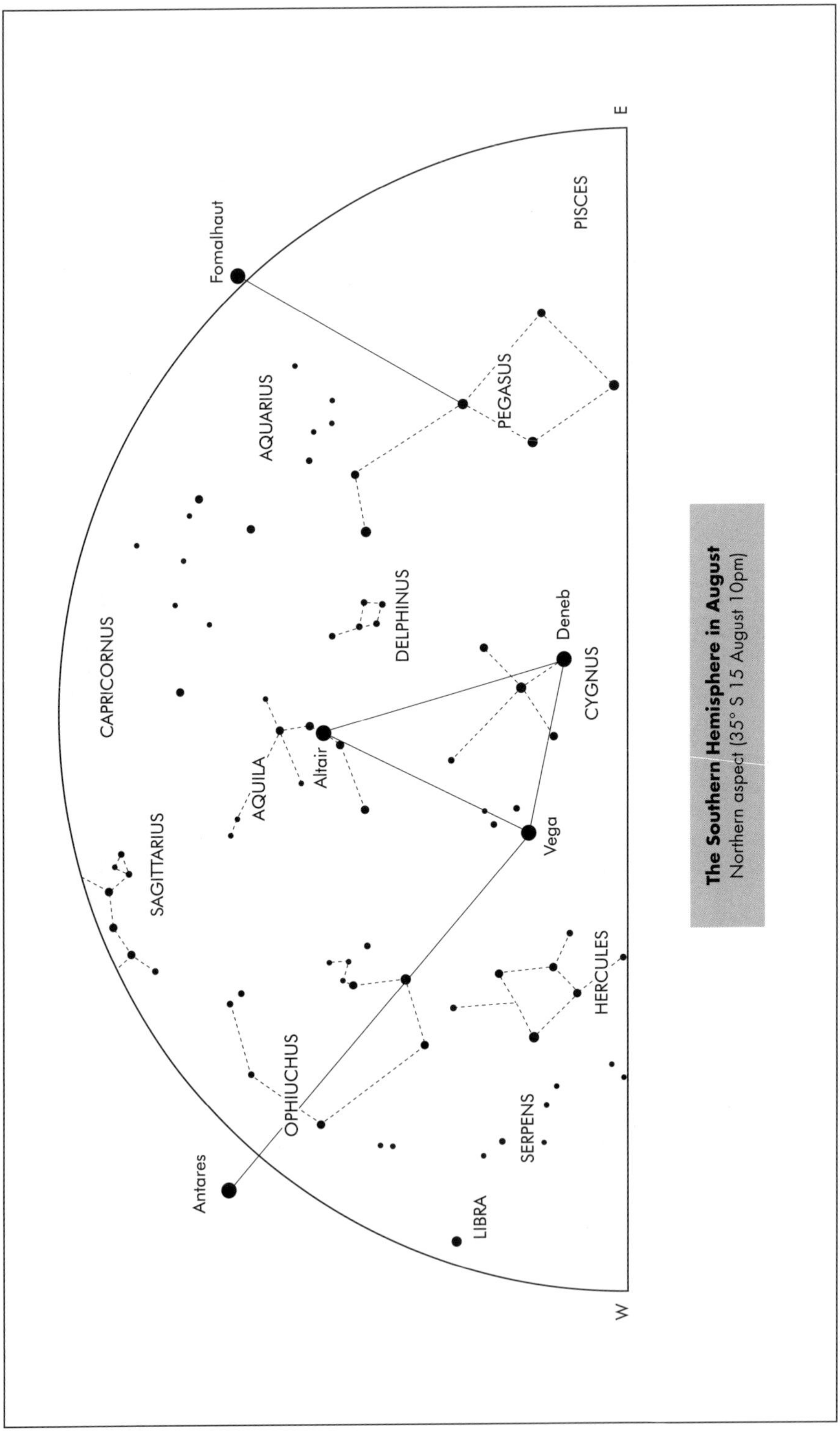

The Southern Hemisphere in August
Northern aspect (35° S 15 August 10pm)

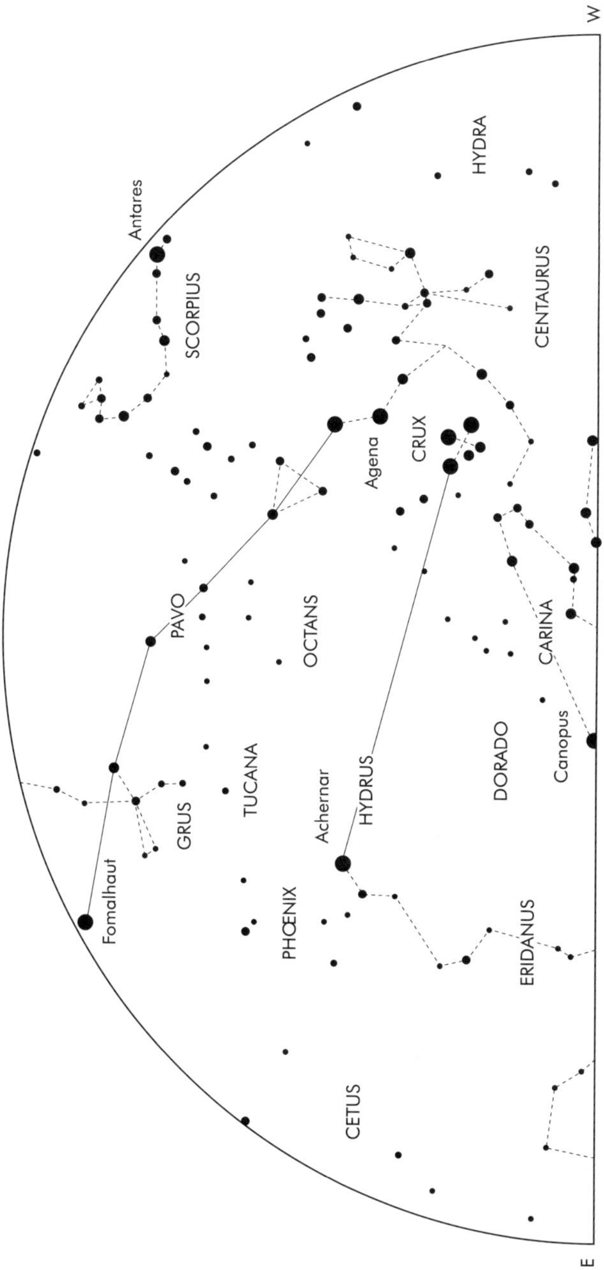

The Southern Hemisphere in August
Southern aspect (35° S 15 August 10pm)

August 1

The August Sky

For the northern observer, the nights are lengthening. The Summer Triangle continues to dominate the scene, while the main autumn constellation, Pegasus, makes its entry in mid-evening. The Great Bear is rather low in the north-west, which means that the W of Cassiopeia is high in the north-east. Arcturus is setting, and Capella is so low over the northern horizon that any mist or light pollution will conceal it. We have lost the Scorpion, though Sagittarius remains on view very low in the south.

In the early hours of the morning, the lovely star-cluster of the Pleiades appears in the east. By the end of August it rises well before midnight – a reminder that the hot days are over, and winter, with its frosts and fogs, lies ahead.

Remember that August is the 'meteor month'. The Perseids, which peak on 12 August, may always be relied upon to give good displays, and even in the presence of the Moon quite a number of 'shooting-stars' can always be expected. The shower has already started at the beginning of the month, and goes on until near the end of the third week.

From the southern hemisphere, the Vega–Deneb– Altair triangle is visible in the east, though only Altair is reasonably high from Australia or South Africa. Scorpius and Sagittarius remain prominent, though Scorpius is sinking in the west. The Southern Cross is low, and so is Canopus, with the Southern Fish high up, and Pegasus appearing in the north-east. Of course the Perseids can be seen, but the radiant lies far in the north, so that for once Britons have the advantage over Australians. The Clouds of Magellan are gaining altitude, and this is a good time to pick out the admittedly rather confusing Southern Birds – Grus, Pavo, Phœnix and Tucana.

August 2

Neptune: the Outermost Giant

Throughout the period we are covering here, Neptune reaches opposition in the late northern summer – from 21 July in 1997 to 4 August in 2003; the synodic period is only 367.5 days. Throughout the period Neptune remains in Capricornus, around declination −19°; at the 2002 opposition it is not far from the star Rho Capricorni, magnitude 4.8.

Future Points of Interest

2002: Neptune at opposition.

Neptune's mean opposition magnitude is 7.7. This means that it is well below naked-eye visibility, but binoculars will show it in the guise of a starlike point. Telescopically it appears as a small, bluish disk, with an apparent diameter of around 2 seconds of arc. Ordinary telescopes will show no surface features, though markings have been recorded very distinctly from the Hubble Space Telescope – and of course from the one space-craft to have passed by it, Voyager 2 in 1989.

There are eight known satellites, of which six were found during the Voyager 2 encounter. Of the two attendants previously known, Triton is an easy telescopic object with adequate equipment, but Nereid is very faint indeed, and the recently-found satellites are beyond the range of Earth-based instruments.

Future Points of Interest

1998: Uranus at opposition.

August 3

The Perseids – and Comet Swift–Tuttle

It is time to start checking on the Perseids, because – unlike the Leonids of November – they never fail us. The Zenithal Hourly Rate can rise to 75 around the 12th of the month.

Meteors, as we have seen, are cometary débris, and many annual showers are known to have parent comets. In the case of the Perseids the comet was discovered in 1862, independently by Lewis Swift and Horace Tuttle, so that it bears both their names. In September of that year the nucleus reached the second magnitude, and there was a tail at least 30° long, so that it was an impressive sight. It was followed until 31 October. The calculated period was 120 years, a nd there could be no doubt of the association with the Perseids.

Searches for the comet began in 1980; but 1982 came and went, without any sign of Swift–Tuttle. Then new calculations were made, giving a period of 130 years, and in 1992 the comet duly came back, even though it was not as bright as it had been in 1862. At the next return it will come very close to the Earth, and will be truly magnificent. There were even alarmist suggestions that it might hit us, though it now seems that it will miss the Earth by a comfortable margin.

The Perseid radiant lies in the northern part of the constellation, not far from Gamma Persei (11 November), but of course the meteors are not confined to Perseus, and streak across the sky.

August 4

Observing Neptune

Let us now return to Neptune. This is a good time to observe it, particularly since it lies in a fairly barren part of the sky; Capricornus is one of the less rich constellations of the Zodiac.

Neptune was discovered in 1846, as a result of some mathematical calculations by the French astronomer U.J.J. Le Verrier. When the planet Uranus was discovered in 1781, by William Herschel (4 August), its movements were worked out, but it persistently wandered away from its predicted path. From these irregularities, Le Verrier calculated the position of a remote planet which was pulling Uranus off course, and in 1846 Neptune turned up very close to the position which Le Verrier had given. Similar calculations had been made in England by John Couch Adams, of Cambridge, but the actual discovery – by Johann Galle and Heinrich D'Arrest, at the Berlin Observatory – was made on the basis of Le Verrier's work.

Neptune is a giant, about 30 000 miles in diameter and 17 times as massive as the Earth. A silicate core may be present, but most of the interior is dominated by 'ices', primarily water, above which comes the cloud-layer which we actually see. The orbital period is nearly 165 years, but the rotation period is only 16 hours 7 minutes; Neptune, like all the giant planets, has a short 'day'. The upper atmosphere consists of 85% hydrogen, 13% helium and 2% methane. Neptune has a marked internal heat source, but the temperature of the upper clouds is very low: around –220 °C.

A three-inch telescope is powerful enough to show that Neptune is not a star; it appears as a tiny disk. The disk is very evident with a telescope of, say, six inches aperture, and this will also show Triton, the senior satellite. Triton is, in fact, brighter than any of the satellites of Uranus, though it was only with the pass of Voyager 2 in 1989 that we found out how extraordinary a world it is.

Nothing else will be made out on Neptune, but it is always satisfying to locate 'the outermost giant', which will be reasonably well placed all through the period covered here.

Anniversary

1969: Mariner 7 passed Mars at a range of 2178 miles. It returned 126 high-quality images, mainly of the planet's southern hemisphere. Mariner 7 is now in solar orbit.

Future Points of Interest

2003: Neptune at opposition.

August 5

Aquila

Our next constellation must be Aquila, the Eagle, which contains the first-magnitude Altair, one of the members of the unofficial Summer Triangle. The usual legend about it states that it represents an eagle sent by Jupiter to collect a Phrygian shepherd-boy, Ganymede, who was to become cup-bearer to the gods – following an unfortunate accident to Hebe, the former holder of the office, who tripped and fell awkwardly during a particularly solemn ceremony.

The leading stars of Aquila are listed below.

Aquila really does conjure up a vague impression of a bird in flight. It contains some interesting objects, and since it is crossed by the Milky Way the whole area is decidedly rich; it is worth sweeping with binoculars. In 1918 a nova flared up here, and for a while outshone every star in the sky apart from Sirius, though it has now become very faint.

August 6

Antinoüs

In many early maps of the sky the Eagle is referred to as 'Aquila et Antinoüs', and in some maps, notably Bayer's of 1603 (in which he allotted Greek letters to the stars) Antinoüs is depicted as an entirely separate constellation. We know who he was; he was a boy favourite of the Roman emperor Hadrian, and when he died it is said that Hadrian persuaded the Alexandrian astronomers to place him in the sky. The boy's head is marked by the stars now known as Eta and Theta Aquilæ, and the body extends down to Lambda Aquilæ. It also included part of the little constellation of Scutum, the Shield (14 August).

In the event Antinoüs did not survive, and has long since been forgotten, so that his leading stars have been returned to the Eagle.

Aquila

Greek letter	Name	Magnitude	Luminosity (Sun=1)	Distance (light-years)	Spectrum
α Alpha	Altair	0.8	10	16.6	A7
β Beta	Alshain	3.7	4.5	36	G8
γ Gamma	Tarazed	2.7	700	186	K3
δ Delta	–	3.3	11	52	F0
ζ Zeta	Dheneb	3.0	60	104	B9
η Eta	–	3.5–4.4	5000	1400	G0
θ Theta	–	3.2	180	199	B9
λ Lambda	Althalimain	3.4	82	98	B8

August 7

Altair

The leader of Aquila is actually the twelfth brightest star in the sky. It is one of the nearest of the first-magnitude stars, at less than 17 light-years; only Alpha Centauri, Sirius and Procyon are closer. It is a white A-type star, ten times as luminous as the Sun. (In astrological lore it was said to be a mischief-maker, and to portend danger from reptiles – which presumably makes as much sense as any other statement made by astrologers!). It is less than 9° north of the celestial equator, which means that it can be seen from every inhabited country. It is flanked to either side by fainter stars (Tarazed and Alshain) which makes it very easy to identify.

Altair seems to be about 1 400 000 miles in diameter, but it is not a perfect sphere. It is in very rapid rotation, with a period of only 6.5 hours, so that the rotational speed at the star's equator is over 150 miles per second; we can tell this from the way in which the dark lines in the spectrum are broadened: one limb is approaching us and the other receding, so that they give opposite Doppler shifts, and the spectral lines are spread out. Therefore Altair must be egg-shaped, with the equatorial diameter nearly twice the polar diameter.

There is a faint companion, of magnitude 9.5, at a separation of 165 seconds of arc, but this is not genuinely associated with Altair, and lies far in the background.

It is worth using binoculars to look at Altair's two neighbours. Gamma (Tarazed) is obviously orange, while Beta (Alshain) is only slightly yellowish. Of the two, Tarazed is much the more remote and more luminous – and indeed, Tarazed is 70 times as powerful as Altair.

Future Points of Interest

1999: Uranus at opposition.

August 8

Eta Aquilæ

Eta Aquilæ is one of the brightest and best-known of the Cepheid variables (28 January). Its variability was discovered only a few weeks after that of Delta Cephei itself, so that if it had been identified first the stars would presumably have become known as Aquilids rather than Cepheids.

It is easy to find. South of the three stars of which Altair is the central member, there is a much longer line of three, all of which are of much the same brightness. The central member is the variable Eta, which has a range of magnitudes from 3.5 to 4.4; to one side is Delta Aquilæ (3.4) and to the other Theta

Future Points of Interest

1998: Penumbral eclipse of the Moon. Mid-eclipse occurs at 02 h 26 m GMT, but as only 12% of the disk enters the penumbra the eclipse will be rather difficult to detect. You will certainly not notice it unless you are deliberately looking out for it.

(3.2), which make excellent comparison stars. You can also use Iota Aquilæ (4.4) when Eta is near minimum.

The period is 7.2 days, and of course the fluctuations are easy to follow with the naked eye; moreover Eta is only one degree off the celestial equator, so that it can be well seen from either the northern or the southern hemispheres. Since its period is longer than that of Delta Cephei, it must be more powerful; on average it is some 5000 times more luminous than the Sun.

August 9

Nova Aquilæ, 1918

(Do not forget the Perseids. The maximum of the shower now lies close ahead, and it is worth keeping a visual and a photographic watch from tonight onward – particularly after midnight.)

Novæ, or 'new stars', can be very spectacular. They are not really 'new' at all; what happens is that a violent outburst occurs in the white dwarf component of a binary system lasting for a few days, weeks or months. The brightest nova of modern times flared up near Theta Aquilæ in 1918, and reached magnitude −1.1.

It was discovered on 8 June, and reached its peak on 9 June, after which it began to fade; it remained a naked-eye object until March 1919. Subsequently it threw off shells of material, so that for a while it looked like a small planetary nebula, but these shells have now disappeared. The distance has been given as 1200 light-years, so that the actual outburst occurred around the year 800 AD; the peak luminosity was of the order of 450 000 times that of the Sun.

Novæ generally appear in or near the Milky Way, and of course cannot be predicted; amateurs, who know the sky well, have the best chance of detecting them. Indeed, the 1918 star was independently found by a boy of 17, who later became one of America's foremost amateur astronomers: Leslie Peltier.

Anniversary

1877: Deimos, the outer satellite of Mars, discovered by Asaph Hall.

August 10

The Moon from Orbiter

We are working up to the maximum of the Perseid meteor shower, and many astronomers will prefer the Moon to be absent! For the period covered here, the yearly situations are:

1998 Moon full on 8 August, so at shower maximum it will not be too obstrusive.
1999 Moon new on 11 August: ideal for Perseids.
2000 First Quarter on 7 August, full on 15 August; tolerable conditions.
2001 Last Quarter on 12 August; reasonably good for meteor watchers.
2002 New moon on 8 August, so the crescent sets at a reasonable hour.
2003 Full moon on 12 August; as unfavourable as it could be.

The five Lunar Orbiters of 1966–1968 provided the first large-scale lunar map, covering both the Earth-turned and the averted hemispheres of the Moon. It is not too much to say that the Orbiter results were essential to the manned landing programme, and made all earlier work more or less obsolete. When each Orbiter had finished its work, it was deliberately impacted on to the surface. No doubt the wreckage will be found and collected by future lunar explorers. Orbiter 5, last in the series, impacted on 31 January 1968.

Anniversary

1966: Launch of Lunar Orbiter 1. This was the first of the five Orbiters, all of which were successful, and which provided high-quality pictures of the whole of the Moon's surface.

1990: The Magellan probe entered orbit round Venus, and began its long programme of radar mapping. It finally burned away in Venus' atmosphere on 11 October 1994.

August 11

The Perseid Shower

Most meteor showers have known parent comets. That of the Perseids is Swift–Tuttle; that of the November Leonids is Tempel–Tuttle (in each case the co-discoverer was an American, Horace Tuttle, who had an extraordinary career as astronomer, US naval officer, and financial embezzler). The Leonid shower is 'bunched up', so that major displays are to be expected only when the comet returns to perihelion, every 33 years or so. The Perseids are spread all round the comet's orbit, so that good displays are seen annually. Records of them go back to the eighth, ninth and tenth centuries, though it was only in 1835 that the Belgian astronomer Quételet realized that we are dealing with a definite and regular shower. Later, between 1864 and 1866, G.V. Schiaparelli proved the association between the Perseids and Comet Swift–Tuttle.

The peak time is not constant, owing to the erratic nature of our calendar, but is usually in the early hours of 12 August, so that meteor observers will be very much on the alert; obviously, there is strong activity for several nights before and after the actual peak. Gaze up into a clear, dark sky for a few

Future Points of Interest

1999: Total eclipse of the Sun. This is the first total eclipse to be visible from England since 1927, and will be the last until 2090. The track crosses Cornwall and Devon, including Padstow, Newquay, Fowey, Kingsbridge, Torquay and Plymouth; it then crosses Alderney, and off into Europe. Mid-eclipse occurs at 11 hours GMT. If you are thinking about booking a hotel room in the West Country, I must tell you that you are already too late!

2000: Uranus at opposition.

minutes around this time of the year, and you will be very unlucky not to see several meteors – most of them Perseids, though, as we have seen, there are also sporadic or non-shower meteors which may appear from any direction at any moment.

Future Points of Interest

Maximum of the Perseid meteor shower.

August 12

Meteor Photography

The time of the Perseid maximum is obviously favourable for meteor photography. The procedure is straightforward enough; use a fairly fast film and simply give a time exposure, of from four or five minutes up to several hours. Obviously it is a good idea to use as many cameras as you can muster.

Remember, too, that the hours after midnight are the best, because the meteors enter the Earth's 'leading' hemisphere; this means that the velocities are greater than for evening meteors, which have to catch the Earth up – and therefore the morning meteors are the brighter.

With clear skies, you should record several meteors, and you will also obtain interesting photographs of star-trails. If you are lucky, you may even 'net' a really brilliant fireball.

The shower does not end until after the middle of August, but after maximum the activity falls off fairly rapidly. A few Perseids may however linger on until at least 20 August.

Anniversary

1898: Discovery of asteroid 433, Eros, by G. Witt from Berlin. A. Charlois, at Nice, independently discovered it on the same night, but Witt's announcement was made first.

August 13

Eros, the Cosmic Lozenge

All the large asteroids, and most of the small ones, keep strictly to the wide gap between the orbits of Mars and Jupiter. It was therefore something of a surprise when, in 1898, Witt discovered an asteroid which came well within the orbit of Mars, and could approach the Earth to a distance of no more than 15 000 000 miles. It was the first asteroid to be given a masculine name. (To be strictly accurate, asteroid 134, Æthra, can dip inside the orbit of Mars but only just, and seldom).

Eros is lozenge-shaped, measuring 24 miles by 8 miles. It has an orbital period of 1.76 years, and its distance from the Sun ranges between 105 000 000 miles and 228 000 000 miles, so that its path crosses that of Mars but not that of the Earth.

It made a close approach in 1930, and was studied by observatories all over the world. It looks like a star, and so its position against the background of real stars can be measured very precisely. A good knowledge of its orbit would lead to a better estimate of the length of the astronomical unit, or Earth–Sun distance. The method is now obsolete (and in any case, the results from 1930 were not so good as had been hoped), and the last close approach, in 1975, passed almost unnoticed, even though at its best the asteroid becomes bright enough to be seen in binoculars. It spins round in a mere 5 hours 18 minutes.

However, Eros is still 'in the news'. On 16 February 1996 a spacecraft, NEAR (*Near Earth Asteroid Rendezvous*) was launched to it. The probe is scheduled to reach Eros in January 1999, and will 'keep pace' with it for at least nine months, sending back assorted data.

Since 1898 many close-approach asteroids have been found, some of which have passed well within the distance of the Moon, but Eros remains one of the largest of these curious wanderers.

August 14

The Shield and the Wild Duck

Let us now return to Aquila. At its southern end, adjoining Lambda (magnitude 3.4) and 12 (4.0) we find the little constellation of Scutum Sobieskii, Sobieski's Shield, usually referred to simply as Scutum (an older name for it was Clypeus). It was formed by Hevelius in 1690 in honour of John Sobieski, King of Poland, famous for his relief of Vienna in 1683 when it had been besieged by the Turks. (Hevelius was himself a Pole. He lived in Danzig – the modern Gdańsk – where he set up an elaborate observatory; it was from here that he drew up one of the earliest useful maps of the Moon.) Scutum has no star brighter than the reddish Alpha, magnitude 3.8, and has no striking outline, but it is crossed by the Milky Way, and is very rich.

In Scutum lies the open cluster M11, which is visible with the naked eye even though it is not too easy to pick out against the star-studded background. The nineteenth-century astronomer Admiral W.H. Smyth wrote that it 'somewhat resembles a flight of wild ducks in shape', and the nickname is still used, though the shape more nearly resembles that of a fan. It is a compact, impressive cluster, about 5500 light-years away, with an 8th-magnitude star at the apex of the fan. Unquestionably it is a magnificent sight. Its position is RA 18 h 51 m, declination −06° 16′.

Future Points of Interest

1999: Mercury at western elongation.

2003: Mercury at eastern elongation.

Future Points of Interest

2001: Uranus at opposition.

August 15

R Scuti

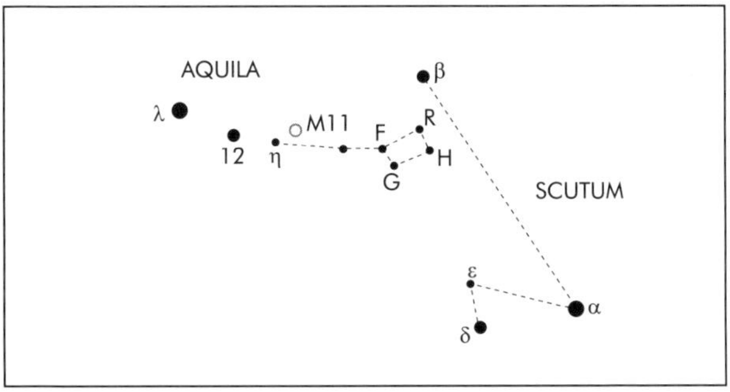

Close to M11 in the Shield lies a very interesting variable star, R Scuti. At its peak it can rise to magnitude 4.4, but the usual range is from about 5.7 to 8.6, so that it can sometimes be seen with the naked eye and seldom drops out of view with binoculars.

It belongs to a very rare type of variable named after the first-discovered member of the class, RV Tauri, which however is not nearly as bright as R Scuti. The stars are yellow supergiants, of immense size but low mass and density. They pulsate, and in general have alternate deep and shallow minima, though there are occasional spells of complete irregularity. With R Scuti the period is of the order of 140 days, but one never quite knows how the star is going to behave – and this is what makes it so interesting to follow.

At its peak R Scuti is about 8000 times as luminous as the Sun, but it is fair to say that there are still some features about these unusual stars which are not yet completely understood.

North of the R Scuti group are two more isolated stars, Epsilon Scuti (magnitude 4.9) and Delta (4.7). Delta Scuti is the prototype of another class of variable. It too pulsates, but in a very short period of only 4.6 hours. The range is less than two-tenths of a magnitude, so that the fluctuations are not detectable except with sensitive measuring equipment. Delta Scuti stars are young, with spectra of types A to F.

Anniversary

1877: Discovery of Phobos, the inner satellite of Mars, by Asaph Hall at Washington. (He had discovered Deimos, the outer satellite, on 10 August).

August 16

Cygnus, the Swan

Our next constellation is Cygnus, the Swan – often known as the Northern Cross, for obvious reasons; unlike the Southern Cross it really does take the form of an X, though one of its members (Albireo) is fainter than the others and further

Cygnus

Greek letter	Name	Magnitude	Luminosity (Sun=1)	Distance (light-years)	Spectrum
α Alpha	Deneb	1.2	70 000	1800	A2
β Beta	Albireo	3.1	700	390	K5
μ Gamma	Sadr	2.2	6000	750	F8
δ Delta	–	2.9	130	160	A0
ϵ Epsilon	Gienah	2.5	60	81	K0
ζ Zeta	–	3.2	600	390	G8
η Eta	–	3.9	60	170	K0
ι Iota	–	3.8	11	134	A5
κ Kappa	–	3.8	105	170	K0
o^1 Omicron^1	–	3.8	650	520	K2
τ Tau	–	3.7	17	68	F0

away from the centre, which rather spoils the symmetry. The leader of the constellation, Deneb, is one of the three members of the Summer Triangle (25 July), and although it looks the faintest of the three it is actually much the most luminous. It lies over 45° north of the celestial equator, and this means that to southern-hemisphere observers it is always rather low down; from Invercargill, at the southern tip of New Zealand, it does not rise at all.

The leaders of Cygnus are listed above.

Cygnus is immersed in the Milky Way, and the whole area is particularly rich, with many variable stars, open clusters and gaseous nebulæ. The last really bright nova flared up north of Deneb in 1976; this is a good 'nova area', and it is always worth making a quick survey.

August 17

Deneb: A Celestial Searchlight

The name Deneb comes from the Arabic *Al Dhanab al Dajajah*, the Hen's Tail, and Deneb does indeed lie in the tail of the Swan. In mythology Cygnus is said to represent a swan into which Jupiter changed himself on a visit to Leda, the wife of the King of Sparta, for the usual discreditable reasons!

Deneb is a particularly powerful supergiant, at least 70 000 times as luminous as the Sun. It has an A-type spectrum, and therefore appears pure white. The surface temperature is almost 10 000 °C, and the diameter may be as much as 60 times that of the Sun (over 50 million miles), though the mass is not likely to be more than 25 times that of the Sun; remember that among the stars the range in mass is much less than in size or luminosity, because large stars are always more

Anniversary

1958: Failure of America's lunar probe Able 1. This was the first US attempt to send a probe to the Moon. Unfortunately the flight came to a sad end after only 77 seconds, at a height of about 12 miles, when the lower stage of the launcher exploded.

rarefied than small ones (15 January). The distance is 1800 light-years, so that we now see Deneb as it used to be during the time of the Roman Occupation of Britain.

Because it is so luminous and energetic, Deneb is squandering its mass at a furious rate, and cannot last for long on the cosmical scale before running out of available 'fuel'. Eventually it will no doubt explode as a supernova, but for the moment its light is almost steady.

It is worth noting that if we could observe from the system of Deneb, our Sun would appear as a very dim star of well below the thirteenth magnitude.

August 18

The North America Nebula

Within 3° of Deneb lies a fascinating nebula, NGC 7000; No. 20 in the Caldwell catalogue. It is visible with the naked eye as a brighter part of the Milky Way, but binoculars bring out its form, and with a wide-field eyepiece on a telescope it is a splendid sight.

It is a huge cloud of dust and gas, mixed in with stars; it earns its nickname because its shape is uncannily like that of the North American continent. Even the dark Gulf of Mexico is very evident.

The distance from us is thought to be about 1800 light-years. This is the same as the distance of Deneb, and in fact the nebula is no more than about 70 light-years from Deneb. This makes it very likely that Deneb is the chief source of illumination; remember that it is exceptionally powerful, so that its influence is marked over a very wide range.

Not surprisingly, the nebula is a favourite target for astrophotographers, and amateur pictures can be truly impressive. The real diameter cannot be less than 50 light-years. The position is RA 20 h 59 m, declination +44° 20′.

August 19

Anniversary

1891: Birth of Milton Humason, the American astronomer who collaborated closely with Hubble during the work on the spectra, nature and movements

(Continued overleaf)

Albireo, the Coloured Double

Albireo or Beta Cygni, the faintest of the stars of the Northern Cross, is a favourite target for the owner of a small telescope – or for that matter a telescope of any kind, because Albireo's lovely contrasting colours put it in a class of its own. The primary, of magnitude 3, is golden yellow, while the fifth-magnitude companion is vivid blue. Firmly-held binoculars will show both components, because the separation is over 34 seconds of arc.

The two components are genuinely associated, but they are a very long way apart – at least 500 000 million miles, and probably much more – so that no orbital motion has been detected. The distance from us is thought to be almost 400 light-years. This means that the primary is some 700 times as luminous as the Sun.

Though most observers describe the companion as blue, some have seen it as green. Just how much of this colour is due to contrast with the primary is not clear, but contrast must play an important rôle; to me, the companion is far 'bluer' than any other star, not excluding Vega. Look at it, and judge for yourself. It may be justifiably claimed that Albireo is the most beautiful double star in the entire sky.

of the galaxies. He was self-taught; most of his career was spent at the Mount Wilson Observatory, where his first post was that of a mule driver and general handyman. By the time of his death, in 1972, he had become one of the most skilful and respected astronomers in the world.

August 20

P Cygni, the Unstable Star

Close to Sadr or Gamma Cygni, the central star of the Northern Cross, lies an unremarkable-looking star, P Cygni, which is in fact a most extraordinary object. It is decidedly variable. It seems to have been first noted in 1600, when it was of third magnitude, and was classed as a nova.

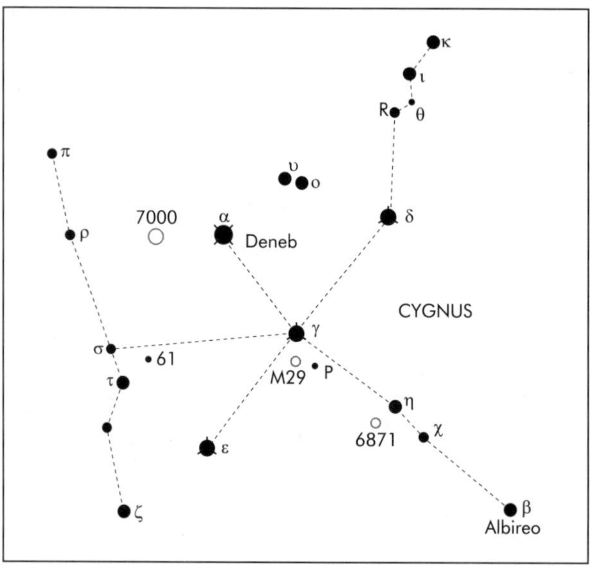

Anniversary

1885: Discovery of S Andromedæ, the supernova in the Andromeda Galaxy (see 27 October).

Future Points of Interest

2002: Uranus at opposition.

Virtual end of the Perseid meteor shower.

It faded slowly, and by 1620 was no longer visible with the naked eye. It reappeared in 1655, and reached magnitude 3.5, but by 1662 it was again absent. It was back once more in 1665, and since about 1715 has fluctuated between magnitudes 4.6 and 5.2. It is easy to estimate; it lies inside the 'bowl' made up of a little semicircle of stars, two of which, 28 Cygni (magnitude 4.8) and 29 Cygni (5.0) are ideal as comparisons.

Spectroscopic observations show that P Cygni is very massive and unstable, and is losing mass at a tremendous rate. If it is around 6000 light-years away, as seems probable, it must be not far short of a million times more luminous than the Sun, dwarfing even Rigel and Deneb. In view of its past history it is well worth monitoring, and binoculars are quite adequate; it has changed little for the past two and a half centuries, but it may brighten up or fade away again at any time. There seems to be no doubt that its final fate will be to explode as a supernova, though perhaps not for at least a million years yet.

August 21

The 'Flying Star'

Still with the Swan, let us look next at a dim star, 61 Cygni, close to Tau; its position is RA 21 h 7 m, declination +38° 45′. It is not bright enough to have been given a Greek letter, so we merely use its Flamsteed number. It is a wide, easy double; the magnitudes of the two components are 5.2 and 6.0. The orbital period is uncertain, but is probably between 600 and 700 years. The present separation is over 30 seconds of arc, so that almost any telescope will split the pair; both components are feeble red dwarfs, and even the brighter member has less than one-tenth the luminosity of the Sun.

The individual or proper motions of the stars are very slight. Very few stars have proper motions of more than one second of arc per year, but 61 Cygni moves by over 4 seconds of arc annually, and has been nicknamed the Flying Star – though it would probably be more appropriate to give this nickname to Barnard's Star, where the annual proper motion is more than twice as great (11 July).

The fact that 61 Cygni is a wide binary, and moves so quickly, led the German astronomer F.W. Bessel to believe that it must be close. He set out to measure its distance, using the method of parallax and in 1838 he was able to show that 61 Cygni lies at a distance of just over 11 light-years. So, obscure though it may look, 61 Cygni has its place in astronomical history.

August 22

Variable Stars in Cygnus

Still with Cygnus, let us discuss some of the many variable stars in the constellation. One of these is Chi Cygni, a long-period variable of the Mira class. It lies between Gamma and Beta, along the 'longest arm' of the Cross, close beside Eta Cygni, a normal K-type star of magnitude 3.9. Chi was one of the very first variables to be identified as such – by Gottfried Kirch, in 1686 – and at its best it is an easy naked-eye object; it has even been known to attain magnitude 3.5, so that it surpasses Eta. Yet when at minimum it drops to below magnitude 14, and is then extremely hard to locate, because it lies in a very crowded region. Only when it is bright does its redness betray it. The period is on average 407 days, so that it comes to maximum 42 days later each year; the 1997 maximum falls on 26 September. It is notable because it has the greatest magnitude range of any known Mira star, and it is one of the strongest infra-red sources in the sky. Its position is RA 19 h 51 m, declination +32° 55′.

A different kind of Mira variable is U Cygni, close to the little pair of Omicron^1 and Omicron^2, roughly between Deneb and Delta Cygni. Here the period is 462 days, and the magnitude range is from 5.9 to 12.1. The position is RA 20 h 20 m, declination +47° 54′: so that it is in the same low-power field as Omicron^2. What singles it out is its intense redness; the spectrum is of type N, and U Cygni is one of the reddest stars in the sky. Telescopically it reminds one of a glowing coal.

Our third variable, SS Cygni, is much fainter, so that a telescope is needed; the star usually hovers around the 12th magnitude, and never becomes brighter than 8.6, so that it is below binocular range. It lies near the fourth-magnitude star Rho Cygni, at RA 21 h 43 m, declination +43° 35′, and it is not hard to identify, because it lies between two members of a distinctive little triangle. For much of the time it remains almost steady, but every 50 days or so it flares up by up to four magnitudes, remaining at maximum for a day or two before fading back to its normal obscurity. The outbursts are not all equally violent, and neither is the interval between them a strict 50 days. SS Cygni is an unpredictable star.

It is called a dwarf nova, because, like true novæ, it is a binary system; one component is a red dwarf and the other a white dwarf. As with novæ, the white dwarf pulls materials away from its less dense companion until conditions become ripe for an explosion – but with SS Cygni everything is on a much milder scale. Many other dwarf novæ are known, but SS Cygni is the brightest of them. The class is known either after SS Cygni or after the much fainter U Geminorum, in the Twins. If you have an adequate telescope, keep a watch on SS Cygni and note when it flares up.

Future Points of Interest

1998: Annular eclipse of the Sun. Mid-eclipse occurs at 2 h GMT; the eclipse will be seen from the Indian Ocean area, the East Indies and the Pacific. Annularity lasts for 3 minutes 14 seconds.

2002: Venus at eastern elongation..

Anniversary

1961: Failure of Ranger 1, first in the series of American probes designed to crash-land on the Moon and send back data before impacting. Ranger 1 never reached orbit, owing to a launch malfunction. The first really successful Ranger was No. 7, which landed on the Moon on 31 July 1964 and returned over 4000 images.

August 23

Clusters in Cygnus

We have spent a long time with Cygnus, but the Swan has so much to offer that there was really no alternative. Before saying adieu, let us glance at some more of the open clusters to be found there.

M29 (RA 20 h 24 m, declination +38° 32′) is in the same wide binocular field as Gamma, which makes it easy to locate, It was discovered by Messier himself in 1764, but is very sparse, and there is nothing at all distinguished about it. Binoculars show it as a dim blur. Nearby, in the direction of Eta, is NGC 6871 or 27 Cygni (RA 20 h 6 m, declination +35° 47′), which is also very sparse, and with low powers hardly looks like a cluster at all. Finally there is M39 (RA 21 h 32 m, declination +48° 26′), in the same binocular field with Pi and Rho Cygni. It too is loose, but it is obvious enough. In the nineteenth century Admiral Smyth described it as 'large and splashy'.

These are only a few of the glories of Cygnus. Fortunately for British observers, the constellation is so far north that it is on view for an appreciable part of every year.

Anniversary

1966: Launch of the Russian probe Luna 11. This was a successful lunar satellite; it approached the Moon to a range of 99 miles, and sent back excellent data. Contact was maintained until 1 October.

Future Points of Interest

2003: Uranus at opposition.

August 24

Uranus

Uranus, the first planet to be discovered in 'telescopic times', has a synodic period of 369.7 days, so that it reaches opposition about four days later in every year. Throughout the period we are covering here it remains in Capricornus. The position at the 1997 opposition was RA 20 h 37 m, declination −19°, south of Beta Capricorni (14 September); by 2003 it will have moved further north, near the boundary between Capricornus and Aquarius.

Uranus is just visible with the naked eye if you know where to look for it, but of course it appears exactly like a star; the magnitude is 5.7, and it is not surprising that it was not known in ancient times. Telescopically it shows up as a small greenish disk, between 3 and 4 seconds of arc in diameter. It is a giant world, with a diameter of 31 770 miles; it takes 84 years to complete one orbit round the Sun, at a mean distance of 1 783 000 000 miles. Almost no detail can be seen on it; we are limited to a view of a virtually featureless cloud-deck.

Perhaps the strangest feature of Uranus is the tilt of its axis. The Earth's axis is tilted to the perpendicular by $23\frac{1}{2}$°, which is why we have our seasons; but the inclination of Uranus is 98° – more than a right angle. The Uranian calendar is most peculiar. First one pole, then the other, has a 'midnight sun' lasting for 21 Earth-years, with a corresponding period of night at the opposite pole. The reason for this extraordinary state of affairs is not known.

Jupiter, Saturn and the outermost giant, Neptune, have strong internal heat-sources, but Uranus has not. There is a definite magnetic field, but here again Uranus is strange; the magnetic poles are nowhere near the poles of rotation, and the magnetic axis does not even pass through the centre of the globe. The axial rotation period is 17 hours 14 minutes, and the mass of the planet is 14 times that of the Earth.

Though Uranus is similar to Jupiter and Saturn in having a gaseous surface, its make-up is different. There is a hydrogen-rich upper atmosphere, with a good deal of helium and some methane, but these gases are mixed with 'ices' – that is to say, substances which would be frozen at the low temperature of the surface. Water is a major constituent; water, ammonia and methane condense in that order to form thick, icy cloud layers. Methane freezes to form the uppermost layer; it absorbs red light, which is why Uranus appears green in colour.

August 25

Giants from Close Range

Most of our knowledge of the two outer giants comes from the one probe, Voyager 2. It was sheer luck that the situation in the late 1970s made it possible to send Voyager on a 'round trip'; it will be more than a century before this can happen again – and if the alignment had occurred in, say, the 1960s we would not have been ready to take advantage of it.

In fact Uranus and Neptune are non-identical twins. When Voyager flew past Uranus it showed very little apart from a few inconspicuous clouds; it was of course approaching the planet 'pole-on', because of the unusual inclination of the axis. Neptune proved to be far more dynamic, and on its lovely blue cloud-layer Voyager recorded a massive storm which was named the Great Dark Spot. There were other spots also, and it was found that Neptune is a windy place. The surface temperature is about the same as that of Uranus; Neptune's internal heat source compensates for the fact that it is so much further from the Sun.

Both giants are radio sources, and both have magnetic fields; with Neptune, as with Uranus, the magnetic axis is offset to the axis of rotation – in Neptune's case, by 47°.

Voyager 2 has long since gone on its way, but the Hubble Space Telescope has been able to record features on Uranus

Anniversary

1981: Voyager 2 passed Saturn, at 63 000 miles.

1981: Voyager 2 passed Neptune, at 18 000 miles. (It had bypassed Uranus at 49 000 miles on 24 January 1986.)

1993: Contact lost with the US probe Mars Observer, which had been launched to the Red Planet on 25 September 1992. The reason for the abrupt loss of contact is not known – and probably never will be.

and Neptune. As expected, Uranus displays a few cloud structures and little else. Neptune seems to have lost its Great Dark Spot, but other features have developed, so that it is far more changeable than had been expected.

August 26

Rings and Moons

In 1977 it was found that Uranus has a system of rings. The discovery was quite unexpected, and one has to admit that it was accidental. Uranus passed in front of a star, and occulted it; both before and after occultation the star 'winked' regularly, showing that it was being briefly hidden by material associated with Uranus.

The rings cannot be seen with ordinary Earth-based telescopes, though they are within the range of the Hubble Space Telescope. They are not in the least like the glorious icy rings of Saturn; they are thin and dark. Our only good views of them have come from Voyager 2.

Before the Voyager mission Uranus was known to have five satellites: Miranda, Ariel, Umbriel, Titania and Oberon, all of which are smaller than our Moon; Titania, the largest, is 980 miles in diameter. Voyager showed them to be icy and cratered; Miranda has an amazingly varied surface, with cliffs, craters and large enclosures which have been nicknamed 'race-tracks'. Voyager discovered another ten satellites, all small and close-in to the planet.

Even the satellites known in pre-Voyager times are not really easy objects, though a 12-inch telescope should be able to show all except Miranda.

Neptune too has rings, but they are very obscure. However, the satellite system is of great interest. Before the Voyager mission two attendants were known; one large (Triton) and one very small (Nereid). Triton has an almost circular orbit, but moves in a retrograde sense – that is to say, opposite to the direction in which Neptune is rotating – and is almost certainly a captured body rather than a bona-fide satellite. Nereid moves in a very eccentric orbit, more like that of a comet than a satellite. Voyager discovered six new small, inner satellites.

Triton proved to be smaller than expected – only 1680 miles across – and to have a thin nitrogen atmosphere. The surface is icy, and the pole is covered with pink nitrogen 'snow'; Triton is so cold that nitrogen freezes out on to the surface. Strange dark streaks indicate the existence of geysers. It is thought that below the surface there is a layer of liquid nitrogen; if this migrates toward the surface the pressure is relaxed, so that the nitrogen explodes in a shower of ice and gas. The outrush sweeps dark débris along with it, and this débris is wafted downwind in the thin atmosphere, producing the dark streaks which we can see. There are no mountains on Triton, and very few craters, so that it is unlike any other world in the Solar System. Unfortunately we cannot hope to find out much more until the dispatch of a new space mission, and this may be delayed for many years.

This is a good time to locate Uranus and Neptune, and they are always worth finding – but do not expect your telescope to show any details on them.

August 27

The Migrating Star

Before turning back to the Solar System, let us pause for a moment to look at a star which, though unremarkable in itself, has changed constellations in recent years – in fact, in 1992.

We have noted the line of three stars of which Altair, the leader of Aquila, is the central member (7 August). Adjoining Aquila is the small, compact constellation of Delphinus, the Dolphin (4 September). Our migrating star is Rho Aquilæ, which used to lie at the extreme edge of the Eagle. Its magnitude is 4.9; it is of type A, about 30 times more luminous than the Sun, and it is 166 light-years away. It is easy enough to identify; it is situated at the boundary of the Milky Way.

Like most stars, Rho Aquilæ has a measurable individual or proper motion. This has been carrying it north and west. The northward trend is only 0.06 of an arc second per year, but this has been enough to carry it over the border – so that it is now in Delphinus. Of course there is nothing significant in this, because the constellation boundaries are quite arbitrary and mean nothing at all; all the same, it provides a good trick question for any astronomical quiz – in which constellation will you find Rho Aquilæ?

No other naked-eye star is scheduled to change constellations in the near future. Apparently the next will be Gamma Cæli, of magnitude 4.6, which will track across the border of the adjacent Columba; but this will not happen until about the year 2400.

August 28

Mars at its Nearest

Mars, the Red Planet, has always held a special fascination for us, because it has always been regarded as a possible abode of life. Certainly it is less unlike the Earth than any other planet

Future Points of Interest

2003: Opposition of Mars.

Mars in Opposition

Date	Minimum distance from Earth (millions of miles)	Maximum apparent diameter (seconds of arc)	Maximum Magnitude	Constellation
1997 March 17	61	14.2	−1.1	Virgo
1999 April 24	54	16.2	−1.5	Virgo
2001 June 13	42	20.8	−2.1	Sagittarius
2003 August 28	35	25.1	−2.7	Capricornus

in the Solar System. Even in the earlier part of the twentieth century there were still astronomers who believed that the straight, artificial-looking lines crossing the red 'deserts' were true waterways, built by the local inhabitants to form a vast irrigation system – drawing water from the polar ice-caps and pumping it through to the centres of population. Alas, the canal-building Martians have long since joined the canals themselves in the realm of myth, but we still cannot be sure that Mars is completely sterile.

Mars is a small world, only 4200 miles in diameter. It is further away from the Sun than we are; its 'year' is 687 Earth-days or 669 Mars days or 'sols', because the axial rotation period is 24 hours 37.5 minutes. The axial inclination is almost the same as that of the Earth, so that the 'seasons' are of the same basic type, though of course they are much longer. The synodic period is 780 days, so that Mars comes to opposition only in alternate years.

Not all oppositions are equally favourable, because Mars has an orbit which is decidedly eccentric; the distance from the Sun ranges between over 154 000 000 miles at aphelion to only 128 500 000 miles at perihelion. Obviously, Mars is at its best when opposition occurs with the planet near perihelion. During the period covered here, the dates are shown above.

The opposition of 2003 is particularly favourable; Mars is in Capricornus, but still attains a respectable height above the horizon from the viewpoint of British observers. It is brighter than Jupiter can ever be, and is indeed an imposing object in the night sky.

August 29

Mars through the Telescope

The small size of Mars means that it can be studied to advantage only for a few months to either side of opposition, so that the observer has to make the most of his limited opportunities. Moreover, a high magnification is needed, and at least a four-inch telescope is essential for any useful work, though the main features – the polar caps and the main dark areas – can be seen with smaller instruments.

Mars has a thin atmosphere, made up chiefly of carbon dioxide, and the ground pressure is below 10 millibars everywhere – too low to support advanced life of our kind. Clouds can form, but in general the atmosphere is transparent unless dust-storms are present.

Most of the surface is 'desert'. There are however important differences between the deserts of Mars and Earth; the Martian areas are covered with reddish minerals rather than sand – Mars is a 'rusty' place – and they are bitterly cold rather than scorching hot. Material whipped up from the deserts by winds in the atmosphere can produce widespread dust storms, some of which may at times cover the entire planet and may persist for weeks.

The polar caps are icy, though they are not the same as ours; the underlying water ice is overlaid by carbon dioxide ice, which disappears when the warmer weather arrives in Martian spring. The dark areas are permanent. Once thought to be old sea-beds filled with vegetation, they are now known to be mere albedo features, where the red material has been scoured away to leave the underlying surface exposed.

Both the satellites, Phobos and Deimos, are real dwarfs; they are irregularly shaped and crater-pitted. Phobos has a longest diameter of less than 20 miles, Deimos less than 10, so that they are beyond the range of small telescopes even when Mars is best placed. Phobos has an orbital period of only 7 hours 39 minutes; this is less than a Martian day, so that to an observer on the planet's surface Phobos would rise in the west, gallop across the sky and set in the east 4.5 hours later. Neither satellite would provide much illumination at night; indeed Deimos would look like nothing more than a large, dim star. Probably both are captured asteroids rather than bona-fide satellites.

August 30

Observing Mars

In 1877 the Italian astronomer G.V. Schiaparelli drew a map of Mars, using a fine nine-inch refractor. The 'canals' which he recorded proved to be non-existent (they were tricks of the eye), but the names which he gave to the bright and dark areas are still in use, though they have been modified in view of the space-craft results.

The most prominent feature on Mars is the dark V-shaped Syrtis Major, which is now known to be a lofty plateau rather than a depression. Further north there is the wedge-shaped Acidalia Planitia (formerly called the Mare Acidalium). One of the most spectacular bright regions is Hellas, a circular area south of Syrtis Major; it was once believed to be a snow-covered

Anniversary

1745: Birth of Johann Hieronymus Schröter, first of the really great observers of the Moon and planets. He established his observatory at Lilienthal, near Bremen in Germany, and equipped it with good telescopes, in cluding one made by William Herschel. He made outstanding observations of Solar

(Continued opposite)

System bodies; for example his drawings of Mars were much the best of their time. Unfortunately many of his observations were destroyed in 1813, when his observatory was sacked by invading French troops. Schröter died three years later.

plateau, but has been found to be a deep basin, often cloud-filled. At times it looks remarkably like an extra polar cap.

If Mars is observed for half an hour or so, the shift of the main features across the disk, by virtue of the planet's rotation, is very obvious. Because the rotation period is 37.5 minutes longer than ours, any feature will reach the central meridian 37.5 minutes later each night.

When setting out to make an observation, it is wise to begin by putting in the main features, using a relatively low magnification. Then change to a higher power and add the finer details, paying special attention to the polar caps and to any clouds or dust-storms which may be present. Mars is always capable of springing surprises, and the well-equipped observer can do valuable work by monitoring it.

Anniversary

1913: Birth of Sir Bernard Lovell, the great radio astronomer who was responsible for building the 250-foot Jodrell Bank telescope now named after him.

1992: Discovery of 1992 QB1, an asteroid-sized body moving in the outer Solar System. It is about 175 miles in diameter; its orbital period is 296 years, and its distance from the Sun ranges between 3800 million miles and 4400 million miles, well beyond Neptune. Other similar bodies have since been found, and it is thought that there is a whole belt of them – now called the Kuiper Belt in honour of the Dutch astronomer G.P. Kuiper, who first suggested its existence. The magnitude of 1992 QB1 is only 22.8.

August 31

Close-Range Views of Mars

Many space-craft have now been sent to Mars; most of the results have come from American vehicles, since as far as Mars is concerned the Russians have been strangely unsuccessful. We now have detailed maps of the entire surface. There are craters, valleys, mountains and towering volcanoes, one of which, Olympus Mons, is three times the height of our Everest. Of course, all these features are beyond the range of ordinary Earth-based instruments, though good results are being obtained by the Hubble Space Telescope.

In 1976 two Viking probes made controlled landings on Mars, and undertook a search for life; no definite traces of living organisms were found. Then, in 1996, it was claimed that some meteorites found in Antarctica had been blasted away from Mars by a massive impact, and that they contained traces of primitive past life. The announcement caused a great deal of interest, but we have to admit that it is simply an interesting, if remote, possibility.

We have seen pictures of what are almost certainly dry riverbeds on Mars, in which case there must once have been running water there together with a fairly dense atmosphere. However, it is not likely that higher life-forms ever developed there, and certainly none exist on Mars now. Future explorers will, alas, have no chance of being met by a Martian welcoming committee. The Pathfinder probe landed on Mars in 1997 (July 4) and the orbiting Mars Global Surveyor reached Mars in September 1997.

Future Points of Interest

1998: Mercury at western elongation.

September

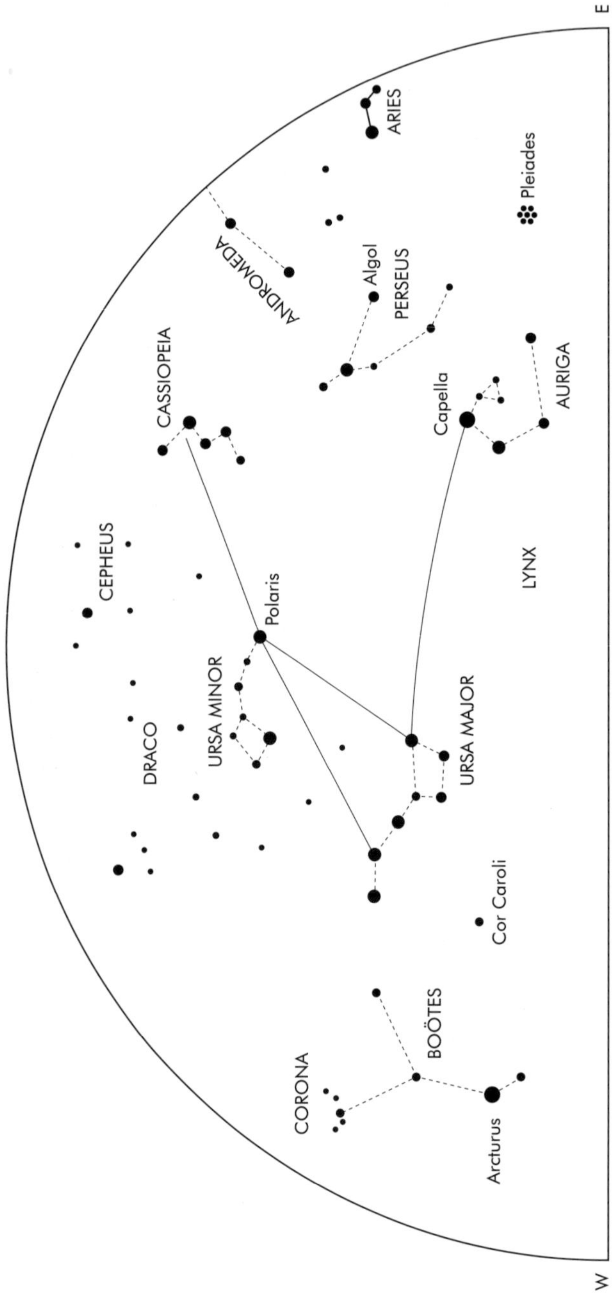

The Northern Hemisphere in September
Northern aspect (50° N 15 September 10pm)

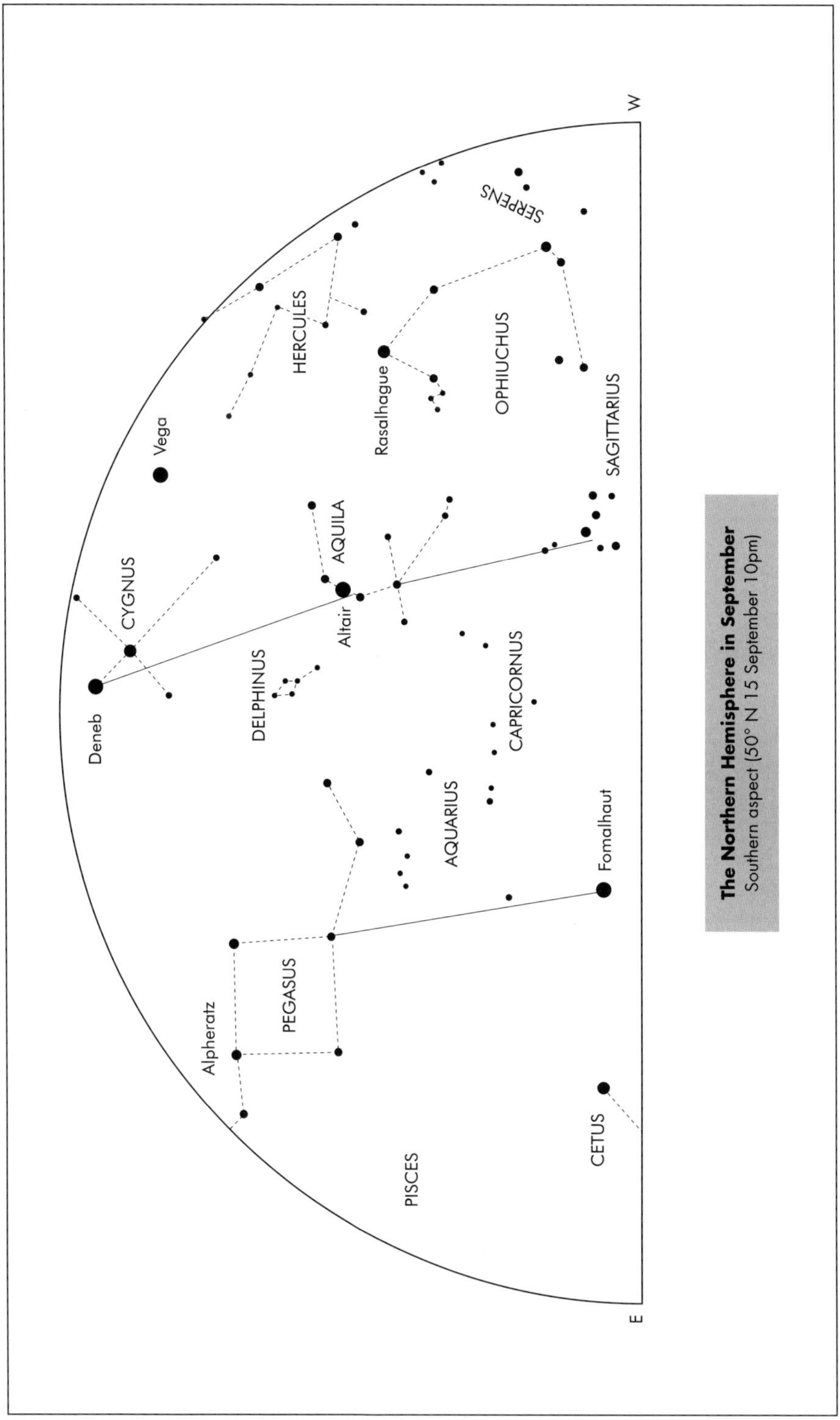

The Northern Hemisphere in September
Southern aspect (50° N 15 September 10pm)

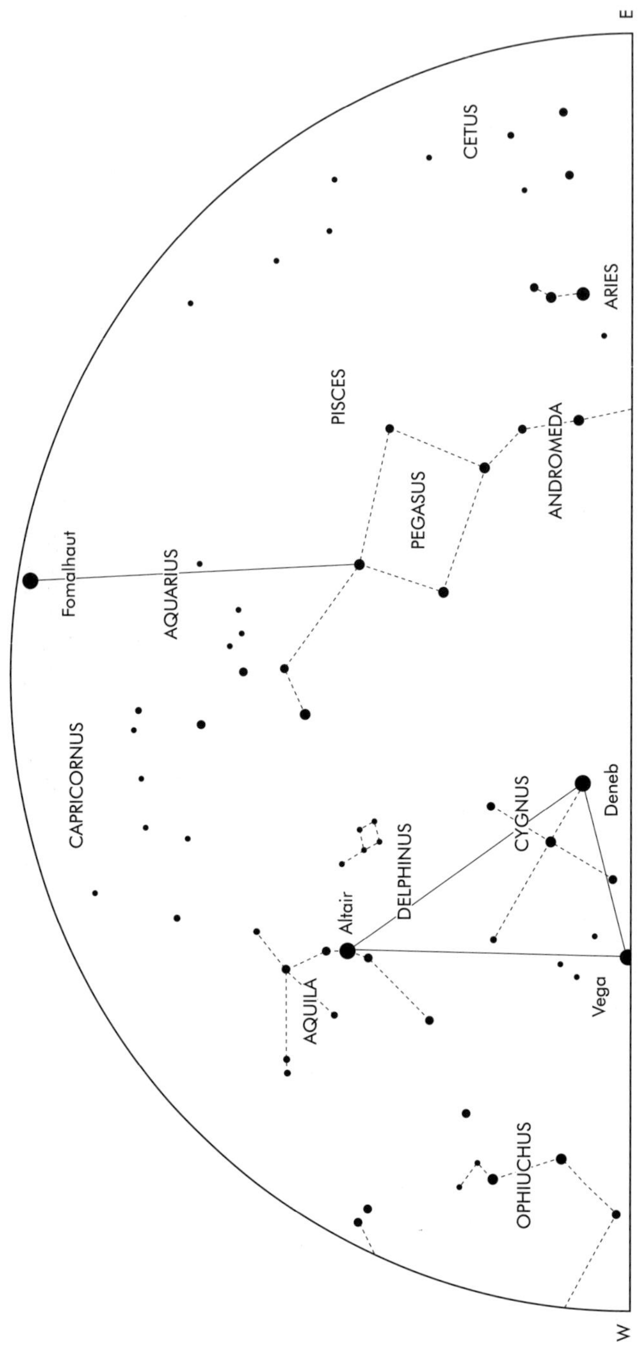

The Southern Hemisphere in September
Northern aspect (35° S 15 September 10pm)

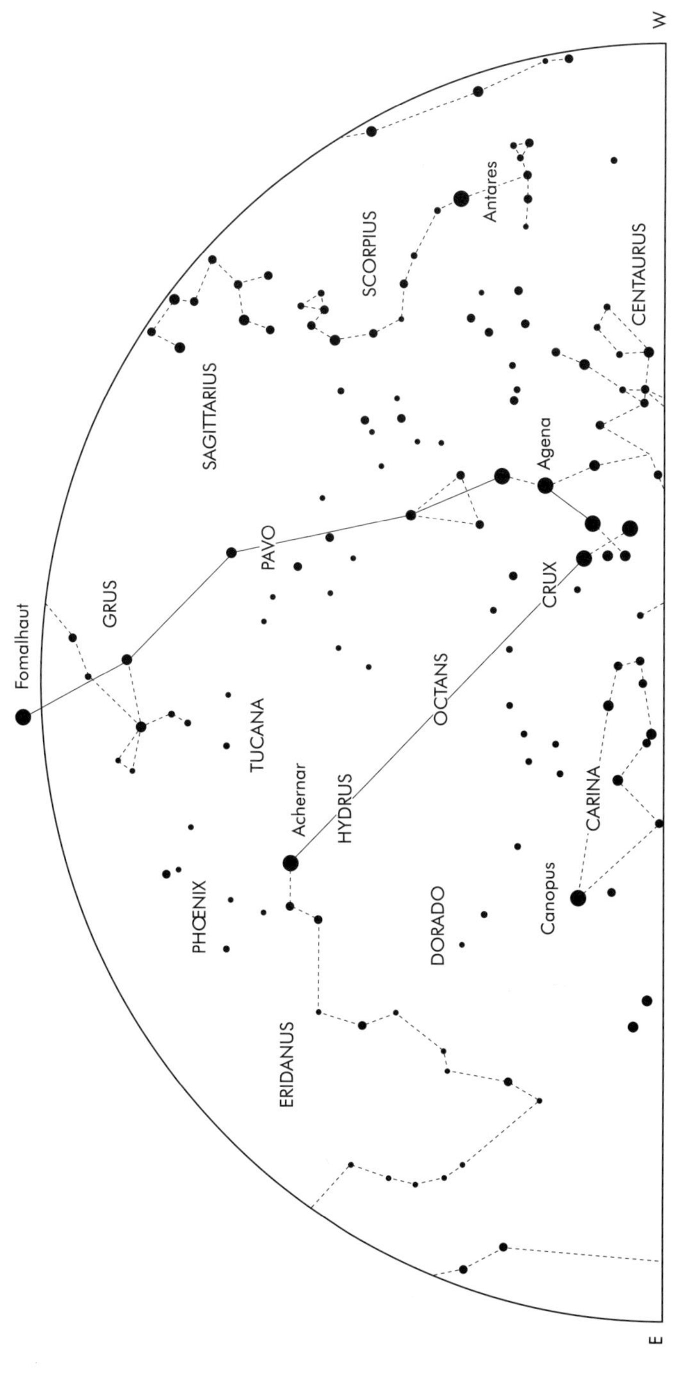

The Southern Hemisphere in September
Southern aspect (35° S 15 September 10pm)

September 1

The September Sky

We are coming to the northern autumn, with its longer nights – and already we can start to see the return of the Orion area; the Hunter himself does not rise until shortly before dawn, but the Pleiades have made their entry in the east and Aldebaran follows, while Capella is gaining altitude. Ursa Major is at its lowest in the north, which means that the W of Cassiopeia is near the zenith. The Summer Triangle is still very much in evidence, and will remain so throughout the month. Arcturus has disappeared into the evening twilight.

The main autumn constellation is Pegasus, the Flying Horse, high in the south; its four main stars make up the famous Square, and cannot be missed, though it is true that maps tend to make the Square look smaller and brighter than it really is. Leading off the Square is Andromeda, and beyond Andromeda is Perseus, the gallant hero of the celebrated legend (11 November). This is the best time to look for Fomalhaut in the Southern Fish, the southernmost of the first-magnitude stars. It is low even from England, and from North Scotland you will be lucky to see it at all.

From southern-hemisphere countries, however, Fomalhaut is very high up, and British observers who see it for the first time will find that it is unexpectedly bright. Pegasus is well above the northern horizon, but to Australians and New Zealanders the Vega–Deneb– Altair triangle is very low in the south-west (as we have noted, not all of it can be seen from the southernmost part of New Zealand). Scorpius is still prominent in the west, followed by Sagittarius with its glorious star-clouds. Crux is low in the south, and so is Canopus. The Clouds of Magellan are very high up, and the Milky Way stretches across the sky from the Cross through to the north-western horizon.

Anniversary

1804: Discovery of the third asteroid, Juno, by Karl Harding from Schröter's observatory at Lilienthal, near Bremen.

1979: Pioneer 11 by-passed Saturn at a range of just under 13 000 miles. This was the first Saturn encounter. Pioneer had been launched on 5 April 1973, and had already flown past Jupiter on 2 December 1974.

Future Points of Interest

2002: Mercury at eastern elongation.

September 2

Juno, the Third Asteroid

Eighteenth-century astronomers were puzzled by the wide gap in the Solar System between the orbits of Mars and Jupiter. They believed that a planet should exist there, and in 1800 a group headed by Johann Schröter and the Baron Franz Xavier von Zach began a systematic hunt for it; they called themselves the Celestial Police, and decided to patrol the Zodiac. Ironically, they were forestalled. In 1801 Piazzi at Palermo, not then a member of the team (though he joined

The Four Largest Asteroids

Name	Diameter (miles)	Mean opposition magnitude	Orbital period (years)	Rotation period (hours)	Mean distance from Sun (millions of miles)
Ceres	584	7.4	4.61	9.08	257.0
Pallas	360 × 292	8.0	4.62	7.81	257.4
Juno	179 × 143	8.7	4.36	7.21	247.8
Vesta	358	6.5	3.63	6.50	219.3

later) discovered the first asteroid, Ceres. The Police were not in the least discouraged. Olbers, one of their leading members, discovered asteroid No. 2 (Pallas) in 1802, and then came Harding's success with Juno, on 1 September 1804. Next came Vesta, another Olbers discovery, in 1807. No more seemed to be forthcoming, and the Police disbanded following the destruction of Schröter's observatory by the French. The next asteroid, Astræa, was not found until 1845.

The four original asteroids are nicknamed the Big Four and it is true that Ceres, Pallas and Vesta are the three largest members of the swarm, but Juno does not even rank with the first dozen. The main characteristics of the original four are given above.

Ceres and Vesta are almost regular in shape, but Pallas and Juno are not. Juno is what is termed an S-type asteroid, and its composition seems to be much the same as that of metal-bearing meteorites known as chondrites.

Juno is never visible with the naked eye; indeed, only Vesta ever reaches naked eye visibility. Binoculars will of course show the 'big four', and many others, but they look like stars.

It was once thought that the asteroids represented the débris of an old planet which either exploded or was broken up by collision, but it is now thought more likely that no planet could form in this part of the Solar System because of the powerful disruptive influence of Jupiter.

Anniversary

1976: Landing of the Viking 2 space-craft on Mars.

September 3

The Martian Scene

Let us return briefly to Mars. This is, after all, an anniversary day; Viking 1 had landed in the 'Golden Plain' of Chryse on 19 June (see 3 July) and Viking 2 followed, coming down in Utopia on 3 September.

Both landers sent back floods of data. In particular, analyses were made of the atmosphere, which proved to be over 95% carbon dioxide; most of the rest was nitrogen, with 0.03% of water vapour. The climate is decidedly chilly, with a temperature ranging from about −30 °C near noon down to a fearsome −86 °C before dawn. The atmospheric pressure is below 10 millibars. The Pathfinder mission of July 1997 was even more successful, and the rover, *Sojourner*, analyzed various rocks in the vicinity, which were given attractive nicknames such as Barnacle Bill, Yogi and Soufflé. It was confirmed that the Ares Vallis had once been a raging torrent of ordinary water. *Sojourner* operated for several Martian days or sols; temperatures ranged from −13°C (−8°F) at Martian noon down to −79°C (−105°F) at night, while winds were light and westerly. The pinkness of the daytime sky enhanced the eerie beauty of the Martian scene.

September 4

The Legend of the Dolphin

We have spent a long time on the Summer Triangle, but we must also pay some attention to the smaller constellations in the area; they are still on view during evenings in the northern hemisphere, and have not yet been completely lost in the twilight from the southern hemisphere.

First there is Delphinus, the Dolphin, which may be small but is far from inconspicuous. It lies near Aquila (5 August). It has no brilliant stars, but its leaders are close together; the overall appearance is not unlike that of a star-cluster, and unwary observers have been known to mistake it for the Pleiades. The leaders are listed below.

These make up the main pattern together with Delta (magnitude 4.4) and Epsilon (4.0). The curious names given to Alpha and Beta have an equally curious origin. They were given by Niccolo Cacciatore, an astronomer at the Palermo Observatory in Sicily. The Latinized version of his name is Nicolaus Venator – spell these names backwards, and you will see why Alpha and Beta Delphini are so called!

Delphinus is an ancient constellation, and there is a legend attached to it. Arion, known as a great singer, lived at the court of Periander, ruler of Corinth. On one occasion he sailed to Tænarusk in Sicily, to take part in a musical contest,

Delphinus

Greek letter	Name	Magnitude	Luminosity (Sun=1)	Distance (light-years)	Spectrum
α Alpha	Svalocin	3.8	60	170	F5
β Beta	Rotanev	3.5	46	108	B9
γ Gamma	–	3.9	4.5	75	G5+F8

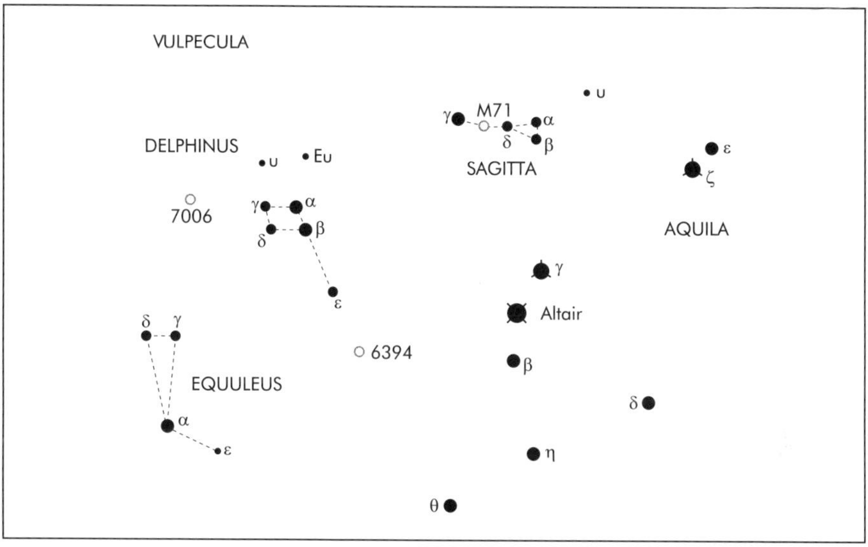

and, predictably, won every prize. He was sailing homeward when the members of the ship's crew decided to murder him and seize the prizes. Arion was thrown into the sea, and would have drowned but for the intervention of a kindly dolphin, which took him on its back and carried him ashore. Years later, when the dolphin died at a great age, it was brought back to life and placed in the sky as a token of gratitude.

Anniversary

1977: Launch of Voyager 1, which bypassed Jupiter in 1977 and Saturn in 1981. It is now on its way out of the Solar System permanently.

September 5

Doubles and Variables in Delphinus

Delphinus does not lack interesting objects. Beta is a close binary; the magnitudes are 4.0 and 4.9, but the separation never exceeds 0.3 of a second of arc, so that this is a severe test for a modest telescope. Gamma is a much easier double; magnitudes 4.5 and 5.5, with a separation of almost 10 seconds of arc. The two share a common motion through space, and are certainly associated, but they are a long way apart – at least 300 astronomical units. Colour estimates of the two components vary, but most people describe them as slightly yellowish.

There are two bright variables in the Dolphin; U Delphini and EU Delphini. Both are of the semi-regular type, so that their periods are very rough – 110 days and 59 days respectively. EU is the brighter of the two, with a magnitude range of 5.8 to 6.9; U fluctuates between 7.6 and 8.9. Both have M-type spectra, so that they are obviously red.

A very interesting nova (HR Delphini) flared up in the constellation in 1967. It was discovered by the English amateur G.E.D. Alcock, who makes a habit of discovering novæ and comets. (Incidentally, he has no telescope; he uses powerful, well-mounted binoculars.) The nova reached magnitude 3.3. Unlike most novæ, it remained visible with the naked eye for many months, and the fading was slow and irregular. It was identified with a star which had been of magnitude 11.9 before the outburst, and by now it has returned to its old state; it is unlikely to fade further, so that it remains within the range of a small telescope. Whether it will again flare up in the future remains to be seen; the variations over the past few years have been slight.

September 6

The Delphinus Globulars

There are two globular clusters in the Dolphin which are worth finding. One is C47 (NGC 6394); RA 20 h 34 m, declination +07° 24′. It lies south of Epsilon; the magnitude is above 9.

The second is C42 (NGC 7006); RA 21 h 05 m, declination +16° 11′ near Gamma. It is fairly faint (magnitude 10.6) and is small, with an apparent diameter of no more than 2.8 minutes of arc, but it is exceptionally remote. According to some estimates it may be as much as 150 000 light-years away, and it may even be one of the few known 'intergalactic tramps'. After all, there is no reason why globular clusters should not be found in intergalactic space; single stars could also be there, though they would be very hard to identify. In every other respect C42 seems to be an ordinary globular.

September 7

Equuleus

Our second small constellation is Equuleus, the Little Horse or the Foal, which adjoins Aquila and Delphinus. Mythologically it is said to represent the foal given by Mercury to Castor, one of the 'Heavenly Twins' (6 February), but there are no well-defined tales attached to it. The three main stars are Alpha (magnitude 3.9), Gamma (4.7) and Delta (4.5); these make up a small triangle, but there is nothing distinctive about them. Kitalpha, the old proper

Future Points of Interest

1998: Occultation of Jupiter by the Moon (5 h GMT). The occultation will be seen from the SE Pacific, South America, the Atlantic, and Western Africa.

name for Alpha, comes from the Arabian name for the whole constellation.

Epsilon Equulei is a wide double; magnitudes 6 and 7.1, separation 10 seconds of arc. The brighter component is itself a close double; it is a binary system with a period of 102 years. The separation is never more than one second of arc.

Delta Equulei used to be known as the closest of all visual binaries, though its record has long since been broken. The two components are almost equal at magnitude 5; the orbital period is only 5.7 years. As the separation never exceeds 0.3 of a second of arc, this is a test object for a large telescope. The real separation is rather less than that between our Sun and the planet Jupiter.

September 8

Sagitta, the Arrow

The next small constellation is Sagitta, due north of Altair. (It really is small; it covers a mere 80 square degrees, and only the Southern Cross is smaller.) The brightest stars are Gamma (magnitude 3.5) and Delta (3.8). Together with Alpha (4.4) and Beta (also 4.4) these make up the 'arrow' shape. Again there are only vague mythological associations, even though Sagitta is one of Ptolemy's original 48 constellations. One version says that Sagitta was an arrow shot by Apollo against the one-eyed Cyclops; later it was identified with Cupid's bow.

There are several variable stars in Sagitta. U Sagittæ is an eclipsing binary of the Algol type; RA 19 h 19 m, declination +19° 37′. The period is 3.4 days, and the magnitude range is from 6.6 to 9.2, so that this is a good subject for the owner of a small telescope. Near it is the red, N-type semiregular X Sagittæ; RA 20 h 05 m, declination +20° 39′; magnitude range 7.9–8.4, with a rough period of 196 days. Of more note is WZ Sagittæ (RA 20 h 08 m, declination +17° 42′), near Gamma; it is usually very faint – below magnitude 15– but occasionally flares up to binocular visibility, as it did in 1913, 1946 and 1978. It is worth keeping a watch here, as recurrent novæ are rather rare.

The cluster M71 (RA 19 h 54 m, declination +18° 47′) is easy to locate, as it lies between Delta and Gamma. It is on the fringe of binocular visibility, and a small telescope shows it easily. Nobody seems to be quite sure whether it is a very compact open cluster or a very loose globular; at any rate it is very remote, with an estimated distance of 18 000 light-years from us.

Future Points of Interest

Maximum of the Piscid meteor shower. This is a weak and diffuse shower, not associated with any known parent comet. It lasts from about 12 August to 7 October; the radiant position is RA 00 h 36 m, declination +07°, but the ZHR never exceeds 10.

September 9

Vulpecula

Our last minor constellation in this area is Vulpecula, the Fox, which lies between Delphinus and Cygnus. It was originally known as Vulpecula et Anser, the Fox and Goose, but the Goose has long since vanished from the sky.

Vulpecula has no star brighter than Alpha, magnitude 4.4. Alpha is of type M, and is very red; it forms a wide optical pair with 8 Vulpeculæ, magnitude 5.8. The pair is worth observing with binoculars simply for the colour contrast.

The most notable object in the constellation is M27, usually regarded as the finest of all planetary nebulæ. It is just visible with binoculars, and is in the same wide field with Gamma Sagittæ, but a telescope is needed to show it properly. Unlike the famous Ring Nebula, M57 in Lyra (30 July), it is not symmetrical; it merits its nickname of the Dumbbell, and the 'waist' can be seen with a telescope as small as six inches aperture. It is 975 light-years away, and the present diameter is 2.5 light-years; like all planetaries, it is expanding (remember that a planetary nebula is simply an old star which has thrown off its outer layers). The illuminating star of M27 is very faint; it is a hot bluish-white dwarf of magnitude 14.

September 10

The Sea of Tranquillity

The lunar Mare Tranquillitatis has been the scene of several landings. The first was the unsuccessful Ranger 6 on 2 February 1964; a TV failure meant that no images were received. Next came Ranger 7 (28 July 1964) and Ranger 8 (February 1965) which succeeded; then Surveyor 5, on 10 September 1967, and then the first manned landing by Neil Armstrong and Buzz Aldrin, on 19 July 1969.

Mare Tranquillitatis is one of the principal seas on the Earth-turned hemisphere of the Moon. It covers 170 000 square miles, and is in the eastern half of the disk, so that it can be seen when the Moon is between about 5 and about 15 days old. It connects with several other maria; Mare Serenitatis, Mare Nectaris, Mare Fœcunditatis and Palus Somnii. The Mare Nectaris is clearly a separate basin, joined to Mare Tranquillitatis by a comparatively narrow 'neck'; the same is true of the Mare Fœcunditatis, while Palus Somnii is a curiously-coloured area leading away in the direction of the separate Mare Crisium. On the strait between Tranquillitatis and Serenitatis, between the two capes of Acherusia and Argæus, the magnificent crater Plinius 'stands sentinel'.

Anniversary

1967: Landing of the US probe Surveyor 5 on the Moon, at latitude 1.4° S, longitude 23.2° E, in the Mare Tranquillitatis. It returned 18 000 images, and undertook soil analysis. Contact was maintained until 16 December.

There are no high, continuous mountain borders to the Mare Tranqillitatis, and neither are there any large craters on the floor, which in general is smooth by lunar standards. One very notable area is that of Arago, in the west of the Mare and not far from its border, where there is a superb system of lunar domes.

With a telescope, it is interesting to compare the Mare Tranquillitatis with the Mare Serenitatis. Tranquillitatis is much the lighter and patchier of the two, and much the less regular in outline.

September 11

Tracing the Zodiac: Continued

Earlier on (see 2 May) we traced the path of the ecliptic across the sky; now, with the approach of northern autumn, it may help if we continue the process.

As we have seen, the ecliptic runs from Scorpius through Ophiuchus and into Sagittarius. Sagittarius is the southernmost of the Zodiacal groups, and the ecliptic runs from it into Capricornus, the Sea-Goat, which has no really distinctive outline and no star brighter than Delta (2.9). The ecliptic cuts right through the constellation, just north of Theta (4.1) and appreciably to the north of Delta.

Next it enters Aquarius, the Water-bearer. Which again covers a wide area but is far from distinctive. The ecliptic just misses the star Lambda Aquarii (3.7), close to a little group of stars all lettered Psi, and then goes into Pisces, the Fishes, an even more obscure group consisting of a long line of dim stars south of the Square of Pegasus. Pisces contains the vernal equinox (29 September), on a barren area north of Omega Piscium (4.0). Before reaching Aries, the ecliptic passes near the edge of Cetus, the Whale; though it does not enter Cetus, the distance is so slight that the constellation can (and often does) contain planets. The ecliptic runs well south of the trio of stars which marks Aries, and then passes into Taurus (10 December), passing between the Pleiades and Aldebaran.

It is always interesting to follow the ecliptic line. In many ways it is a pity that there are no bright stars to make it more readily identifiable.

September 12

The Sea-Goat

To northern observers Capricornus is now in the south-west after sunset; from the southern hemisphere it is not very far

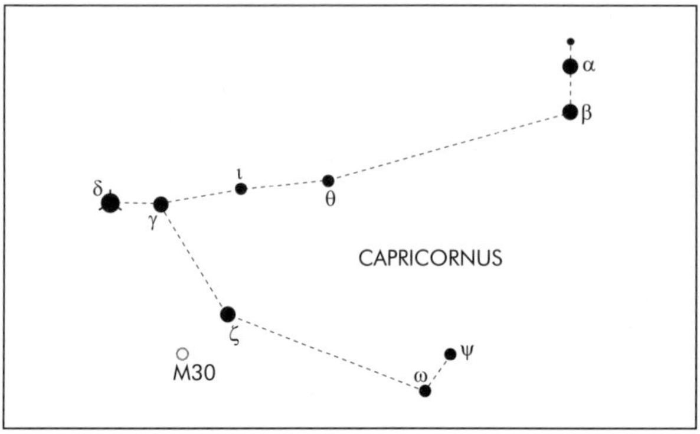

from the zenith. It covers over 400 square degrees, but is not at all prominent. It was of course one of the original groups listed by Ptolemy, but its mythological associations are not at all clear. It represents a 'sea-goat', but there seems to be no definite story of a marine goat anywhere. There is a suggestion that Capricornus may have some link with the demigod Pan, but opinions differ.

The brightest star, Delta, is only of magnitude 2.9, and there is no obvious pattern. However, it is not hard to find; it lies more or less between Altair (7 August) and Fomalhaut in the Southern Fish (2 October), and the line of three stars of which Altair is the central member points to it. The leaders are shown below.

Delta Capricorni is actually an eclipsing binary system. The range in magnitude is less than 0.1, and so cannot be detected without special instruments known as photometers; the period is only just over one day, and the components are no more than about 2 000 000 miles apart.

It is interesting to note that the point where Neptune was first identified, by Galle and D'Arrest in 1846 (4 August), was about 4° north-east of Delta Capricorni.

Capricornus

Greek letter	Name	Magnitude	Luminosity (Sun=1)	Distance (light-years)	Spectrum
α Alpha^2	Al Giedi	3.6	60	117	G9
β Beta	Dabih	3.1	2	104	F8
γ Gamma	Nashira	3.7	28	65	F0
δ Delta	Deneb al Giedi	2.9	13	49	A5
ζ Zeta	Yen	3.7	5200	1470	G4

Anniversary

1959: Landing of Luna 2 on the Moon.

September 13

The Flight of Luna 2

Look at the Moon (if the sky is clear, and the phase is right!) and you will see the huge Mare Imbrium or Sea of Showers, which is well defined and fairly regular; it is bounded in part by the lunar Apennines and Alps, and in area it is equal to Great Britain and France combined. On it are some majestic craters, including the 50-mile, dark-floored Archimedes, which makes up a superb trio with its smaller neighbours Aristillus and Eudoxus. It was near here, on 13 September 1959, that Luna 2 landed.

It was not the first attempt by the Soviet Union to send a probe to the Moon. Earlier, Luna 1 had bypassed our satellite and sent back useful data. But Luna 2 actually landed, though it did not transmit from the surface and was destroyed on impact. It was justifiably regarded as a success, and it ushered in a new era.

We do not know whether or not it was meant to come down gently; in those days the Russians were decidedly secretive. In any case, it crashed, and we cannot be sure of the landing site, though it is thought to have been near Archimedes. Signals from it were picked up by the radio telescope at Jodrell Bank as well as in what was then the USSR, and ceased abruptly as the probe hit the Moon. No doubt the broken fragments are still there, awaiting collection by museum authorities. Certainly Luna 2 has its place in history; it was the first space-craft ever to land on another world.

September 14

Beta and Alpha Capricorni

Two of the leading stars of Capricornus are double. Beta has a 6th-magnitude companion at a separation of 205 seconds of arc, so that virtually any telescope will show it; the two share a common motion through space, so that there is a genuine association.

Alpha, or Al Giedi, is different. It is a naked-eye pair; the magnitudes are 3.6 and 4.2 (rather confusingly the brighter star is officially Alpha^2, the fainter Alpha^1) and the separation is 378 seconds of arc, but here we are dealing with a purely optical alignment. Alpha^2 is 117 light-years from us, and 60 times as luminous as the Sun, but Alpha^1, which is only just over half a magnitude dimmer, could match over 5000 Suns and is 1600 light-years away. Like the Sun, it is yellowish, but unlike the Sun it is a giant and not a dwarf.

Both components have faint companions, but these are not to be seen except with fairly large telescopes.

In ancient times the two Alphas were closer together than they are now. The proper motion of the nearer star has moved it away from its neighbour, but apparently it was not until the seventeenth century that the separation was great enough for the two to be seen separately with the naked eye.

September 15

Messier 30

It must be said that as well as being rather dim and formless, Capricornus is decidedly lacking in interesting objects. One, however, is the globular cluster M30, which lies near Zeta at RA 21 h 40 m, declination –23° 11′. Less than a degree from Zeta is a fifth-magnitude star, 41 Capricorni, and M30 is in the same binocular field as 41.

The cluster was discovered by Messier himself, in 1764; he described it as 'Round; contains no star'. Certainly the nucleus of the cluster is small and compact, and all in all this is not an easy globular to resolve. It is around 41 000 light-years away.

It is at the limit of visibility with binoculars; the integrated magnitude is 8.5, so that if you can see it with binoculars you are doing very well. With a modest telescope, of course, it is easy.

September 16

Future Points of Interest

1998: Opposition of Jupiter. The planet is now only 4° south of the celestial equator.

Pegasus

It is time to turn our attention to Pegasus, which is, as we have said, the main (northern) autumn constellation. It is not very far from the celestial equator, and so is well seen from Australia or even New Zealand, but Andromeda, which leads away from it, is always very low from southern countries.

There is a celebrated legend associated with Pegasus. It represents a flying horse, upon which rode the dauntless hero

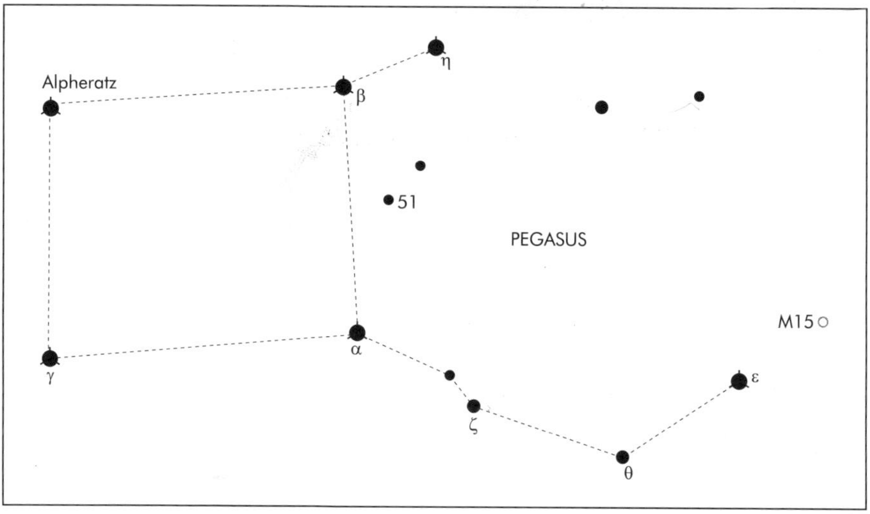

Bellerophon in a mission to dispose of a particularly ferocious fire-breathing monster, the Chimæra. However, it is fair to say that there is nothing equine about the celestial Pegasus, and the four stars which make up the Square form the main pattern. One of the four, Alpheratz, used to be known as Delta Pegasi, but has now been transferred to Andromeda, as Alpha Andromedæ. Since it so obviously belongs to Pegasus, this seems totally illogical; one has the feeling that the decision must have been made by a committee!

The leaders of Pegasus are listed below.

Alpheratz, Markab, Scheat and Algenib form the Square, with Enif some way away in the general direction of Aquila. As we have noted, maps do tend to give the impression that the Square is smaller and brighter than it really is in the sky.

Pegasus

Greek letter	Name	Magnitude	Luminosity (Sun=1)	Distance (light-years)	Spectrum
(α Andromedæ	Alpheratz	2.1	96	72	A0)
α Alpha	Markab	2.5	60	100	B9
β Beta	Scheat	2.3–2.9	310	176	M2
γ Gamma	Algenib	2.8	1320	490	B2
ϵ Epsilon	Enif	2.4	5000	520	K2
ζ Zeta	Homan	3.4	82	156	B8
η Eta	Matar	2.9	200	173	G2
θ Theta	Biham	3.5	26	82	A2
ι Iota	–	3.8	3.8	39	F5
μ Mu	Sadalbari	3.5	60	147	K0

September 17

The Light-curve of Beta Pegasi

If you look at the four stars of the Square of Pegasus one after the other, with binoculars or even with the naked eye, you will notice something at once. While three of the stars are white, the fourth – Scheat or Beta Pegasi – is decidedly orange. It has an M-type spectrum, and is therefore an evolved star which has used up much of its 'fuel' (8 October).

Like so many red giants, Beta Pegasi is variable. The range is not great – about magnitude 2.3 to 2.9 – but it is enough to be noticeable, particularly as there are excellent comparison stars available in Alpheratz (2.1), Alpha Pegasi (2.5) and Gamma Pegasi (2.8). Beta has a period of around 38 days, but it is a semiregular star, so that no two cycles are identical.

It is interesting to draw up a light-curve, plotting the star's magnitude against the date. If you make observations over a period of a few weeks, you will soon see that there is a definite rhythm to the way in which the star is behaving. (Remember, of course, to allow for extinction, see 15 January.) With a little practice it is not difficult to make estimates accurate to at least a tenth of a magnitude.

September 18

Future Points of Interest

2001: Mercury at eastern elongation

Epsilon Pegasi: a Suspected Variable

There are a few bright stars which are not officially recognized as being variable, but who have been strongly suspected of fluctuations. One of these is Enif or Epsilon Pegasi, the brightest star in the Flying Horse (if we exclude the stolen Alpheratz). Its name, Enif, comes from the Arab *Al Anf*, 'the Nose'. It is highly luminous, and with its K-type spectrum is decidedly orange.

It has been claimed that it shows short-term changes in magnitude. A nineteenth-century German observer, Schwabe, went so far as to maintain that it had a period of 25.7 days. This has never been confirmed, but we have an excellent project here for the naked-eye observer: compare Epsilon with the stars in the Square. It should be marginally brighter than Alpha Pegasi, much brighter than Gamma Pegasi, and a little fainter than Alpha Andromedæ. Once allowance has been made for extinction (15 January) it should be possible to get a reliable value.

Probably there are no real changes, but it is worth checking.

Future Points of Interest

1998: Occultation of Venus by the Moon, 18 h GMT. Visible from the Pacific and South America.

September 19

Double Star Separations

While we are with Pegasus, there is a chance to make a useful test. There are two double stars not far from the Square. One of these, Eta, has components of magnitudes 2.9 and 9.9, and as the separation is 108 seconds of arc this is an easy pair even if the secondary is rather faint. Kappa Pegasi, has magnitudes of 4.7 and 5.0, but the separation is only 0.3 of a second of arc, so what aperture telescope will be needed to split it?

It is always difficult to give exact answers to questions of this sort, but there is at least a rule which is better than nothing. The limiting magnitudes and minimum separations attainable are as follows, assuming that with a double star the components are not very unequal:

Limiting Magnitudes and Separation

Telescope aperture (inches)	Faintest magnitude	Smallest separation (seconds of arc)
2	9.1	2.5
3	9.9	1.8
4	10.7	1.3
5	11.2	1.0
6	11.6	0.8
10	12.8	0.5
12	13.2	0.4
15	13.8	0.3

Here, then, are some test objects on view this evening.

Double Star Tests

Star	Separation (secs)	Magnitudes	
Zeta-80 Ursæ Majoris (Mizar/Alcor)	708	2.3, 4.0	
Alpha-8 Vulpeculæ	414	4.4, 5.8	
Alpha Capricorni	371	3.5, 4.2	
Epsilon Lyræ	207	4.7, 5.1	
Beta Capricorni	205	3.1, 5.0	
Eta Pegasi	91	2.9, 9.9	
Nu Draconis	62	4.8, 4.9	
Zeta Lyræ	44	4.3, 5.9	
Beta Cygni	34.4	3.1, 5.1	
Kappa Herculis	28.4	5.3, 6.5	
Zeta Piscium	23	5.6, 6.5	
Alpha Canum Venaticorum	19.4	2.9, 5.5	
Alpha Ursæ Minoris	18.4	2.0, 9.0	Polaris
Zeta Ursæ Majoris	14.4	2.1, 4.8	Mizar

Double Star Test *(Continued)*

Star	Separation (secs)	Magnitudes	
Eta Cassiopeiæ	12.2	3.4, 7.5	
Gamma Andromedæ	9.8	2.3, 4.8	
Gamma Arietis	7.8	4.7, 4.8	
Zeta Cancri	5.7	5.0, 6.2	
Rho Herculis	4.2	4.6, 5.6	
Eta Sagittarii	3.6	3.2, 7.7	
Epsilon Draconis	3.1	3.8, 7.4	
Rho Ophiuchii	3.1	5.3, 6.0	
Gamma Virginis	3.0	3.5, 3.5	
Epsilon Boötis	2.8	2.5, 4.9	
Iota Trianguli	2.3	5.4, 7.0	
Zeta Aquarii	2.0	4.3, 4.5	
Alpha Piscium	1.9	4.2, 5.1	
Epsilon Arietis	1.4	5.2, 5.5	
Zeta Herculis	1.1	2.9, 5.5	
Zeta Boötis	1.0	4.5, 4.6	
Nu Scorpii	0.9	4.2, 6.8	
Alpha Ursæ Majoris	0.7	1.9, 4.8	Dubhe
Lambda Cassiopeiæ	0.5	5.3, 5.6	
Beta Delphini	0.3	4.0, 4.9	
Delta Equulei	0.3	5.2, 5.3	
Kappa Pegasi	0.3	4.7, 5.0	
Zeta Sagittarii	0.2	3.3, 3.4	
Epsilon Ceti	0.1	5.7, 5.8	

Try them, and see how far down the list you can get!

September 20

The Planet of 51 Pegasi?

Let us next turn to a very ordinary-looking star, 51 Pegasi. It is of magnitude 5.5; it lies at RA 22 h 57 m, declination +20°45′, just outside the Square, and roughly half-way between Alpha and Beta. Its mass and luminosity are not very different from the Sun's; it is of spectral type G3, and the distance has been given as 42 light-years, though with some uncertainty.

51 Pegasi became 'news' in October 1995, when two astronomers at Geneva Observatory, M. Mayor and G. Queloz, announced that by measuring slight changes in its velocity they had found it to be attended by a secondary body – not a star, but a planet rather less massive than Jupiter. They claimed that the orbit of the planet was virtually circular, with an orbital period of only 4.2 days. This would give a distance between the star and the planet of a mere 4 300 000 miles, and the surface temperature of the planet would be a torrid 1000 °C.

Future Points of Interest

1998: Occultation of Mercury by the Moon, 7 h GMT. Visible from Italy and Greece through to Indonesia and Australia.

The idea of a massive planet moving so close to a star seemed implausible, and later observations seemed to indicate that the observed effects were due to features of the star itself rather than to any orbiting planet. Few astronomers now believe that 51 Pegasi does, indeed, have a planet moving round it, and the same may be true of other stars reported to be planetary centres, such as 70 Virginis and 47 Ursæ Majoris. However, opinions differ, and although the real existence of these extra-solar planets may be rather unlikely, it is certainly not impossible.

September 21

Remote Objects in Pegasus

It is worth spending one more evening with Pegasus before moving on. The most interesting object we have not so far discussed is the globular cluster M15, which lies in line with Theta and Epsilon at RA 21 h 30 m, declination +12° 10′. It is a fine, bright globular, with an integrated magnitude of 6.3, so that it is not far below naked-eye visibility, and is very easy with binoculars. It has a condensed centre, and the outer parts are not hard to resolve; the distance is about 50 000 light-years. It was discovered on 7 September 1746 by the Italian astronomer G. Maraldi, who was not searching for nebulæ at the time; he was hunting for a comet.

Also in Pegasus, though much too faint to be seen with small telescopes, are five galaxies making up what is known as Stephan's Quintet. They seem to be 'lined up' and associated with each other, but spectroscopic observations show that one of them is much more remote than the rest, which is decidedly odd. As we have seen, distances are measured by the shifts in the spectral lines. Some astronomers, notably H.C. Arp and Sir Fred Hoyle, believe that this method is unreliable. If they are right, then many of our present theories will have to be drastically revised, and certainly Stephan's Quintet gives 'cause for concern'.

Finally, see how many stars you can count inside the boundaries of the Square, first with the naked eye and then with binoculars. You may well be surprised!

September 22

Future Points of Interest

Autumnal Equinox (22 or 23 September).

Equal Day and Night

In March the Sun crosses the celestial equator, moving from south to north (20 March). Now, six months later, it is back to the equator, this time travelling from north to south, so we have the autumnal equinox, or First Point of Libra – which, thanks to precession, is now situated in Virgo, between Beta and Eta Virginis (22 January).

Again the date is not quite constant, because of the irregular nature of our calendar (it is inconvenient of the Earth to take 365.25 days to complete one orbit round the Sun, instead of exactly 365). Therefore the date of the autumnal equinox changes slightly. During the period covered in this book, the dates are:

1998 September 23, 5 h 39 m
1999 September 23, 11 h 33 m
2000 September 22, 17 h 29 m
2001 September 22, 23 h 6 m
2002 September 23, 4 h 57 m
2003 September 23, 10 h 48 m
2004 September 22, 16 h 31 m
2005 September 22, 22 h 25 m

The South Pole now begins its six months' night. This is important, because a major observatory is being set up there, and there will be considerable advantages in having a prolonged period of darkness followed by an equal period of daylight. Seeing conditions are good, and the South Pole Observatory will be in full operation early in the twenty-first century.

September 23

Aquarius, the Water-Bearer

From prominent Pegasus we turn to a large but rather faint Zodiacal constellation: Aquarius. Of course it is one of Ptolemy's originals, and probably represents Ganymede, cupbearer to the Olympian gods. One old description of it, by the Roman writer Manilius, reads: 'He holds the cup or little urn in his hand, inclined downward; and he is always pouring out of it, as indeed he ought to be, to be able from so small a source to form that river, which you see running from his feet, and making so large a tour over all this part of the globe.' The proper name of Alpha Aquarii, Sadalmelik, comes from the Arabic *Al Sa'd al Malik*, the Lucky One of the King.

The leading stars are listed opposite.

There is no particular shape to Aquarius. It is cut by the celestial equator, but almost all of it lies in the southern hemisphere of the sky. It occupies much of the space between Pegasus and Fomalhaut, the first-magnitude star in the Southern Fish (2 October).

Anniversary

1846: Discovery of Neptune, by J. Galle and H. D'Arrest (see 4 August).

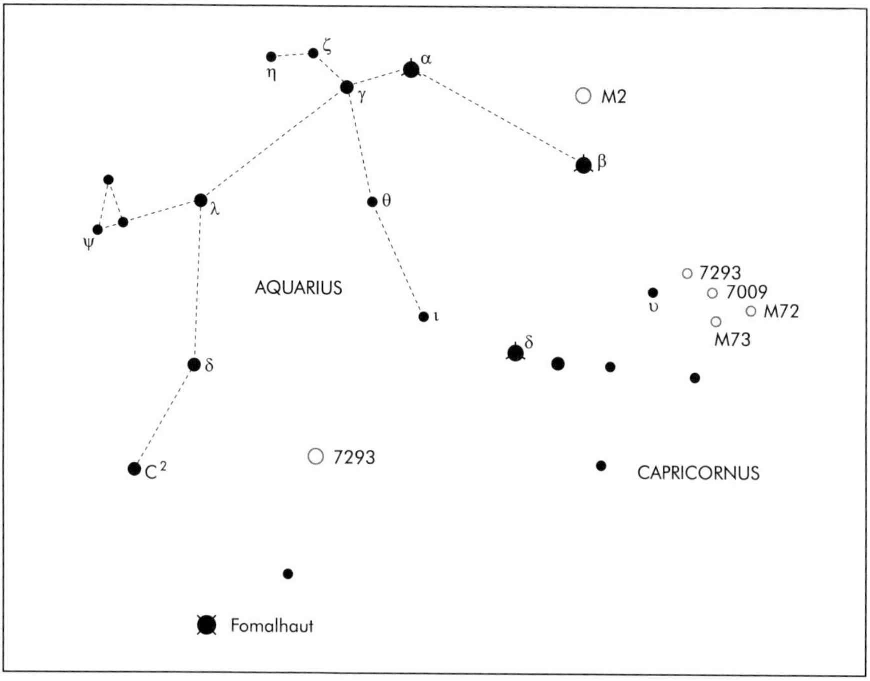

Aquarius					
Greek letter	Name	Magnitude	Luminosity (Sun=1)	Distance (light-years)	Spectrum
α Alpha	Sadalmelik	3.0	5250	945	G2
β Beta	Sadalsuud	2.9	5250	980	G0
γ Gamma	Sadachiba	3.8	50	91	A0
δ Delta	Scheat	3.3	105	99	A2
ε Epsilon	Albali	3.8	28	107	A1
ζ Zeta		3.6	50	98	F2+F2
λ Lambda		3.7	120	256	M2
c^2		3.7	60	107	K0

September 24

Zeta Aquarii

Come now to a very fine binary in the Water-bearer: Zeta Aquarii. It is a member of a well-marked little asterism including Gamma and Eta as well as Zeta; it is often called the Water-Jar.

Zeta shines as a star of magnitude 3.6, but a telescope shows it to be double. The components are almost equal in magnitude (4.3 and 4.5) and are separated by 2 seconds of arc. Both stars are F-type subgiants; the true separation is about 100 times the distance between the Earth and the Sun.

The orbital period is 856 years. At present the separation is gradually decreasing, because we are seeing the pair from a less favourable angle – in 1781 William Herschel had the separation as over 4 seconds of arc – but the long orbital period means that the alteration is very slow, and Zeta Aquarii will remain an excellent object for viewing in the foreseeable future. It lies only 00° 01′ 13″ from the celestial equator, and so is equally well seen from either hemisphere of the Earth.

September 25

Harvest Moon

Everyone knows that the Moon rises later every night. This is because it is moving eastward against the stars. The time-interval between successive moonrises is known as the 'retardation', but its value is not constant throughout the year. It is least around the time of the equinox, because the ecliptic is then at its least 'upright' position with respect to the horizon. The diagram should make this clear. The Moon takes the same time to travel from A to B in each case, but you can see why the retardation is least in the right-hand diagram. It is not true to say (as some books do) that the full moon then rises at the same time for several successive nights, but the retardation may often be reduced to as little as a quarter of an hour.

The full moon nearest the autumnal equinox is called Harvest Moon, because the extra light was said to help farmers to gather in the harvests. The Harvest Moons for the period covered here are:

1997 September 16
1998 October 5
1999 September 25
2000 September 13
2001 October 2
2002 September 21
2003 September 10

Anniversary

1644: Birth of Ole Rømer, the Danish astronomer who was the first to measure the velocity of light (by observations of the eclipses of Jupiter's satellites by the shadow of the planet). In 1681 he became Director of the Copenhagen Observatory, and made many important contributions; in particular he invented the transit circle and the meridian circle. He died in 1710.

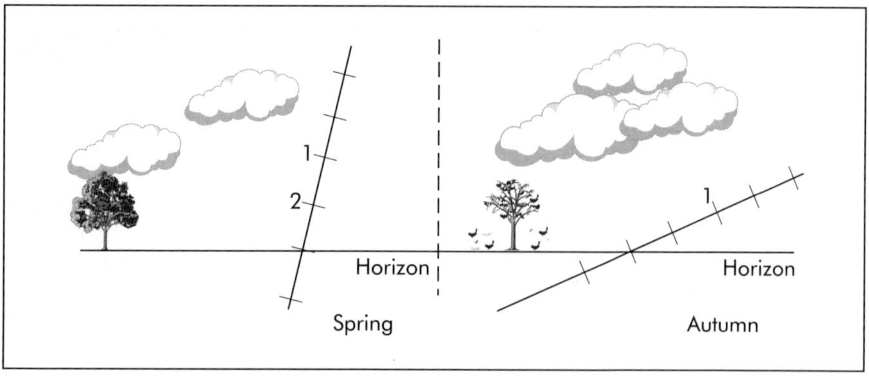

The full moon following Harvest Moon is known as the Hunter's Moon.

Anniversary

The famous 'Blue Moon' of 1950.

September 26

Blue Moons

A very strange sight was seen on 26 September 1950. For a few hours there really was a blue moon. The Moon shone with what I described at the time as 'a pale electric blue radiance, utterly unlike its normal self'.

Of course this had nothing directly to do with the Moon. The cause was dust in the upper atmosphere of the Earth, sent there by major forest fires raging in Canada. Other blue moons have been seen now and then, all for the same basic reason.

There is another sort of 'blue moon', though there is no obvious reason for the name. Since the interval between one full moon and the next is 29.5 days, it is possible to have two full moons in the same calendar month – for instance in January 1999, when there will be full moons on the 2nd and the 31st. In this case the second full moon is termed a blue moon.

It has also been claimed that the full moon looks larger when close to the horizon than it does when high in the sky. This is a well-known and powerful illusion, but it really *is* an illusion and nothing more, as measurements will show. The cause has been hotly debated, but hinges on the fact that the low moon can be compared with objects on or near the ground (trees and houses, for example) while the high moon cannot.

Anniversary

1918: Birth of Sir Martin Ryle, one of the great pioneers of radio astronomy. He spent much of his career at Cambridge; he developed the technique of aperture synthesis, and much of the later work in this field was based upon his researches. In 1972 he succeeded Sir Richard Woolley as Astronomer Royal. He died in 1984.

September 27

Messier Objects in Aquarius

Let us now return to the Water-bearer.

There are three Messier objects in the constellation; Numbers 2, 72 and 73. M2, north of Beta and forming a triangle with Beta and Alpha, is a fine globular cluster (RA 21 h 33 m, declination −00° 49′) which some people can see with the naked eye; it is striking in binoculars. It was discovered by Maraldi in 1746, during a comet-hunt. He wrote that it was 'round, well terminated and brighter in the centre ... not a single star around it for a pretty large distance. I took, at first, this nebula for a comet.' It is not too easy to resolve into stars; its distance is 55 000 light-years.

M72, another globular, was found in 1780 by Pierre Méchain. It is not nearly as bright as M2; it lies more or less between Epsilon Aquarii and Theta Capricorni, at RA 20 h 53 m, declination −12° 32′. It is not impressive, and is not easy to resolve; the distance is 62 000 light-years. The integrated magnitude is below 9.

Close beside M72, in the direction of Nu Aquarii, is M73, which is not nebular at all, but merely consists of a few faint stars. Messier said that it contained 'a little nebulosity', but this is incorrect, and the stars in it are not associated with each other. M73 is always included in the catalogue, but we must agree with Admiral Smyth, who said that he included it 'out of respect to Messier's memory'. Its position is RA 21 h 59 m, declination −12° 38′.

With the naked eye or binoculars it is worth looking at the little group of stars which includes Psi^1, magnitude 4.2. They are close together, and several of them are orange. They can even give the impression of being a very loose cluster, but this is not the case; they are not genuinely associated with each other.

Future Points of Interest

2003: Mercury at western elongation.

September 28

The Saturn Nebula and the Helix

Aquarius contains two of the finest planetary nebulæ in the sky, and it is surprising that neither is in Messier's list. C55 – NGC 7009 – lies near Nu Aquarii; it was discovered by William Herschel in 1792, and its nickname of the Saturn Nebula was given to it by Lord Rosse in 1848; he commented that the dark band of obscuring matter across it does make it look superficially a little like the Ringed Planet. It is strikingly beautiful in large telescopes, though small apertures do not show its form well. The position is RA 21 h 04 m, declination −11° 22′. As the integrated magnitude is only just below 8, it is a relatively easy object to locate.

The Helix Nebula C63 – NGC 7293 – is the nearest and brightest of all the planetaries; its apparent diameter is about half that of the Moon. Its integrated magnitude is between 6 and 7, so that binoculars will show it even though a large telescope is needed to bring out the 'ring' form, not too unlike M57 in Lyra (30 July). The central star is of magnitude 13.5. Splendid views of it have come from the Hubble Space Telescope; the images show giant tadpole-shaped objects which have been ejected from the dying star. The distance from Earth is only 450 light-years. The position is RA 22 h 30 m, declination −20° 48′, roughly half-way between Fomalhaut and the fourth-magnitude star Iota Aquarii.

September 29

Pisces

We have here yet another fairly large but decidedly dim Zodiacal constellation: Pisces, the Fishes, south of the Square of Pegasus and occupying much of the area between Pegasus and Aquarius. We can in fact use Pegasus as a guide. The two western stars of the Square, Beta and Alpha, point southward through Pisces and Aquarius until they arrive at Fomalhaut in Piscis Australis (the Southern Fish), which is the southernmost of the first-magnitude stars to be visible from Britain; it is now very low over the southern horizon, and viewers in northern Scotland will be lucky to see it at all. (To Australian observers, it is almost overhead.) We will return to Fomalhaut later (2 October). Meanwhile, let us concentrate on Pisces, which contains the Vernal Equinox. There are two fishes, joined by a knot, possibly representing the fishes into which Venus and Cupid once changed themselves to avoid the unwelcome attentions of the monster Typhon.

Pisces consists essentially of a long line of faint stars running south of Pegasus. The leaders are listed below.

Alpha Piscium is a fine double; the components are of magnitudes 4.2 and 5.1, and the separation is 1.9 seconds of

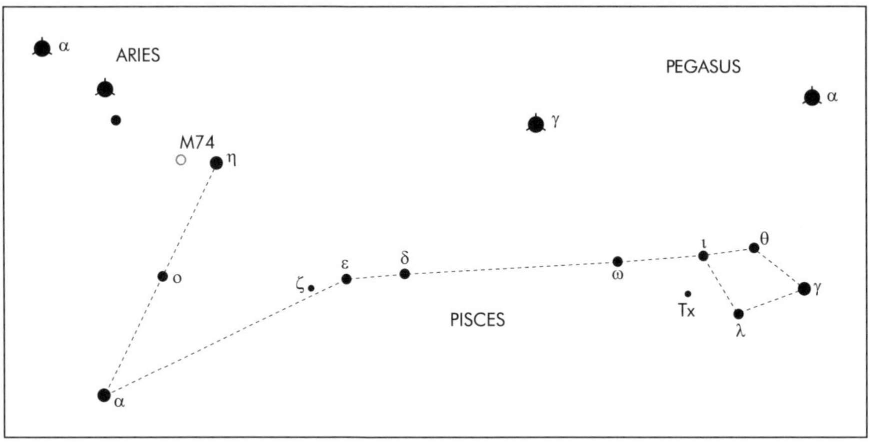

Pisces					
Greek letter	Name	Magnitude	Luminosity (Sun=1)	Distance (light-years)	Spectrum
α Alpha	Al Rischa	3.8	26	98	A2
γ Gamma	–	3.7	58	156	A7
η Eta	Alpherg	3.6	58	133	G8
ω Omega	–	4.0	45	85	F4

arc. It is a binary, with a long orbital period (933 years). Alpha Piscium is a star with several names; Al Rischa, Kaïtain or Okda. Take your pick!

Another easy double is Zeta (RA 01 h 24 m, declination +07° 35′). The magnitudes are 5.6 and 6.5, and the separation is a full 23 seconds of arc, so that virtually any telescope will show the components separately.

September 30

TX Piscium and Van Maanen's Star

One characteristic of Pisces is the little quadrilateral of stars not far from Alpha Pegasi in the Square: Gamma, Iota, Theta and Lambda. Near them, and actually in the same binocular field with Iota and Lambda, is one of the reddest of all stars, TX Piscium (RA 23 h 46 m, declination +03° 29′). It is variable, with a magnitude range of 6.9–7.7; there seems to be no period at all. The spectrum is of type N.

N-stars are very cool. Their surface temperatures are only of the order of 2500 °C, and this of course accounts for their red colour. TX Piscium is one of the best examples, because it is bright enough to be seen in binoculars, and its colour stands out at once. It is several hundreds of light-years away; if it were as close as, for instance, Sirius, it would be a truly magnificent spectacle.

Near Delta Piscium, at RA 00 h 46 m, declination +06° 09′, there is a very different type of star – Van Maanen's Star, a white dwarf probably no larger than the Earth but around a million times as dense as water. White dwarfs are bankrupt stars (21 February) and Van Maanen's Star is one of the best examples. Do not waste time in searching for it; the apparent magnitude is well below 12.

There is one Messier object in Pisces, the spiral galaxy M74: RA 01 h 37 m, declination +15° 47′). It is one of the most elusive of the Messier objects, and its integrated magnitude is no more than 8; it lies near Eta. There is nothing very special about it. The distance is of the order of 78 000 light-years.

October

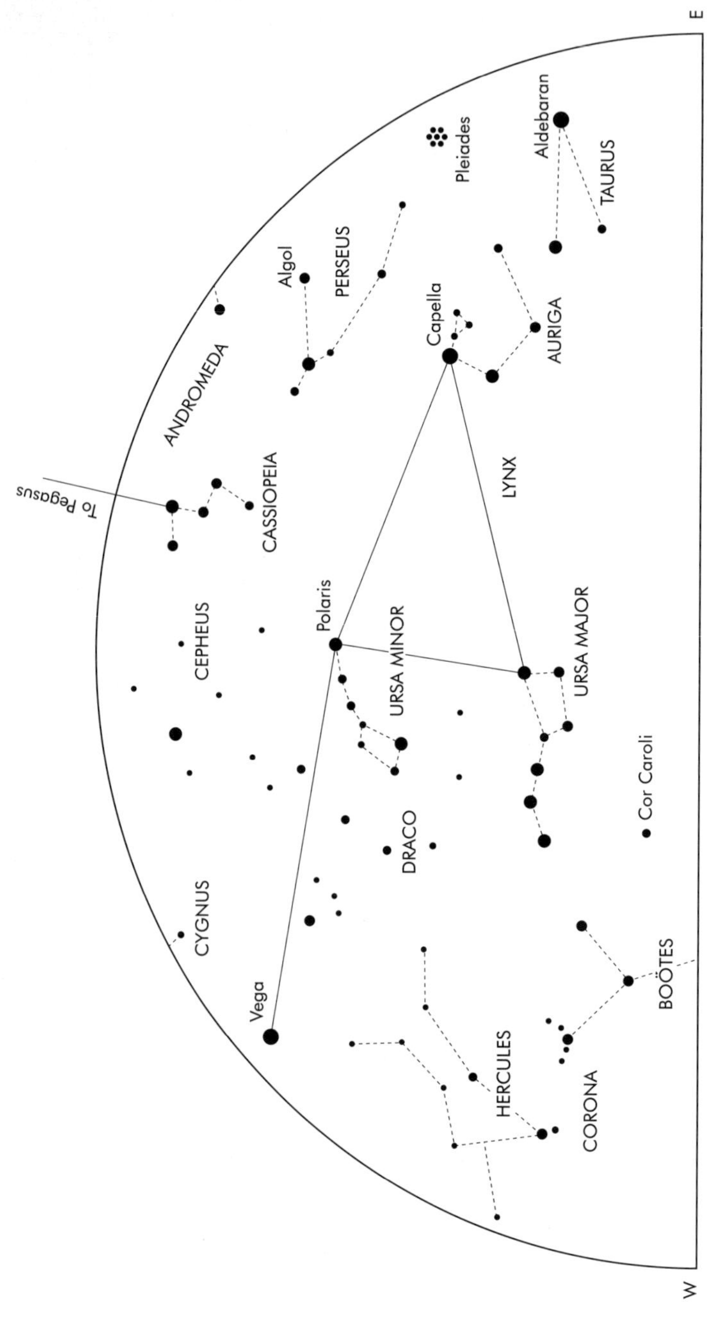

The Northern Hemisphere in October
Northern aspect (50° N 15 October 10pm)

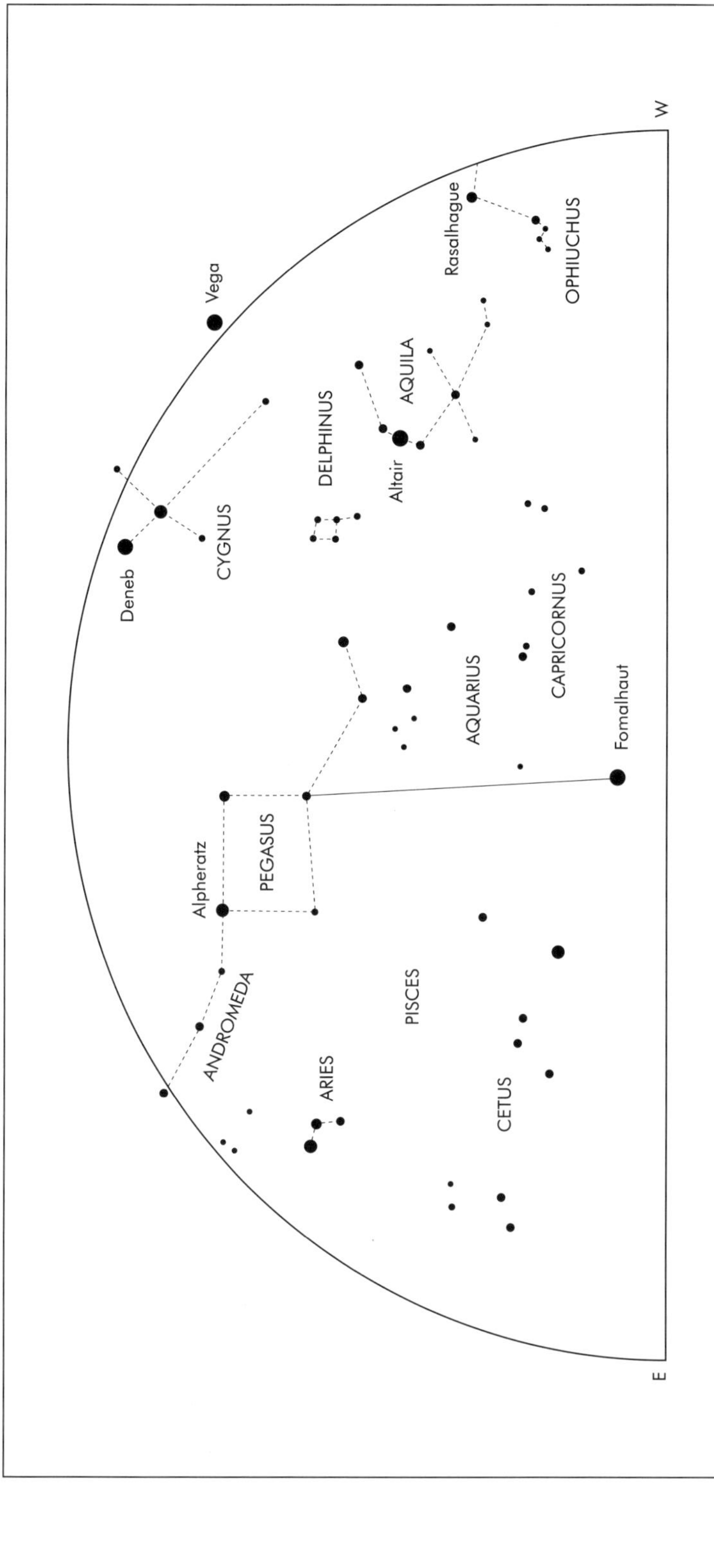

The Northern Hemisphere in October
Southern aspect (50° N 15 October 10pm)

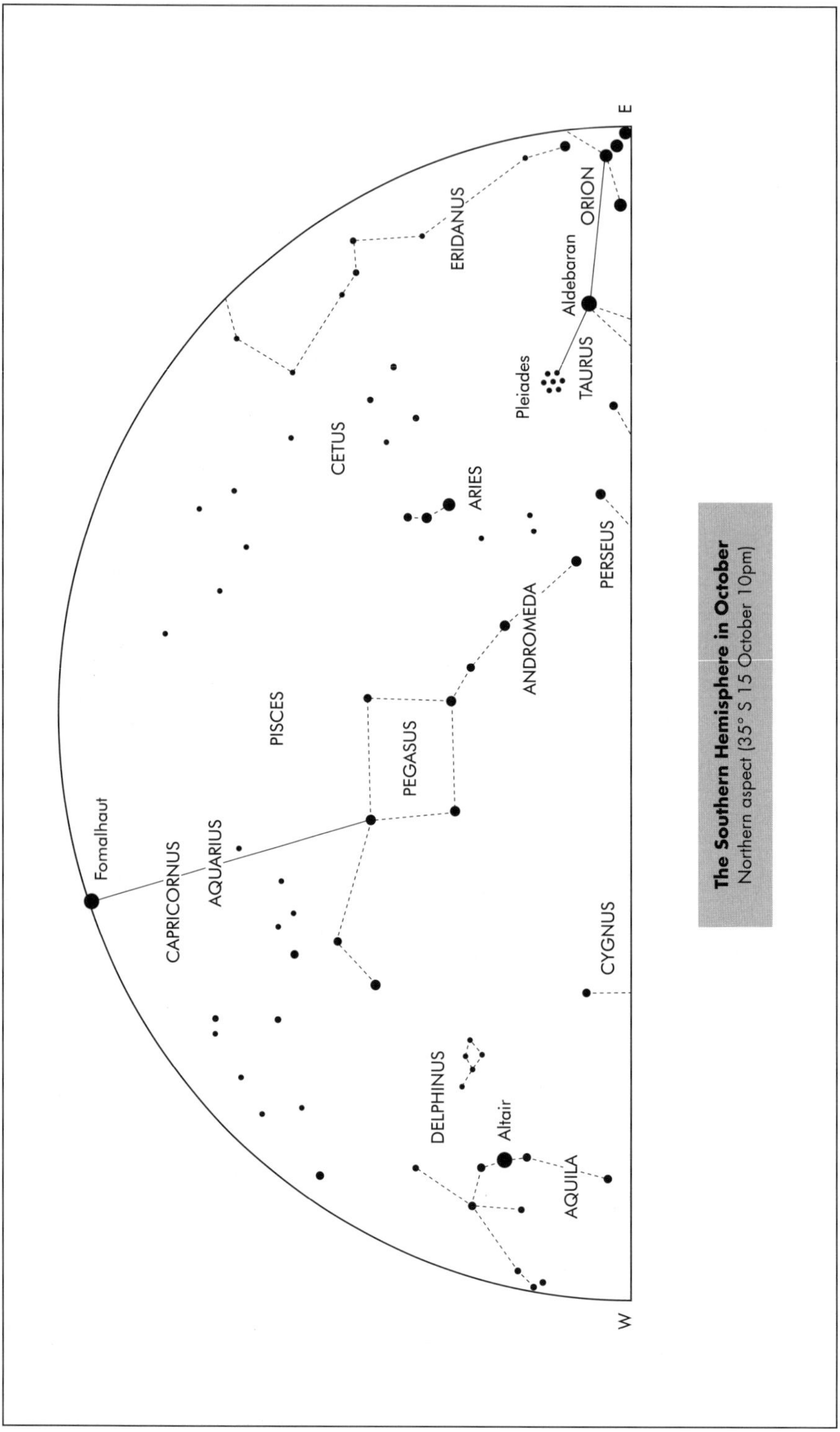

The Southern Hemisphere in October
Northern aspect (35° S 15 October 10pm)

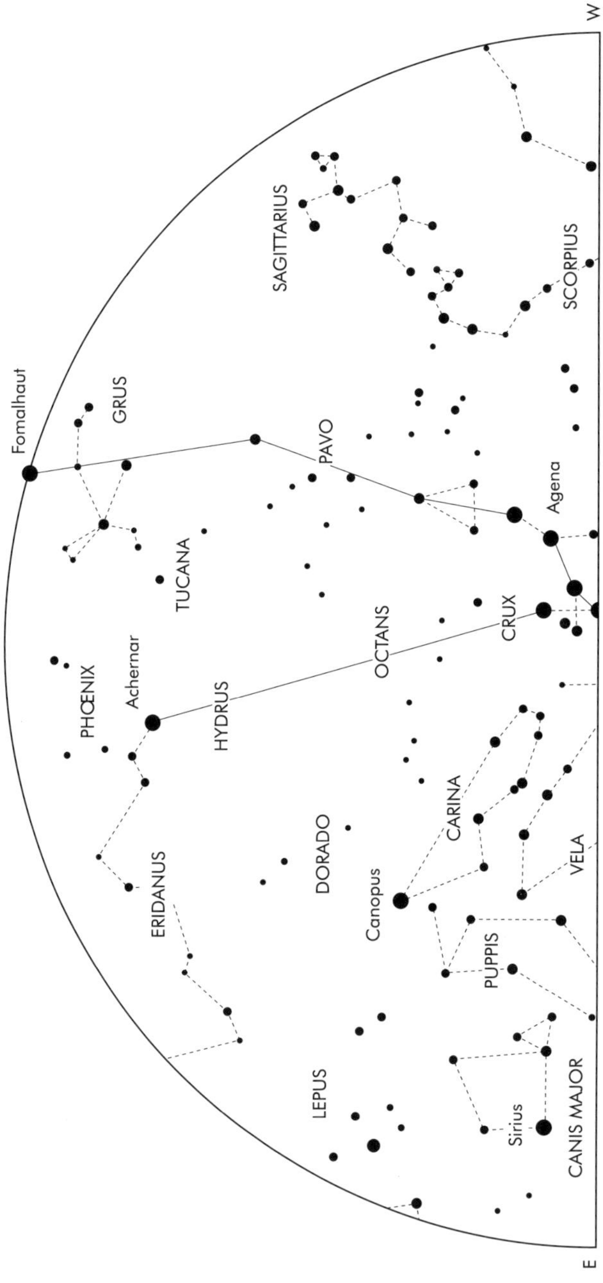

The Southern Hemisphere in October
Southern aspect (35° S 15 October 10pm)

October 1

The October Sky

Northern observers are now preparing for the return of Orion. It rises in the east very late in the evening, and well before that some of the members of the Hunter's retinue – Capella, Aldebaran, the Pleiades – are well in view. The Square of Pegasus is high in the south, with Andromeda leading off toward Perseus and thence to Capella; lower down is the large, sprawling constellation of Cetus, the Whale. Fomalhaut is still on view, and Aquarius and Capricornus are becoming low in the south-west.

The Great Bear is at its lowest in the north, which means, of course that the W of Cassiopeia is almost overhead. We still have the Summer Triangle, and in the evening we will not lose any of it until late in October, though of the three brilliant stars only Altair actually sets over British latitudes. The Milky Way is excellently placed, running from Aquila through Cygnus, Cassiopeia, Perseus and Auriga through to Gemini in the east.

Southern-hemisphere observers have lost the magnificent Scorpion, but Orion can be seen at a reasonable hour; Pegasus remains on view in the north-west. The Southern Cross is at its lowest, and even Canopus is not much in evidence. Achernar, leader of the celestial River, is almost overhead. This is the best time to study the Southern Birds, and both the Clouds of Magellan are well placed. Northern observers never cease to mourn the fact that these magnificent galaxies are so far south of the celestial equator.

October 2

The Southern Fish

Let us return to the Square of Pegasus, this time to use it as a guide. Well to its south lies the Southern Fish, known as Piscis Australis or Piscis Austrinus – occasionally as Piscis Solitarus, the solitary fish, to distinguish it from Pisces. It has one bright star, Fomalhaut.

It is an ancient constellation; and although no definite legends are attached to it, it has sometimes been associated with the Syrian fish-god Oannes, who was said to have come to earth as a teacher at a time when men were still uncivilized barbarians. (Looking at the world situation at the end of the twentieth century, it does not seem that Oannes had much success.) Fomalhaut can be found by using Beta and Alpha

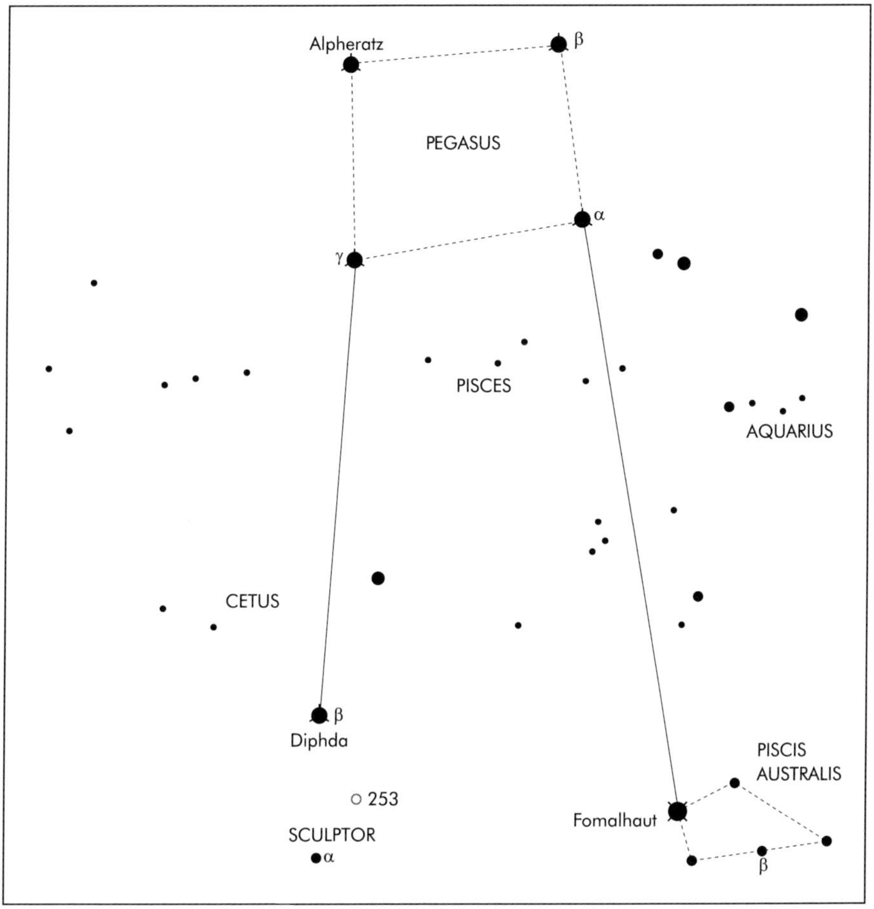

Pegasi as pointers; the line crosses Pisces and Aquarius and leads directly to the Southern Fish. The only possible confusion is with Diphda or Beta Ceti, which is roughly aligned with the other two stars of the Square, Alpheratz and Gamma Pegasi; but Diphda is always higher up than Fomalhaut as seen from Britain, and is not nearly so bright (5 October).

As we have noted, Fomalhaut is the southernmost of the first-magnitude stars to be visible from Britain; its declination is, in round figures, –30°, so that from anywhere north of latitude +60° it will never rise, and from anywhere south of –60° it will never set. It is a white A-type star, 13 times as luminous as the Sun and only 22 light-years away. It is of special interest because in 1983 it was studied from IRAS, the Infra-Red Astronomical Satellite, and was found to be surrounded by a cloud of cool, possibly planet-forming material in the same way as Vega (27 July).

It would certainly be premature to claim that Fomalhaut is the centre of a planetary system, and of this we have no proof at all; we can, however, say that it is possible – and we cannot entirely discount the idea that an intelligent astronomer in the Fomalhaut system may at this moment be looking up at our Sun and making similar speculations. But from Fomalhaut, the Sun would look considerably fainter than Fomalhaut does to us.

There is nothing else of special interest in Piscis Australis, though Beta (Fum el Samarkah), of magnitude 4.4, has a companion of magnitude 7.9 at a separation of 30.3 seconds of arc. This is an optical pair, not a binary.

October 3

Sculptor

This is a good time to look for another southern constellation: Sculptor, the Sculptor (originally Apparatus Sculptoris, the Sculptor's Apparatus). It was formed by Lacaille in 1752, though for no obvious reason, because it is extremely barren; its brightest star, Alpha, is only of magnitude 4.3. From Britain it is always very low indeed; it adjoins Piscis Australis, and lies below Beta Ceti (5 October). To southern-hemisphere observers it is high up, and can be found because it lies inside the large triangle formed by Fomalhaut, Beta Ceti, and Ankaa or Alpha Phœnicis (27 November). It contains the south galactic pole.

The finest object in Sculptor is the edge-on spiral galaxy C65 (NGC 253), near the border with Cetus and only 7.5° south of Beta Ceti. It was discovered by Caroline Herschel, William Herschel's sister, in 1783 during one of her routine searches for comets. When at a reasonable altitude, it is an easy binocular object, and telescopically it is magnificent. John Herschel called it 'a superb object ... very bright and large. Its light is somewhat streaky.' Its distance is between 7 and 8 million light-years, so that it is not very far beyond our Local Group. Also in Sculptor is a second edgewise-on galaxy, C72 (NGC 55), near the border with Phœnix and less than 4° from Ankaa. It too can be seen with binoculars under good conditions, but its declination is over 39° south, so that from Britain it is inaccessible.

October 4

Anniversary

1957: Launch of Russia's artificial satellite Sputnik 1, thereby opening the Space Age.

The Start of the Space Age

Today is one of the most important of all scientific anniversaries. On 4 October 1957 the Space Age began – not with a

> *Anniversary*
>
> 1959: Launch of Luna 3, the Russian space-craft which went round the Moon and sent back the first pictures of the averted hemisphere (26 October).

whimper, but with a very pronounced bang. Scientists in what was then the USSR launched Sputnik 1, the pioneer artificial satellite. It was small, and it carried nothing apart from a radio transmitter which sent out the never-to-be-forgotten 'Bleep', bleep! signals, but its significance was profound.

It was not entirely unexpected. It had been known that the Soviets were working toward such a launch, and so for that matter had the Americans; the White House had announced plans to launch an artificial satellite during the so-called International Geophysical Year, when scientists of all nations collaborated in studies of the Earth and its environment. However, the timing came as something of a shock in the West, and there were some ill-timed comments from the United States; one bemedalled general described Sputnik as 'a hunk of old iron that almost anyone could launch'. In fact, the American programme was in difficulties, and it was not until full rein was given to the team headed by Wernher von Braun, late of Peenemünde, that the first American satellite soared aloft – Explorer 1, on 31 January 1958.

Since then many thousands of satellites have been launched, but Sputnik 1 will never be forgotten. It remained in orbit for months, finally burning away in the atmosphere in January 1958.

October 5

Cetus

We have already mentioned Diphda or Beta Ceti, which is of second magnitude and lies in line with two of the stars in the Square of Pegasus, Alpheratz and Algenib. Cetus itself is a very large constellation; it is usually taken to represent a harmless whale, though it has also been identified with the sea-monster of the Perseus legend (11 November). It extends as far as the boundary of Taurus, and is so near the ecliptic that planets may pass through it. It is also cut by the celestial equator, though most of it lies in the southern hemisphere of the sky.

The leading stars are shown in the table overleaf.

The Whale's head, not far from A1 Rischa in Pisces (29 September) is made up of Alpha, Gamma, Mu (magnitude 4.3) and Xi (4.4); the 'body' runs from here down to Diphda. It includes the long-period variable Mira (6 October), which has been known to rise to magnitude 1.7, but which spends much of its time below naked-eye visibility. Here too we find Tau Ceti, one of our nearest neighbours and a promising candidate as a planetary centre, and also one of the very few naked-eye stars which is less luminous than the Sun.

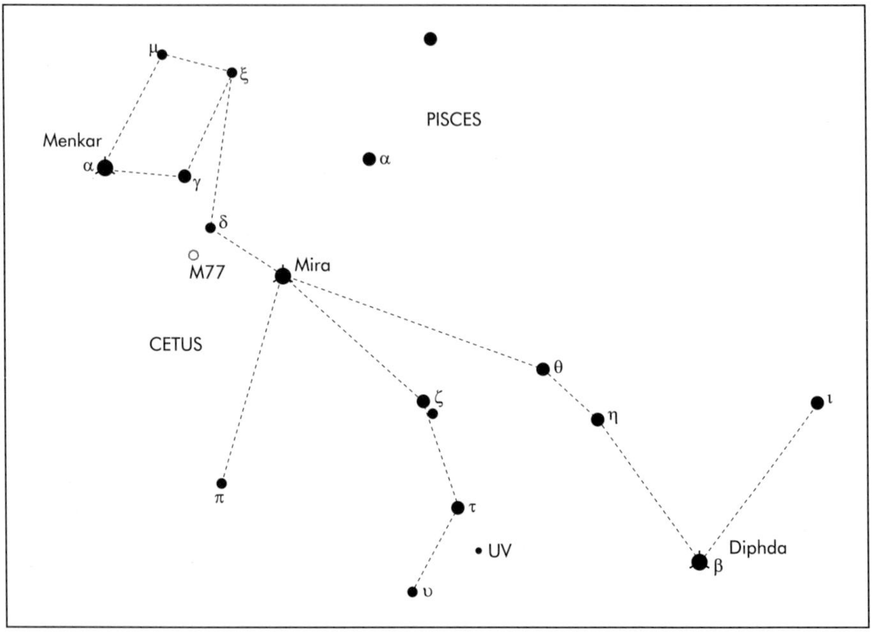

Cetus					
Greek letter	Name	Magnitude	Luminosity (Sun=1)	Distance (light-years)	Spectrum
α Alpha	Menkar	2.5	120	130	M2
β Beta	Diphda	2.0	60	68	K0
γ Gamma	Alkaffaljidhina	3.5	26	75	A2
δ Delta	–	4.1	1320	850	B2
ζ Zeta	Baten Kaitos	3.7	96	190	K2
η Eta	–	3.4	96	117	K2
θ Theta	–	3.6	60	114	K0
ι Iota	Baten Kaitos Shemali	3.6	96	160	K2
τ Tau	–	3.5	0.45	11.8	G8

October 6

The 'Wonderful Star'

Cetus contains one of the most celebrated stars in the sky: Mira, 'the Wonderful' (a nickname suggested by Hevelius). It has given its name to a whole class of variables, and it also has the distinction of being the first variable star to be recognized as such.

The magnitude range is from naked-eye visibility down to magnitude 10, so that it is visible without optical, aid only

Anniversary

1732: Birth of Nevil Maskelyne, who was Astronomer Royal from 1765 to his death in 1811. He was particularly concerned with navigational problems, and was the founder of the *Nautical Almanac*.

Anniversary

1990: Launch of the Ulysses space craft to study the poles of the Sun.

Future Points of Interest

2000: Mercury at eastern elongation.

for a few weeks in every year. Yet we can never be sure. At some maxima it has been known to reach magnitude 1.7; in February 1987 I estimated it as being of magnitude 2.3, but at other maxima it barely exceeds 4. It is, of course, very red, with an M-type spectrum; the distance is thought to be about 220 light-years. If this is correct, then the diameter of Mira is of the order of 350 000 000 miles, so that its huge globe could comfortably accommodate the whole orbit of Mars round the Sun. The position is RA 02 h 19 m, declination −02° 59′.

Because the mean period is less than eleven months, it sometimes happens that Mira reaches maximum twice in a calendar year – for instance, on 13 January and 11 December 1998. Mira has a binary companion, discovered in 1923 by R.G. Aitken with the 36-inch refractor at the Lick Observatory in America. The separation was then 0.9 seconds of arc, but then decreased; it was 0″.8 in 1940, 0″.7 in 1950, 0″.4 in 1980, and will be only 0″.1 in 2000, after which it will increase again. At the moment, therefore, Mira is a difficult test. The orbital period is 400 years. The companion is a dwarf flare star, and sometimes brightens up abruptly from its usual magnitude of 9.5; it even has a variable star designation – VZ Ceti.

A good piece of advice is to check the position of Mira when it is bright, so that you can find it again when it falls below naked-eye range. Near it is a neat little double, 66 Ceti.

October 7

Tau Ceti: a Near Neighbour

It is worth taking more than a passing look at Tau Ceti, in the southern part of the Whale. It looks ordinary enough; it is of magnitude 3.5, and so is easy to identify. Its interest stems from the fact that it is very near, by stellar standards – a mere 11.8 light-years away – and is a dwarf, with only 45% of the luminosity of the Sun. Its spectrum is of type G, and therefore it is not too unlike the Sun even though it is much feebler.

Tau Ceti has been regarded as a possible planetary centre, and was one of the two stars selected as targets for the 1960 Project Ozma, when the large radio telescope at Green Bank, West Virginia, was used to 'listen out' at selected wavelengths in the slim hope of picking up a signal rhythmical enough to be classed as artificial. (Ozma comes from the famous Frank Baum novel, *The Wizard of Oz*, though most people preferred to call it Project Little Green Men!) Not surprisingly, the results were negative. Further attempts have been made since, with a similar lack of success; but one never knows what may happen in the future.

October 8

Giants and Dwarfs of the Sky

Ejnar Hertzsprung was deeply interested in stellar classification and evolution, and in 1905 he came to an interesting conclusion. Red stars were divided into two definite types; giants, of great power and luminosity, and small, feeble dwarfs. Red stars similar in output to the Sun did not exist. The giant-and-dwarf division was still evident for orange stars, and to a lesser extent for yellow stars, but not for white or bluish. Later, the same conclusion was reached by H.N. Russell, in America, and the result was the construction of what is termed the Hertzsprung–Russell or HR Diagram, in which stars are plotted according to their spectral types and their luminosities. A typical HR Diagram is given here, and the division into giants and dwarfs is striking. Most stars lie on the Main Sequence, which runs from the top left of the diagram through to the bottom right. Originally it was

Anniversary

1873: Birth of Ejnar Hertzsprung, the Danish astronomer who established the existence of giant and dwarf stars. He worked at Copenhagen, Göttingen and Mount Wilson before becoming Director of the Leiden Observatory in 1935. He died in 1967.

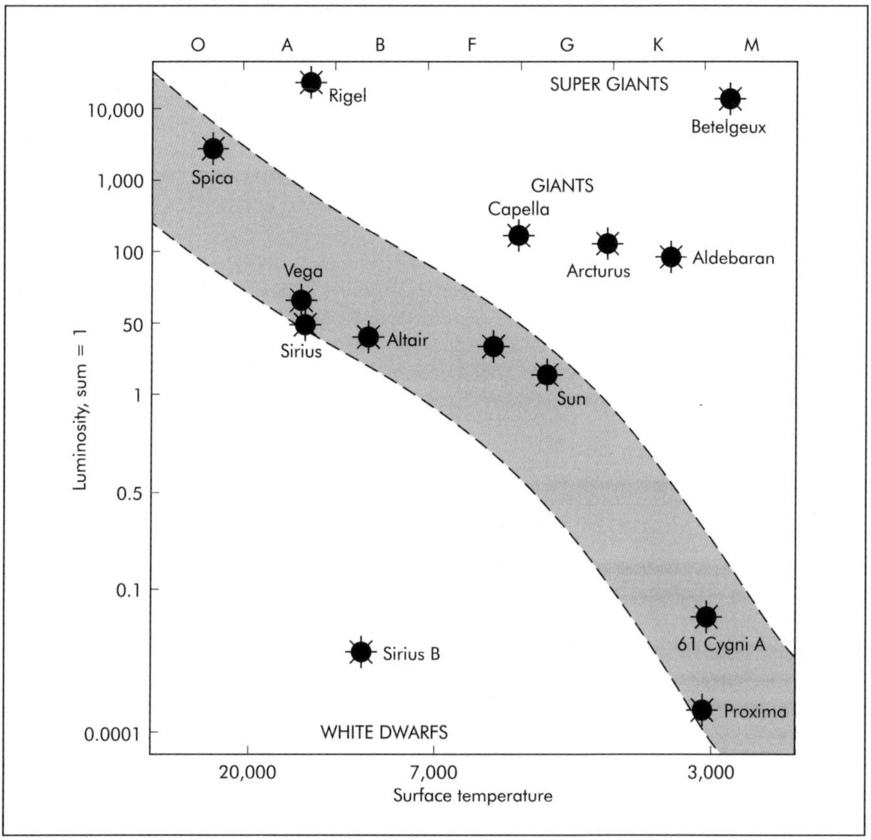

assumed that the diagram represented an evolutionary sequence, with a star beginning at the top right, joining the Main Sequence, and sliding down to extinction at the bottom right. This is now known to be wrong, but the diagram retains its importance.

Both giants and dwarfs are on view this evening. Look for example at a red giant (Alpha Ceti), a red dwarf (61 Cygni), an orange giant (Beta Ceti), an orange dwarf (Tau Ceti), a yellow giant (Capella) and a yellow dwarf (Gamma Delphini); also a white star such as Altair. White dwarfs, discovered much later, come into an entirely different category, and are not merely ordinary white stars of low luminosity (21 February).

Anniversary

1873: Birth of Karl Schwarzschild, a distinguished German astronomer who was a pioneer in stellar photometry and theoretical astrophysics. He was Director of the Göttingen Observatory (1901–9) and then of Potsdam Observatory. He died in 1916 as a result of military service during the First World War.

October 9

Flare Stars

Not far from Tau Ceti, at RA 01 h 39 m, declination $-17° 58'$, lies the remarkable UV Ceti. It is a very feeble red dwarf – or rather a pair of red dwarfs; their luminosities are respectively 0.000 06 and 0.000 04 that of the Sun, so that they are true cosmic glowworms. Even from a distance of a mere 8.4 light-years, they are still faint: well below magnitude 12.

The secondary member, UV Ceti B, is a flare star. It may suddenly brighten abruptly by several magnitudes, and then fade away back to normal brilliancy. For example, on 24 September 1952 it brightened from magnitude 12.3 to 6.8 in about 21 seconds. Only flare activity could account for this sort of behaviour. Our Sun shows flares, but they are very mild, and do not affect the Sun's output to a measurable extent. Things are very different with UV Ceti B. Many other flare stars are known, including Proxima, the nearest of all stars beyond the Sun; all are red dwarfs, and we can never predict when they are going to perform.

October 10

The Draconid Meteors

Tonight is the maximum of the Draconid meteor shower – sometimes called the Giacobinid shower, because the parent comet is Giacobini–Zinner. The comet takes 6.5 years to go round the Sun; it was originally discovered in 1900 by Giacobini and recovered in 1913 by Zinner – hence the double name. It is never bright, though occasionally it develops a tail.

The Draconids are unpredictable. Very often the display is so poor that it cannot be noticed at all, but now and then there are major 'storms', as in 1933, 1946 and 1972; during the 1946 shower the maximum zenithal hourly rate was over 30 000 for a brief period. For once northern observers have the advantage, since the shower radiant lies close to Gamma Draconis in the Dragon's head (15 October).

Watch for the Draconids tonight – and also on the 9th and 11th. Very probably you will see few shower meteors, or even none at all, but one never knows.

Future Points of Interest

1997: Saturn at opposition. It is now 4° north of the celestial equator; the rings are opening, and the magnitude is +0.4.

Maximum of the Draconid meteor shower.

Anniversary

1858: Donati's Comet at its best. This is said to have been the most beautiful comet ever observed, with a long, curved dust tail and a straight ion tail. It will be some time before we see it again; the orbital period is thought to be about 2000 years.

1989: First close-range image of an asteroid (Gaspra), from the Galileo space-craft en route to Jupiter. Gaspra proved to be an irregularly shaped body, pitted with craters. Its longest diameter is 12 miles.

October 11

Objects in Cetus

Returning to Cetus, let us look at the Whale's head. The brightest star in it is Alpha or Menkar, an M-type red giant. (In the same low-power field with it is a bluish fifth-magnitude star, 93 Ceti; there is no genuine association, but the pair is worth looking at because of the colour contrast.) Gamma Ceti is a fine double; magnitudes 3.5 and 7.3, and the separation is 2.2 seconds of arc. The components share a common motion in space, but the orbital period must be immensely long. It has been claimed that the primary is yellow and the secondary blue, but I admit that I have never been able to see any colour in either of them.

Near Delta, at RA 02 h 43 m, declination -00° 01′, is the spiral galaxy M77, with a condensed nucleus and inconspicuous arms; spirals of this type are known as Seyfert systems, and are often very active. M77 is over 50 000 000 light-years away; studies of the movements of stars inside it have indicated that it may contain a massive central black hole. Telescopically M77 is not very conspicuous, with an integrated magnitude of only 8.8, but it is an exceptionally massive galaxy.

Before taking our leave of Cetus, look at Diphda, south of the Square of Pegasus. The magnitude is 2.0, but there have been suggestions that it is somewhat variable. Unfortunately it is a difficult star to check with the naked eye, because there are no comparison stars anywhere near it.

Anniversary

1968: Launch of Apollo 7, carrying Astronauts Schirra, Eisele and Cunningham. It was an Earth orbiter, part of the manned landing programme. It remained aloft for 10 days 20 hours.

1983: Discovery of the asteroid 3200 Phæthon. It is about 3 miles in diameter, and is darkish, with a rotation period of 4 hours. Its orbital distance from the Sun ranges between 13 million and 242 million miles, so that at perihelion it is well inside the orbit of Mercury, and at aphelion well beyond the orbit of Mars. It moves in much the same orbit as the Geminid meteor stream (11 December) and it has been suggested that it is the parent of the stream, in which case it would have to be classed as an extinct comet. Its orbital period is 1.43 years.

October 12

The Chemical Furnace

While in this area, those who feel inclined can seek out another obscure constellation: Fornax, the Furnace, originally Fornax Chemica, the Chemical Furnace. This is yet another of the constellations added to the sky by Lacaille in 1752.

It adjoins Cetus, and also Eridanus (30 January, 2 November), and is well south of the equator. It's brightest star, Alpha, is of magnitude 3.9, and lies at declination −29°. This is much the same as with Fomalhaut, but from Britain Fomalhaut is easy to locate because of its brightness; Alpha Fornacis is much more difficult, particularly since there are no obvious guides to it. It is 46 light-years away, and is an F8-type star four times as luminous as the Sun.

Fornax is crowded with faint galaxies, and is therefore of great interest to deep-sky observers, but there is not much available to the owner of a small telescope. There is, however, a notable galaxy, the 'Fornax Dwarf', which is 420 000 light-years from us – much closer than the Andromeda Spiral – but is a mere 7000 light-years in diameter. Were it further away it would not have been identified as a separate galaxy, and it is very poor compared with the senior members of the Local Group; it contains very little nebulosity, if any. These dwarf galaxies are very common, but obviously we can only see those which are relatively close to us.

October 13

Crater Linné

It is some time since we discussed the Moon, so let us briefly turn out attention back to our faithful satellite, and say something about the strange case of the vanishing crater.

First, where is the Moon tonight? On 13 October during our period:

1997	Three days before full, so a good gibbous moon after dark.
1998	One day after Last Quarter: a thick crescent in the morning sky.
1999	Four days after new; an evening crescent.
2000	Full.
2001	Three days after Last Quarter; a thin crescent before dawn.

2002 First Quarter; half-moon, and ideally placed.
2003 Three days after full; gibbous in the morning sky.

It is usually said that the Moon is a changeless world, and in general this is true enough; no major craters can have been formed for at least at thousand million years. But in the nineteenth century there was a report which caused a great deal of interest. It was claimed that a small crater, Linné on the Mare Serenitatis, had disappeared.

The Mare Serenitatis (Sea of Serenity) is one of the Moon's most prominent maria, and is regular in outline, with a relatively smooth floor. On it there is one conspicuous crater, Bessel, 12 miles across. (It was named after F.W. Bessel, the first man to measure the distance of a star.) In 1838 two German observers, Wilhelm Beer and Johann von Mädler, drew a splendid map of the Moon. They showed Bessel, and also another crater, Linné, which they described as deep and distinct.

In 1866 another German, Julius Schmidt, observed the area, and reported that instead of being a crater, Linné was now a white patch. All sorts of theories were proposed to account for this – moonquakes, rock falls and the like – but it now seems definite that nothing happened; Mädler himself said that Linné looked the same in 1868 as it had done in 1838. It is today a craterlet surrounded by a white nimbus, as the space-probe pictures have shown; but its appearance can alter considerably according to the angle at which sunlight strikes it. So we can discount any real change in Linné – but at least the episode diverted observers' attention back to the Moon.

October 14

William Lassell and his Telescope

Anniversary

1846: Discovery of Triton, Neptune's major satellite, by William Lassell.

William Lassell was one of the great amateur astronomers of Victorian times. He was born in 1799, and took control of the family business when he came of age, but his main interest was astronomical; he built a fine 24-inch reflector, and set it up at his home in Liverpool. He concentrated upon planetary observations; as soon as the discovery of Neptune was announced (4 August) he turned his telescope toward the area, and identified it at once. A few weeks later – on 14 October – he discovered Triton. He also discovered two of the satellites of Uranus (Ariel and Umbriel) and, independently, Hyperion, the seventh satellite of Saturn.

Lassell died in 1880. His telescope had been dismantled, but the optics and plans survived, and in 1996 the telescope was reconstructed, with the original optics. (I was honoured at being invited to perform the opening ceremony). It is now in full working order at the Liverpool Museum, and is used for practical work as well as a museum exhibit. No doubt William Lassell would approve.

October 15

Draco, the Dragon

Draco is one of the largest of all the constellations, occupying over 1000 square degrees of the sky, but it has no outstandingly bright stars; the leader, Gamma, is below the second magnitude. Thuban, the star lettered Alpha, is almost 1.5 magnitudes fainter than this (it used to be the north pole star in ancient times).

Draco sprawls from a point between the Pointers and the Pole Star through to the neighbourhood of Vega. It is circumpolar from British latitudes; the declination of Gamma is +51°. It is not hard to make out the twisted line of stars which forms the reptile's body; the most prominent part of the constellation is the 'head', made up of Gamma, Beta, Xi (magnitude 3.7) and Nu, which is a very wide double; both components are white, and of magmitude 4.9. As they are separated by over 30 seconds of arc, they can be seen individually with good binoculars, and they can even be distinguished by very keen-eyed people with no optical aid at all.

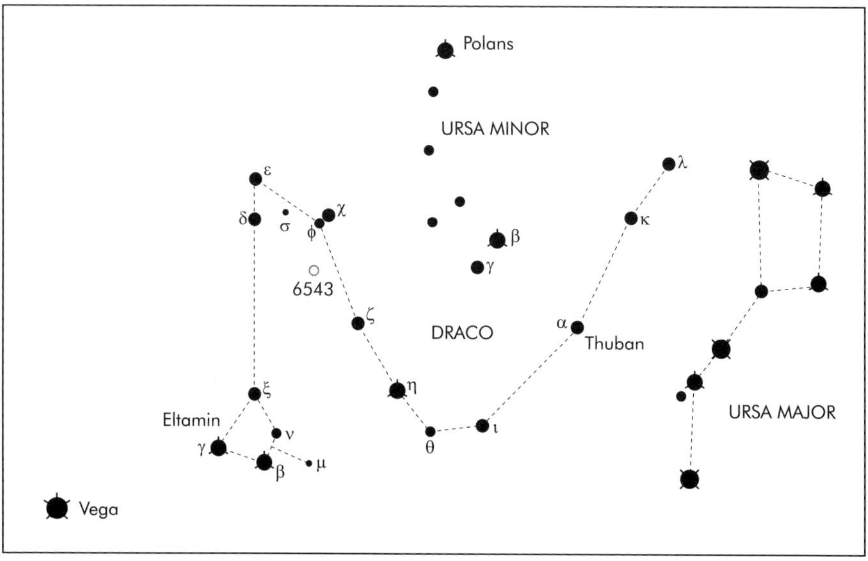

Draco					
Greek letter	Name	Magnitude	Luminosity (Sun=1)	Distance (light-years)	Spectrum
α Alpha	Thuban	3.6	130	230	A0
β Beta	Alwaid	2.8	600	270	G2
γ Gamma	Eltamin	2.2	110	101	K5
δ Delta	Taïs	3.1	60	117	G9
ϵ Epsilon	Tyl	3.8	58	166	G8
ζ Zeta	Aldibah	3.2	500	316	B6
η Eta	Aldhibain	2.7	110	81	G8
ι Iota	Edasich	3.3	96	156	K2
χ Chi	–	3.6	2	25	F7

Mu Draconis is a closer double; the components are equal at magnitude 5.7, but the separation is only 1.9 seconds of arc. This is a slow binary, with an orbital period not far short of 500 years. Both Eta and Epsilon have faint companions which are easy telescopic objects.

The planetary nebula C6 (NGC 6543) lies between Delta and Zeta. Its integrated magnitude is 9, so that it is far from difficult to find with a small telescope though it is probably beyond the range of ordinary binoculars. It was, incidentally, the first nebulous object to have its spectrum observed – by Sir William Huggins, in 1864. Its true diameter seems to be about one-third of a light-year.

October 16

Nearby Stars in Draco

Draco contains two naked-eye stars which are among our closer stellar neighbours. These are Sigma (Alrakis) and Chi, which has never been honoured with a proper name. So let us seek them out.

Chi is easy; it forms a wide pair with Phi, which is a normal A-type star of magnitude 4.2; it is 100 light-years away. Chi's distance, however, is a mere 25 light-years, and it is no more than twice as luminous as the Sun. Its spectrum is of type F, so that it should be slightly yellowish, though the colour is by no means pronounced.

Alrakis is much closer, at 18.5 light-years, and is only about one-third as luminous as the Sun, so that on the cosmical scale it is decidedly feeble; it is orange, and of type K. It could be a promising candidate as the centre of a planetary system, though whether there really are any planets moving round it we do not yet know. It is easy to locate, because it lies almost directly between Epsilon and the Chi–Phi pair.

Future Points of Interest

1998: Occultation of Mars by the Moon (3 h GMT)

October 17

The Old Pole Star

When the Egyptian Pyramids were being built, the north pole star was not Polaris, but Thuban – Alpha Draconis. Before discussing its loss of polar status, let us look at Thuban itself.

It lies between Alkaid, in the Plough, and Kocab in Ursa Minor. Its magnitude is 3.65, and as it is rather 'on its own' it is very easy to identify. It is a spectroscopic binary with an orbital period of 51.4 days; the real distance between the components is of the order of 200 000 000 miles, and the brighter star has an unusual spectrum, with very strong lines due to the element silicon.

Though lettered Alpha, Thuban is only the eighth brightest star in Draco, and – as we have seen – it is almost $1\frac{1}{2}$ magnitudes fainter than Gamma, in the dragon's head. It has been suspected of variability; Bayer, in 1603, ranked it as brighter than Gamma, which could be why he allotted it the first letter of the Greek alphabet. It is not likely that there has been any real change, but it may be worthwhile making occasional comparisons with Lambda Draconis (Giansar), which is an orange-red star of type M. Lambda's magnitude is 3.8, so that it is only marginally fainter than Thuban.

Thuban is 230 light-years away, and is at present approaching us at a rate of 10 miles per second, though of course this will not continue indefinitely – and there is certainly no fear of an eventual collision.

October 18

The Moving Pole

Around 4800 years ago, during the Old Kingdom period of Ancient Egypt, the north pole star was Thuban in Draco. Thuban was at its closest to the pole – less than 10 minutes of arc – in 2830 BC, and there have been many discussions about the celestial pole and the alignment of the Pyramids.

The poles of the sky do not remain in exactly the same positions; each pole describes a small circle in the sky, with a period of around 26 000 years. By now the pole has moved to a position near Polaris; the distance between the polar point and Polaris is still decreasing, and on 24 March 2100 will be less than half a degree. By 4000 AD the pole will be fairly near Alrai or Gamma Cephei (27 January) and by 10 000 AD it will be in the region of the brilliant Deneb, in Cygnus (17 August). It will reach the neighbourhood of Vega in 14 000 AD – and then we will have a really brilliant pole star. After that it will track through Draco, again passing by Thuban before returning to Polaris in 26 000 AD. Meanwhile the south celestial pole will perform a similar manœuvre, and when the northern hemisphere has Vega as its pole star the south will have the even more splendid Canopus.

The reason for this shift – termed precession – is that the Earth is not a perfect sphere. The diameter as measured through the equator is 7927 miles, but only 7900 miles as measured through the poles. The Sun and Moon pull on the equatorial bulge, and make the Earth 'wobble' slightly, in the manner of a boy's gyroscope which is running down and is starting to topple. The gyroscope precesses in a few seconds and the Earth takes thousands of years, but the underlying force is the same. Of course the shift of the pole also affects the equator, which is why the First Point of Aries has now shifted into Pisces (22 January).

October 19

Gamma Draconis and Aberration

Eltamin or Gamma Draconis, in the Dragon's head, looks like – and is – a normal K-type star rather below second magnitude; but it has a place in astronomical history, because observations of it led to a very important discovery.

One of the leading English astronomers of the eighteenth century was James Bradley. He was anxious to measure the distances of the stars, using the method of parallax, and as a start he concentrated upon Gamma Draconis, because it passes directly overhead from the latitude of Kew in outer London – where Bradley set up his equipment – and was thus particularly easy to measure. Bradley knew that the parallax shifts would be very small, but his results puzzled him. There were shifts indeed, but they were not due to parallax; Gamma Draconis appeared to be moving in a tiny circle, returning to its original position after a period of one year. Bradley checked with other stars, and found that they all behaved in the same way.

According to a famous story which may well be true, Bradley found the answer one day when he was out sailing on the Thames. He noticed that when the direction of the boat was altered, the vane on the mast-head shifted slightly, even though the wind remained the same as before. At once he realized that this was the principle which accounted for the behaviour of Gamma Draconis. Light does not travel instantaneously; as the Danish astronomer Ole Rømer had found, it flashes along at a speed of 186 000 miles per second. The Earth also is in motion, moving round the Sun at an average

Anniversary

1910: Birth of Subrahmanyan Chandrasekhar, outstanding Indian astrophysicist who pioneered studies of white dwarf stars and compact objects generally. He worked in England and then in America; in 1983 he was awarded the Nobel Prize for Physics. He died in August 1995.

speed of 18.5 miles per second, and so the Earth's 'direction' is changing all the time. Let the Earth be represented by the boat, and the incoming light be represented by the wind, and the position becomes clear; there is always an apparent displacement of a star toward the direction in which the Earth is moving at that particular moment. This is termed aberration.

You can demonstrate aberration next time you are walking in the rain, holding an umbrella. If you want to keep dry, you will have to slant the umbrella ahead of you as you move along.

Anniversary

1811: Closest approach of the Great Comet of 1811, at a distance of just over 100 000 000 miles. It had been discovered by Honoré Flaugergues on 25 March, and was in view for a (then) record of 512 days; it was one of the brightest comets on record, rivalling that of 1843. In 1811 the wine crop in Portugal was surprisingly good, and this was attributed to the influence of the comet, and 'Comet Wine' was produced. Bottles of it turned up at wine auctions for many years afterwards – very occasionally, they still do.

October 20

Andromeda

To northern observers, the Square of Pegasus remains high up during the evening all through October, and this is a good time to follow the variations of Beta Pegasi, Scheat (17 September). If you observe it on every clear night, you will be able to draw up a light-curve. Remember to allow for extinction when comparing it with Alpha or Gamma Pegasi; Alpheratz is rather too bright to be used as a convenient comparison star.

Alpheratz is, of course, the star 'stolen' from Pegasus; Delta Pegasi has become Alpha Andromedæ. Andromeda consists mainly of a chain of brightish stars leading away from the Square. The principal members are listed below.

Alpheratz is a spectroscopic binary. The orbital period is only 97 days, so that the true distance between the components is of the order of 40 000 000 miles. Beta, or Mirach, is clearly orange; Gamma, or Almaak, is one of the finest of all doubles. The magnitudes are 2.3 and 4.8; the primary is orange, and it has been said that the secondary looks green by contrast, though most observers will call it white. The secondary is itself a close binary, with an orbital period of 561 years, but as the separation is never more than 0.5 of a second of arc this is a reasonably difficult test object. The brighter member of the pair is itself a spectroscopic binary, so that altogether the Gamma Andromedæ system is quadruple.

Andromeda

Greek letter	Name	Magnitude	Luminosity (Sun=1)	Distance (light-years)	Spectrum
α Alpha	Alpheratz	2.1	96	72	A0
β Beta	Mirach	2.1	115	88	M0
γ Gamma	Almaak	2.1	95	121	K2+A0
δ Delta	–	3.3	105	160	K3

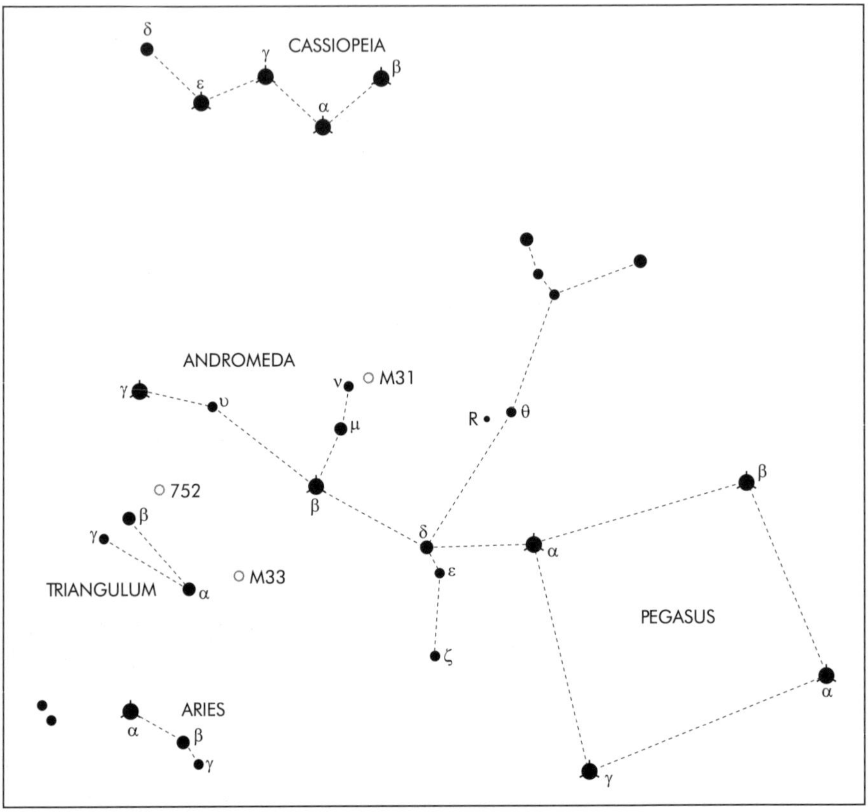

Some way to the north of Alpheratz is Theta Andromedæ, of magnitude 4.6. Close beside it, at RA 00 h 24 m, declination +38° 35′, is the Mira variable R Andromedæ, which has a large magnitude range – from 5.8 to 14.9 – so that at minimum it is a very dim object. At maximum it can attain naked-eye visibility, and is easy with binoculars; the period is 409 days.

October 21

Future Points of Interest

Maximum of the Orionid meteor shower.

The Orionids – and Halley's Comet

We have come to the maximum activity of the Orionid meteor shower. It begins around 16 October and ends on the 27th, with a peak on the 21st, but it is relatively rich, and there should be many meteors for several nights in succession. The position of the radiant is in the north of Orion, at RA 6 h 24 m, declination +15°. The usual Zenithal Hourly Rate is around 25, and is normally exceeded only by the January

Quadrantids (4 January), the May Eta Aquarids (5 May), the August Perseids (11 August) and the December Geminids (13 December).

The Orionid meteors are usually swift, with fine trains. We know the parent comet; it is none other than Halley's.

Halley's Comet – named after Edmond Halley, the second Astronomer Royal, who was the first to realise that it comes back regularly – has a period of 76 years. At its greatest distance from the Sun it recedes well beyond the orbit of Neptune, but it has spread débris all along its path, and the Eta Aquarids as well as the Orionids come from it. Halley's Comet last returned in 1835, 1910 and 1986; in 1986 it was badly placed, and never became bright, but at previous returns it has been brilliant, and at the return of 837 it is said to have cast shadows. It will return in 2061, but will again be badly placed; observe it in 2137, and it will be magnificent. At least we know what it is like; in 1986 the European space-craft Giotto went right through its head, and sent back pictures of the icy, dark-coated nucleus.

October 22

The Great Spiral

The most famous object in Andromeda is M31, the Great Spiral, the closest of the really large galaxies. It also has the distinction of being the most remote object which is clearly visible with the naked eye.

It has been known since very early times; in 964 the Arab astronomer Al-Sûfi described it as 'a little cloud'. Curiously it is not mentioned by Tycho Brahe, greatest of the pre-telescopic observers; the first telescopic record of it is due to Simon Marius, who looked at it on 15 December 1612 and described it as being 'like the flame of a candle seen through horn', or 'a cloud consisting of three rays; whitish, irregular and faint; brighter toward the centre'. Lord Rosse, in 1845, considered – correctly – that it could be resolved into stars, while Sir William Huggins (7 February) found the spectrum to be quite unlike those of gaseous nebulæ such as M42 in Orion's Sword.

To locate M31, start at Beta and Gamma Andromedæ (Mirach and Almaak) and find two fainter stars, Mu and Nu, of magnitudes 3.9 and 4.5 respectively. M31 lies close to Nu. To see it with the naked eye requires a clear, dark night, and even then it is very dim; but binoculars bring it out well, and it cannot be mistaken, though it will show no form and will appear simply as a ghostly patch of light. A small telescope will show its companion galaxies, M32 and NGC 205.

October 23

Anniversary

1822: Birth of F.G.W. Spörer, a pioneer solar observer.

Spörer and his Law

So far we have said little about the Sun, and today is a good moment to return to it, because it is the anniversary of the birth of Friedrich Wilhelm Gustav Spörer, one of the great pioneers of solar studies. He was born in Berlin on 23 October 1822, and educated at Berlin University; he became a teacher, and erected a private observatory, devoting himself mainly to studies of the Sun. In 1874 he became assistant astronomer at Potsdam Observatory, and continued his researches. He died on 7 July 1895.

As we have seen, sunspots are darker patches on the Sun's bright surface or photosphere; they are some 2000°C cooler than adjacent areas – which is why they appear blackish (they are not really so). Obviously they are not permanent. Every 11 years or so the Sun reaches maximum activity, with many spots and flares, while at times of minimum (as in 1996) there may be many consecutive spotless days.

Spörer found that the first spots of a new cycle of activity appear at high solar latitudes; as the cycle progresses new spots form closer and closer to the equator, though they never actually break out at the equator itself. When the spots of the old cycle cease to appear, the first spots of the new cycle break out at higher latitudes. This is known as Spörer's Law; it was discovered independently by Richard Carrington (26 May). We have to admit that even today our knowledge of the detailed workings of the Sun is far from complete, but there is no doubt that Spörer's Law is of fundamental importance.

Remember **never** look directly at the Sun through any optical appliance. I do not apologize for repeating this warning; it is vitally important.

Future Points of Interest

1998: Saturn at opposition; magnitude 0.2, declination +9°.

1999: Jupiter at opposition; magnitude –2.5, declination +10°.

October 24

The Distance of M31

Let us go back to the Andromeda Spiral. Small telescopes will show no structure in it, but larger instruments show that it is resolvable into stars. So what is its nature? Early observers were not sure; they could not tell whether it was part of our Galaxy, or something far more important. The great observer William Herschel believed it to be 'the nearest of all the nebulæ', and no more than two hundred times as far away as Sirius. Lord Rosse disagreed, and wondered whether it might be an external system.

Future Points of Interest

1999: Mercury at eastern elongation.

The question was tackled in the early 1920s by Edwin Hubble, who had the advantage of being able to use the 100-inch telescope at Mount Wilson – then much the most powerful in the world. His method was to use the Cepheid variable stars, which, as we have seen (28 January) 'give away' their distances by the way in which they behave. Hubble found Cepheids in M31, and realized at once that they were far too remote to belong to our Galaxy. He gave the distance as 900 000 light-years, later revised to 750 000 light-years; it was only in 1952 that Walter Baade found an error in the Cepheid scale, and increased the distance of M31 to 2 200 000 light-years.

One light-year is equal to 5.8 million million miles, so that M31 is indeed a long way from us. Even so, it is the very nearest of the large galaxies, and only the Magellanic Clouds and a few dwarf systems are closer. Moreover M31 is considerably larger than our Galaxy, and contains more than our quota of 100 000 million stars.

October 25

The Form of M31

Some galaxies are beautiful objects, shaped like Catherine-wheels; such is M51, the Whirlpool Galaxy in Canes Venatici (23 March), not far from Alkaid in the Great Bear. Unfortunately the Andromeda Spiral is placed at a narrow angle to us, so that its full beauty is lost.

Edwin Hubble, who was the first to establish that the 'starry nebulæ' really are external galaxies, drew up a system of classification which is still in use, though more elaborate systems have been produced since. For the spirals, there were three classes: Sa (conspicuous, often tightly wound arms issuing from a well-defined nucleus), Sb (looser arms, nucleus less condensed), and Sc (arms very loose, nucleus inconspicuous). We now know that our Galaxy is of type Sb. This is also true of M31, but of course M31 is much the larger, with a mass 1.5 times that of our Galaxy; it is surrounded by more than 300 globular clusters. Predictably, it is a radio source (though not a strong one), and is rotating round its nucleus; as with all spirals, the arms 'trail'.

Other galaxies are the so-called barred spirals, with the arms extending from the ends of a bar-like structure through the nucleus; again we have three classes, SBa, SBb and SBc. There are also elliptical galaxies, with no indications of spirality; they range from E7 (highly flattened) down to E0 (virtually spherical, looking superficially like globular clusters even though there is a vast difference in mass). Finally, there are some galaxies which are quite irregular in form.

October 26

The Other Side of the Moon

Anniversary

1959: First pictures received of the other side of the Moon.

This is another lunar anniversary day. For the first time we received definite information about those parts of the surface which we can never see.

To recapitulate: the Moon goes round the Earth, or more precisely round the barycentre, in 27.3 days. It takes exactly the same time to spin on its axis, so that it seems to keep the same face turned toward us all the time, allowing for slight wobbles termed librations; all in all there is 41% of the Moon which is permanently unobservable from Earth. Moreover, the libration areas are very foreshortened and hard to map. (I spent many years in studying these regions; it was gratifying to know that my results were used by the Russians to link the averted regions with the visible hemisphere.)

The pictures came back from the unmanned Luna 3 (or Lunik 3), which was launched on 4 October 1959, exactly two years after the epic flight of Sputnik 1 (4 October). Luna 3 went on a round trip. When over the far side it took pictures, and on 26 October sent them back to Earth by television techniques. I must have been among the first to see them; the Russians sent them to me, and they arrived when I was actually making a live television broadcast on the BBC.

The far side proved to be just as mountainous and crater-scarred as the regions we have always known, but there are differences in detail; in particular there are fewer of the broad maria, and only one large 'sea', the Mare Orientale, extends from the visible on to the far side.

The final fate of Luna 3 is unknown; contact with it was soon lost. Probably it burned away in the upper atmosphare, though there is a chance that it is still orbiting, silent and dead. In any case, it had accomplished its task well. Its pictures look very crude by modern standards, and by now we have detailed maps of the entire surface, but it was Luna 3 which showed the way.

October 27

The Baroness and the Supernova

Many novæ have been seen in the Andromeda spiral. There have also been two supernovæ, one in 1885 and the other in 1987. Supernovæ, as we have seen (24 February), are the most colossal explosions known in nature, and may shine more brightly than all the stars of a normal galaxy combined. The story of the 1885 supernova, now called S Andromedæ, is rather interesting, and involves a rather improbable Hungarian baroness.

In 1885 the Baroness de Podmaniczky was having a house-party at her castle. A small telescope was available, and with it the Baroness and her guests looked at M31; they saw a 'star' there, but naturally did not realize what it was, even though an astronomer, Dr de Koveslígethy, was among the guests. It was of course the supernova, and reached magnitude 6, so that it was on the fringe of naked-eye visibility; it soon faded, and by early 1890 had faded markedly. Eventually it was lost, though its remnant has now been identified. We have to admit that the Baroness was not the first to see it; it had been detected on 20 August by Hartwig at the Dorpat Observatory in Estonia, and it was he who first realised that it was something unusual.

In 1885 supernovæ had not been recognized, and nobody seemed to be sure whether or not S Andromedæ really belonged to M31; it is a pity that it appeared before astronomers had learned enough to take advantage of it. However, things were different when another supernova flared up in 1987 (24 February). We cannot tell when M31 will provide us with a third supernova; it may be this week, or it may not be for a very long time indeed.

October 28

Companions of the Andromeda Spiral

Look at the Andromeda Galaxy with good binoculars, and you will see a much fainter patch of light beside it. This is another galaxy, No. 32 in Messier's list, which is a true companion of the Great Spiral.

It shows little structure, and is not easy to resolve; it is an elliptical system, of type E2 (25 October). It was discovered by the French astronomer G. Legentil in 1749; in 1764 Messier called it 'a small round nebula without a star'. Its real diameter is of the order of 8000 light-years.

Telescopes show another companion galaxy, NGC 205. It was not included in Messier's catalogue, presumably because there was no danger of confusing it with a comet; it is No.225 in the NGC list. Some later observers have 'added' it to the Messier catalogue, as M110, but this does not seem to have received general acceptance. It is an E6-type elliptical, distinctly fainter than M32, but is quite impressive as seen through a telescope of 4 inches aperture or more. It seems that Messier first saw it in 1773; ten years later it was discovered independently by Caroline Herschel.

October 29

Aries, the Ram

Some way south of the line of stars marking Andromeda you will find Aries, the Ram, still regarded as the first constellation of the Zodiac even though the vernal equinox has now shifted into Pisces (22 January).

The leading stars of Aries are listed below.

Alpha, Beta and Gamma make a conspicuous little trio; the reddish hue of Alpha is very evident. Gamma is a fine, wide double; the components are equal, or virtually so, and the separation is 7.8 seconds of arc, so that the pair is very easy to split. There can be little doubt that the two are physically associated, but the distance between them is very great. There has been little relative shift since the first observation of the star was made, by Robert Hooke in 1664; Hooke discovered its duplicity accidentally, while searching for a comet.

There is a third star, of magnitude 9, at a distance of 221 seconds of arc. It is a very close binary, but does not appear to be genuinely linked with the Mesartim pair, because it has a different motion through space – even though catalogues still refer to it as Gamma Arietis C.

Aries

Greek letter	Name	Magnitude	Luminosity (Sun=1)	Distance (light-years)	Spectrum
α Alpha	Hamal	2.0	96	85	K2
β Beta	Sheratan	2.6	11	46	A5
γ Gamma	Mesartim	3.9	60+56	117	A0+B9
c	Nair al Botein	3.6	105	117	B8

Anniversary

1656: Birth of Edmond Halley, one of the greatest of all British astronomers. Though he is best remembered in the popular mind for his prediction of the return of the comet that bears his name, this was only one of his many achievements, and by no means the most important one. He catalogued the southern stars from St Helena; he studied star-clusters and nebulæ, and discovered the proper motions of some bright stars. He was Astronomer Royal at Greenwich from 1720 to his death in 1742, and carried out an extensive and very important series of observations of the movements of the Moon. His scientific experiments extended into many other fields; he even experimented with a primitive submarine.

Future Points of Interest

2001: Mercury at western elongation.

October 30

Triangulum

Between Andromeda and Aries lies the little constellation of the Triangle – one of the few groups whose appearance justifies its name. Look for it directly between Almaak in

Anniversary

1937: Close pass of the asteroid Hermes, at 485 000 miles – a close-approach record at that time, though often

(Continued opposite)

Triangulum					
Greek letter	Name	Magnitude	Luminosity (Sun=1)	Distance (light-years)	Spectrum
α Alpha	Rasalmothallah	3.4	10	59	F6
β Beta	–	3.0	58	114	A5
γ Gamma	–	4.0	45	150	A0

broken since. Hermes was small, only a mile or two across. It has not been seen again, and is unlikely to be recovered; it has the unenviable distinction of being the only asteroid to have been given a name but no number.

Andromeda and Hamal in Aries. The three triangle stars are listed above.

It is an original constellation, despite its small size, and there is a rather tenuous legend about it. This involves Proserpina, daughter of the earth-goddess Ceres, who was waylaid by Pluto, King of the Underworld, who was anxious to marry her. Finally an agreement was reached; Proserpina was to spend six months of the year with Pluto, and the other six 'above ground'. As a wedding present she was given the island of Sicily, which is represented in the sky by Triangulum.

Closely south of Triangulum are three stars of around the fifth magnitude (6, 10 and 12 Trianguli) which were formed by Hevelius, in 1690, into a separate constellation, Triangulum Minor, which, however, has not survived the revisions made in more modern times.

Future Points of Interest

1999: Venus at western elongation.

Future Points of Interest

1998: Occultation of Jupiter by the Moon (16 h GMT).

October 31

The Pinwheel Galaxy

Triangulum contains one important object: M33, a spiral galaxy known as the Pinwheel. It is a member of our Local Group, and is around 2 300 000 light- years away, slightly further than the great spiral in Andromeda.

M33 is not hard to locate with binoculars; it has been said to be visible with the naked eye, and no doubt keen-sighted people can glimpse it, though I certainly cannot. With normal × 7 binoculars it is in the same field with Alpha Trianguli. Look past the fainter star 1 Trianguli, and at about the same distance beyond it, in the direction of Mirach or Beta Andromedæ, you will see the dim haze of M33. Oddly enough it can be much harder to locate with a telescope, because of its low surface brightness. It is of type Sc, which means that as a spiral it is much looser than our Galaxy or M31, but it is much more 'face-on' than M31, and is therefore a beautiful object when photographed or when seen through a really powerful telescope.

It is relatively small, with a diameter of about 60 000 light-years; its mass is no more than 1/15 that of our Galaxy, but it contains objects of all kinds, and several novæ have been seen in it. Predictably, it has a surround of globular clusters. When searching for it, do not use too high a power – and it is a good idea to locate it first by using binoculars.

November

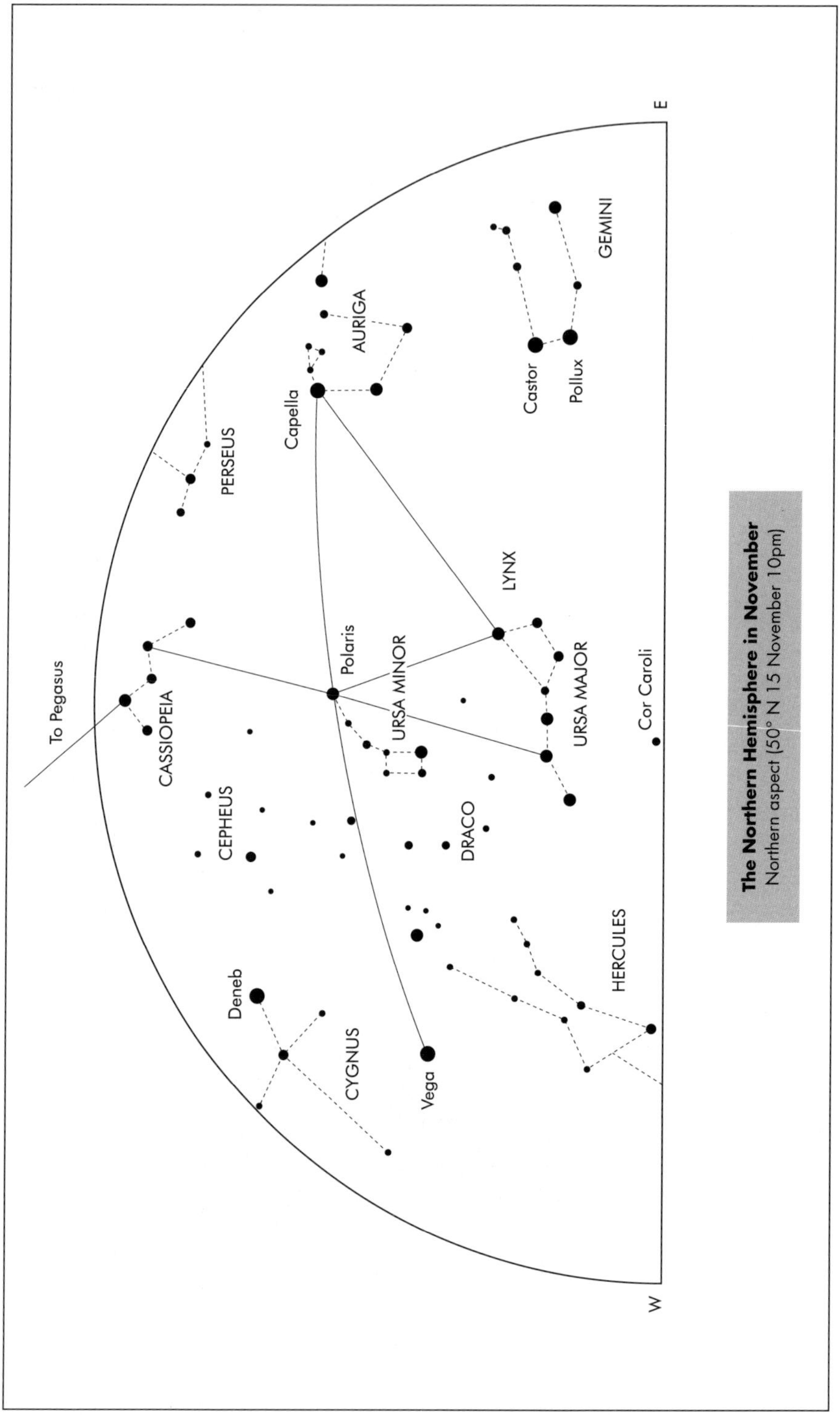

The Northern Hemisphere in November
Northern aspect (50° N 15 November 10pm)

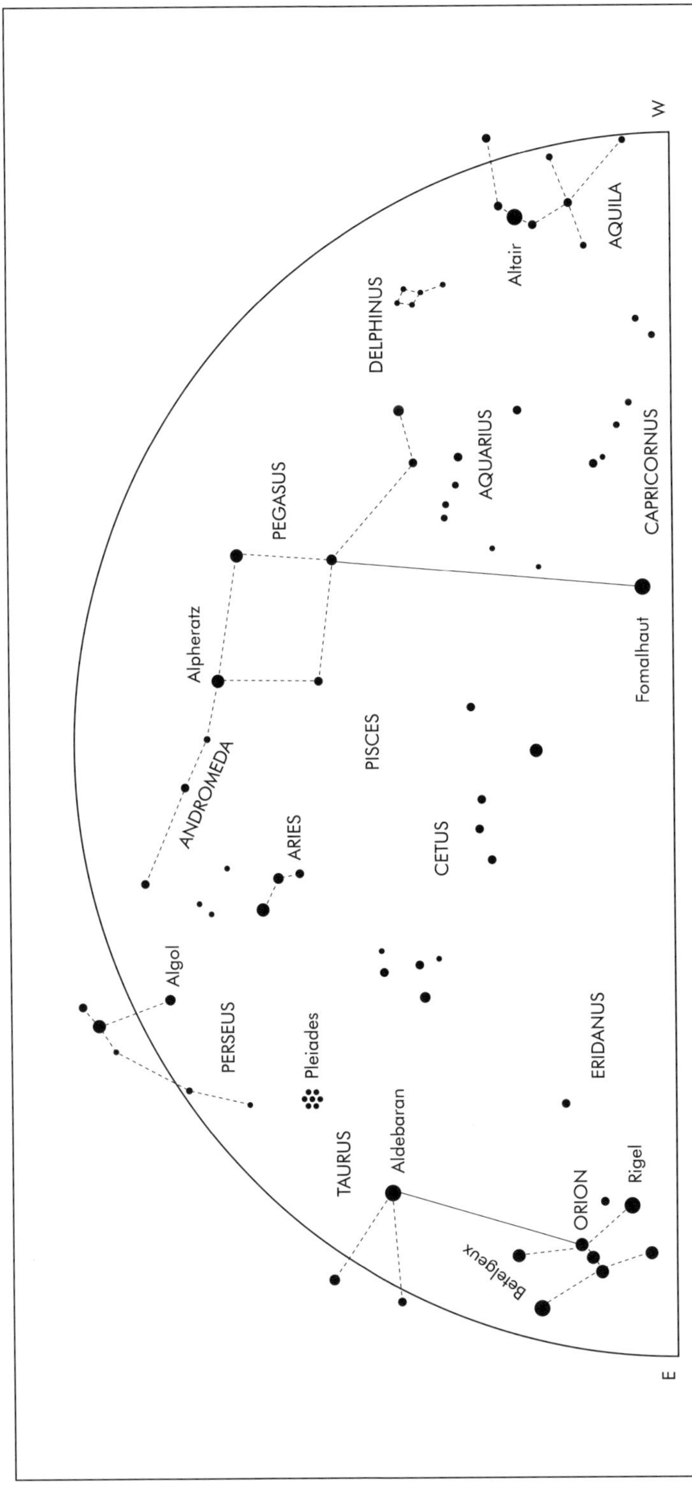

The Northern Hemisphere in November
Southern aspect (50° N 15 November 10pm)

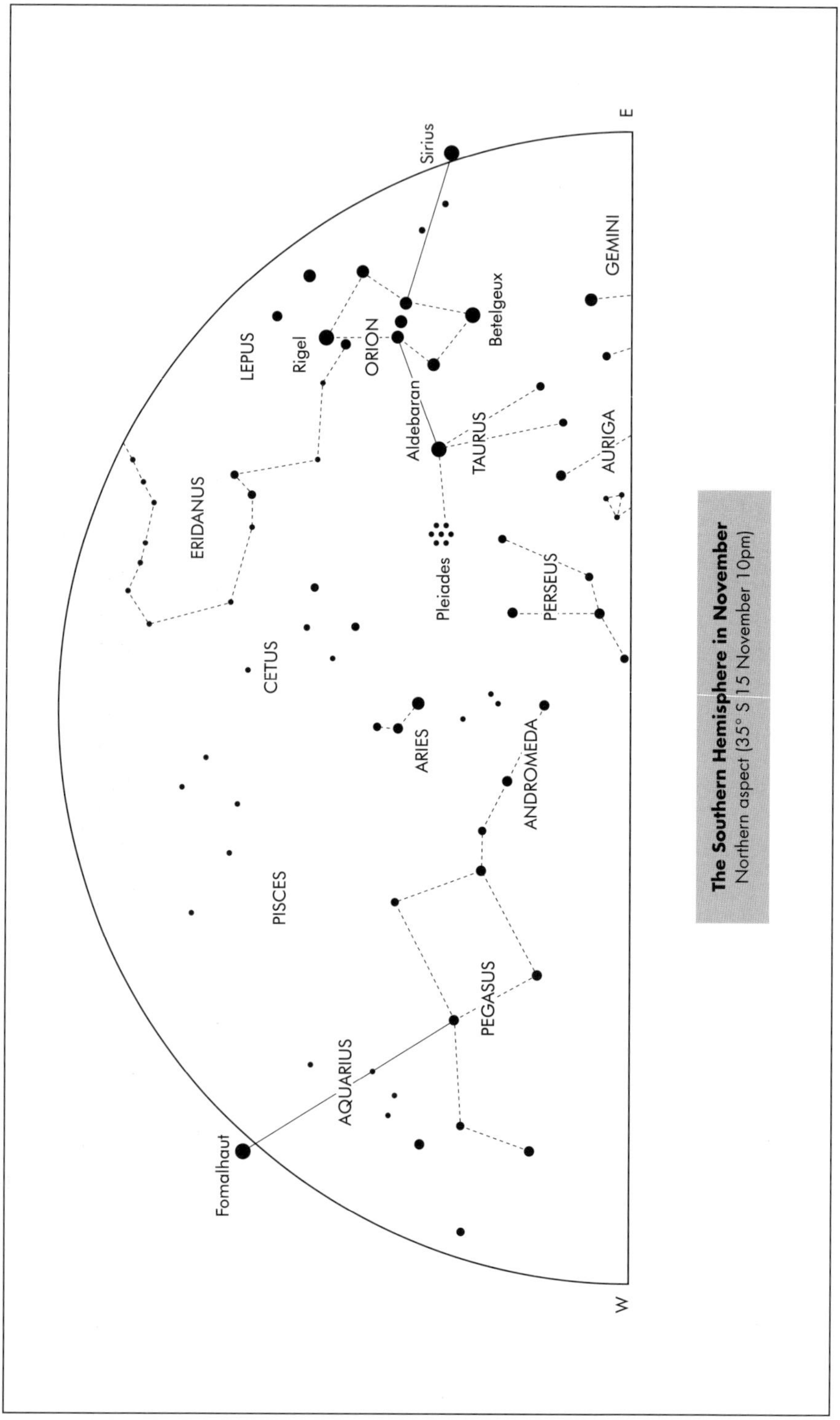

The Southern Hemisphere in November
Northern aspect (35° S 15 November 10pm)

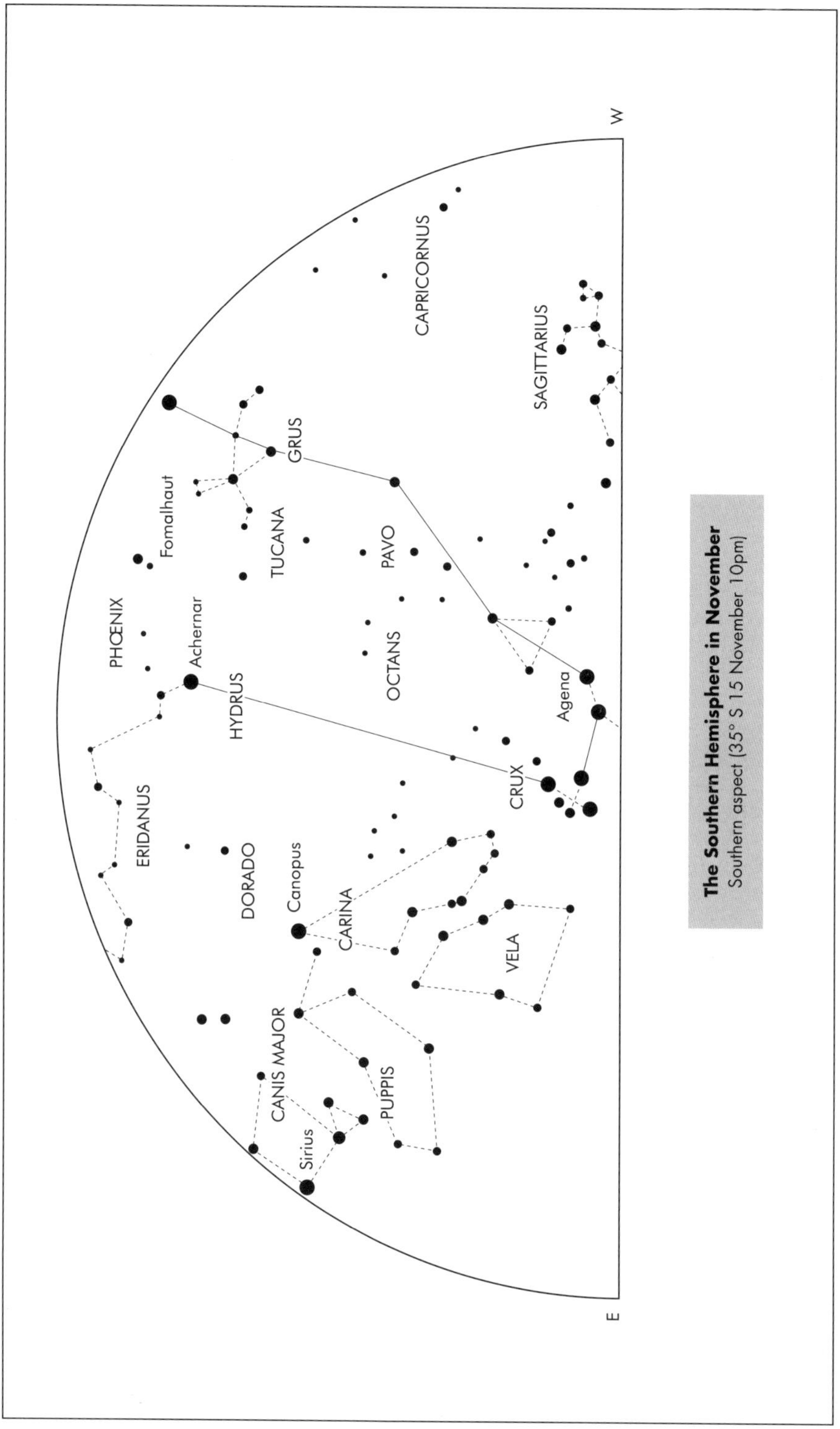

The Southern Hemisphere in November
Southern aspect (35° S 15 November 10pm)

November 1

Anniversary

1962: Launch of the first attempted interplanetary probe – Mars 1, sent up by the Russians. Contact with it was lost at 66 000 miles from Earth, so we will never know what happened to it.

The November Sky

Northern hemisphere observers can once again use Orion, which rises in mid-evening though it is not at its very best until after midnight. Much of the Hunter's retinue is on view – Capella, Aldebaran, the Twins – though Sirius does not appear until later. Ursa Major is still low in the north, and Arcturus has disappeared; the W of Cassiopeia is almost at the zenith. We are losing the Summer Triangle as a dominant feature, and Altair sets before midnight. Pegasus is still there, with Andromeda and Perseus; Cetus and Eridanus sprawl across the southern aspect, but we have to all intents and purposes lost Fomalhaut in the evening twilight. This is a good time of the year to track the Milky Way, from Cygnus right across the zenith and down to Gemini in the east.

From southern countries, this is an ideal time to study the Clouds of Magellan, which, with the Southern Birds, are almost overhead. Orion is with us once more, and Sirius shines brilliantly in the east; Canopus is high up, and it is interesting to compare the two. Sirius looks much the brighter, and one has to use one's imagination to realize that compared with Canopus it is puny; according to the figures in the Cambridge catalogue, it would take more than 7500 stars of the luminosity of Sirius to equal the power of Canopus. Achernar is high, and the Cross still rather low in the south. The Square of Pegasus is setting in the north-east, but Andromeda remains visible low over the horizon. Cetus is well displayed, and we can see the whole of the River Eridanus, from the area of Orion through to the far south.

November 2

Anniversary

1885: Birth of Harlow Shapley, the American astronomer who was the first to give an accurate value for the size of the Galaxy, and also made many important contributions to theoretical astrophysics. From 1921 he was Director of the Harvard College Observatory. He died in 1972.

Eridanus, the River

For our first November constellation it will be convenient to begin with Eridanus, the celestial River, which is immensely long, and runs from the Orion area down to the 'deep south'. From Australia or New Zealand it is visible in its entirety, but from Britain much of it is cut off by the horizon. In giving a list of leading stars, I have therefore added their declinations. To see whether a star is accessible, subtract your latitude from 90°. Thus assume that you are at latitude 52° N; 90–52 = 38; therefore Epsilon Eridani (declination –09°) rises, while Theta (declination –40°) does not.

Achernar is the nearest brilliant star to the south celestial pole. Its name means 'the Last in the River', but there have

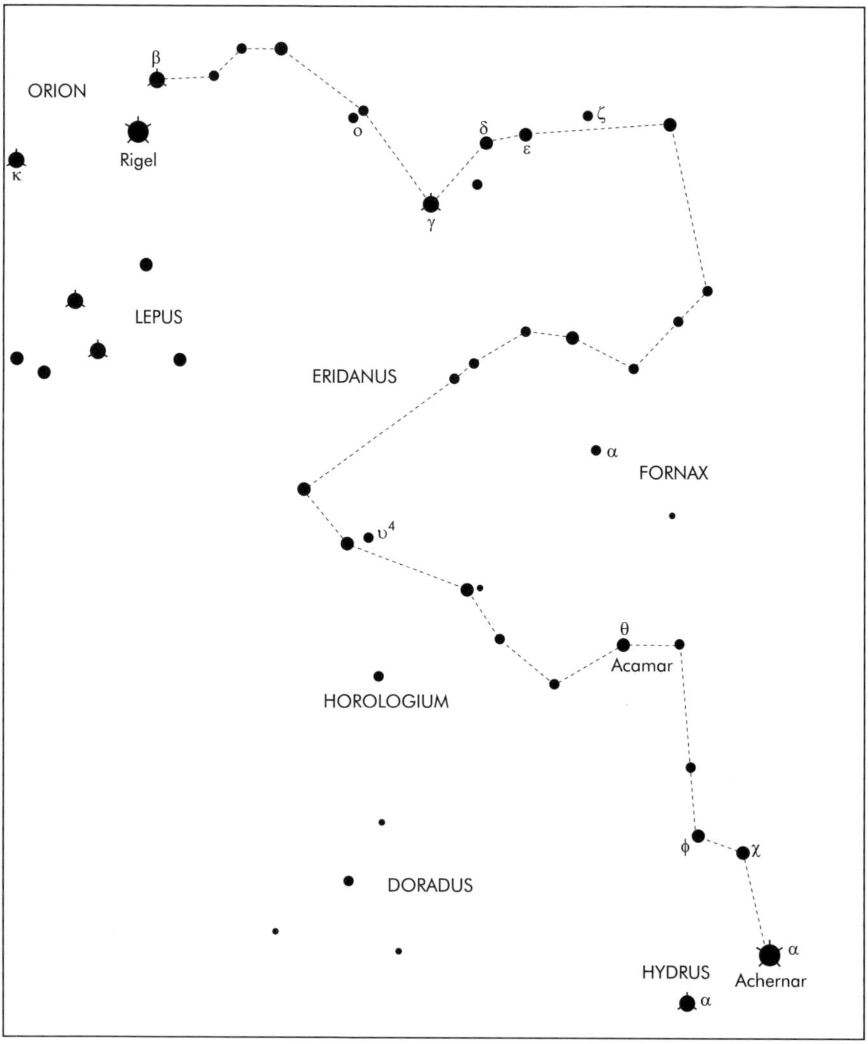

been suggestions that this name should really apply to Theta, which is a fine double; magnitudes 3.4 and 4.5, separation 8.2 seconds of arc. In the northern part of the River, note Epsilon Eridani, which is one of the two nearby stars which are not too unlike the Sun, and are regarded as possible planetary centres; Tau Ceti is the other (7 October). Note also the little pair of Omicron^1 (Beid), magnitude 4.0, and Omicron^2 (Keid), 4.4 They are not connected. Beid is a normal F-type star, over 250 light-years away; Keid is a double, made up of a red dwarf together with a white dwarf binary companion. This pair is a mere 16 light-years away, and in fact Omicron^{2}B appears as the brightest of all white dwarfs in our sky. The separation is over 7 seconds of arc, and the orbital period is 248 years.

Eridanus						
Greek letter	Name	Magnitude	Luminosity (Sun=1)	Distance (light-years)	Spectrum	Declination (degrees)
α Alpha	Achernar	0.5	400	85	B5	-57
β Beta	Kursa	2.8	82	100	A3	-05
γ Gamma	Zaurak	2.9	110	144	M0	-13
δ Delta	Rana	3.5	3	29	K0	-10
ε Epsilon	–	3.7	0.3	10.7	K2	-10
θ Theta	Acamar	2.9	50+17	55	A3+A2	-40
τ Tau	–	3.7	120	225	M3	-21
υ Upsilon[4]	–	3.6	82	130	B8	-33
φ Phi	–	3.6	105	120	B8	-52
χ Chi	–	3.7	4.5	49	G5	-52

November 3

The Taurids

Tonight is the official maximum of the Taurid meteor shower, but the shower is not generally rich; the ZHR seldom exceeds 10, though there was a fine display in 1988. There have been suggestions that the Taurids were much more spectacular in the past. The shower is active between 20 October and 30 November; the maximum is rather ill-defined. In general the meteors are slow but fairly bright.

The parent comet is Encke's, which has a period of 3.3 years (the shortest known). It was discovered by Pierre Méchain in 1786; its periodicity was established by J.F. Encke, who correctly predicted the return of 1822. Since then the comet has been seen at every return except that of 1944, when it was badly placed and most astronomers had other things on their minds.

The aphelion distance lies within the orbit of Jupiter, and the comet can now be followed all round its orbit. It seldom reaches naked-eye visibility, though in 1829 it rose to magnitude 3.5; at some returns it develops a pronounced tail. Perihelion years are 1997 (May), 2000 and 2003.

Future Points of Interest

Maximum of the Taurid meteor shower.

November 4

The South Celestial Pole

Northern observers have an excellent Pole Star – Polaris in Ursa Minor (7 January). Southerners are not so lucky. The south pole lies in a very barren region, and there is no bright

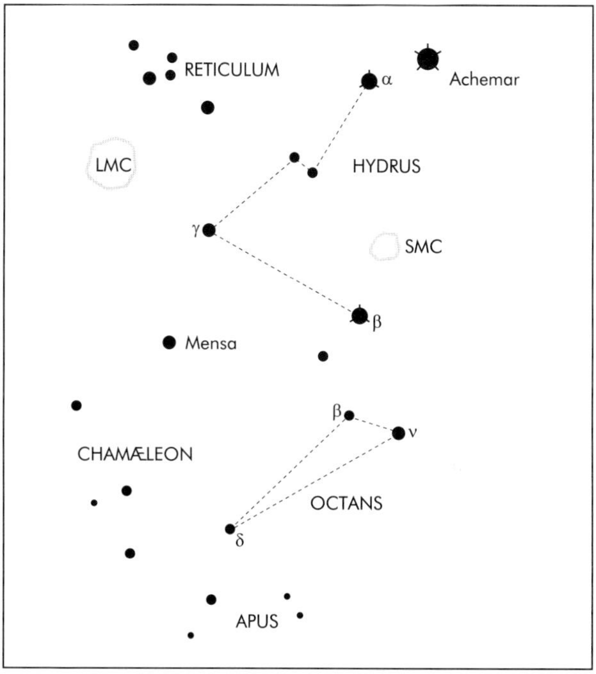

star anywhere near it. The best way to locate the general area is to look midway between Achernar (2 November) and the Southern Cross.

The polar constellation is Octans, the Octant, which is very obscure. The sequence of Greek letters has gone badly wrong here; the leader of the constellation is the orange Nu (magnitude 3.8), followed by Beta (4.2) and Delta (4.3). Alpha Octantis is below the fifth magnitude.

The south pole star, Sigma Octantis, is of magnitude 5.5; its position is RA 21 h 8 m 45 s, declination –88° 57′ 21″, so that it is further from the actual pole than Polaris is from the north pole. It is an A-type star, 121 light-years away and seven times as luminous as the Sun. Unfortunately its faintness makes it far from easy to identify; it is one of a triangle of fifth-magnitude stars (Chi, Sigma and Tau) more or less between Delta Octantis and Beta Hydri.

The nearest reasonably bright star to the south pole is Beta Hydri. Hydrus, the Little Snake (not to be confused with Hydra) contains three stars above the fourth magnitude.

Hydrus

Greek letter	Name	Magnitude	Luminosity (Sun=1)	Distance (light-years)	Spectrum
α Alpha		2.8	7	36	F
β Beta		2.8	3	20	G
γ Gamma		3.2	115	160	M

Hydrus contains little of interest, but is not too hard to locate, because Alpha Hydri lies close to Achernar. Gamma is clearly orange-red in colour.

November 5

Anniversary

1964: Launch of Mariner 3, America's first Mars probe. Unfortunately contact was lost soon after launch from Cape Canaveral. No doubt Mariner is still in solar orbit, but it sent back no data. Mariner 4, following on November 28, was however a great success.

The Clouds of Magellan

Before leaving the far south, I must ask northern observers to bear with me while I say something about the Clouds of Magellan, which are now high during evenings as seen from the latitudes of Australia or New Zealand.

These are satellite galaxies of our own, but are still of considerable size. Both are very prominent naked-eye objects; even moonlight will not drown the Large Magellanic Cloud (LMC). Originally they were classed as irregular systems, but there are obvious indications of spirality with at least the LMC.

Their main importance to astronomers is that they contain objects of all kinds, and these may for practical purposes be regarded as being at the same distance from us: 169 000 light-years for the LMC (very recent work may reduce this to 166 000) and rather further for the Small Magellanic Cloud (SMC). The Tarantula Nebula in the LMC is a magnificent sight with binoculars or a telescope; novæ have been seen, and in 1987 a supernova flared up in the LMC, thereby providing astronomers with a chance to study one of these colossal outbursts from what is, cosmically, close range. The globular cluster 47 Tucanæ is more or less silhouetted against the SMC, but is of course in the foreground.

November 6

Anniversary

1993: Transit of Mercury, the last in the century apart from that of 15 November 1999. The next will be in 2003 and 2006.

Venus in the Evening Sky

In November 1997 Venus is near eastern elongation, and is a magnificent object in the evening sky (this will also be the case in November 2005). It appears well before sunset, and far outshines any other star or planet.

In 1997 elongation occurs on 6 November, and at this time the phase should be exactly 50% (dichotomy). However, because of the effects of Venus' dense, extensive atmosphere, theoretical and observed dichotomy usually differ by a few days – at evening elongations dichotomy is early, in morning

Future Points of Interest

1997: Venus at eastern elongation.

1999: Saturn at opposition. The declination is +13°, and the magnitude 0.0.

elongations it is late. This was first noted by J.H. Schröter, and many years ago I nicknamed it the 'Schröter effect' – a term which has now come into general use.

It has to be admitted that telescopically Venus is not a very inspiring object. In size and mass it is almost a twin of the Earth, but there the resemblance ends. The atmosphere is dense and cloud-laden, hiding the surface so completely that before the Space Age our knowledge of the surface conditions was minimal. Some astronomers pictured a raging, scorching-hot dust-desert, while others maintained that there might be broad oceans supporting aquatic life. It was only with the flight of Mariner 2, in 1962, that temperature measurements showed Venus to be far too hot for life of our kind to exist there. A thermometer would show a temperature not far short of 1000 degrees Fahrenheit. Moreover, the atmosphere is made up chiefly of carbon dioxide, and the clouds are rich in sulphuric acid.

November 7

The Rotation of Venus

Before the space-probes flew, many attempts were made to measure the rotation period of Venus by observing the surface markings. This is easy enough with Mars, or for that matter Jupiter, but not with Venus, simply because the cloudy shadings seen there from time to time are so vague and impermanent that they can provide no real information. Estimates ranged from less than 24 hours up to 225 days. This latter value is the same as Venus' orbital period, so that it would mean a synchronous rotation with Venus keeping the same face turned permanenrly toward the Sun.

Radar showed that the rotation must be slow, but infrared measurements then proved that the night side of the planet was much warmer than it would be if it never received any sunlight. The favoured period was 'about an Earth month'. Finally, probe-carried radar solved the problem. Venus rotates in 243.1 Earth days, so that on Venus a sidereal day is longer than the planet's 'year' of 224.7 days. (The 'solar day' is equal to 118 Earth days.) Moreover Venus spins from east to west, not west to east, so that if you could observe the Sun from the surface it would rise in the west and set in the east. Not that this would ever be possible; there is no such thing as a clear day on Venus.

The reason for Venus' retrograde rotation is not known and the favoured theory, that in its early evolution the planet was hit by a massive impactor and knocked over, does not sound plausible; we have come across the same sort of problem with Uranus (24 August).

The orbital inclination of 3° 23′ 40″ means that transits are rare; those of 2004 and 2012 will be followed by another pair in 2117 and 2125. Mercury, where the orbital inclination is over 7°, transits much more often (see 7 May). However, Mercury is too small to be seen with the naked eye during transit, while Venus is said to appear as a conspicuous black disk. It is interesting to reflect that as I write these words (August 1997) there can be no living person who can recall a transit of Venus; the last was in 1882.

November 8

Venus: the Phantom Satellite

Venus and the Earth differ in another important respect: Venus has no satellite. Yet little more than a century ago it was still widely believed that a satellite existed.

In 1686 the famous Italian astronomer G.D. Cassini, then at Paris, announced the discovery of a satellite with about one-quarter the diameter of Venus itself. In 1730 a satellite was recorded by James Short, an experienced telescope-maker, and during the transit of 1761 a German observer, A. Scheuten, reported a small black disk following Venus across the face of the Sun. There were also reports in 1761 by Montaigne at Limoges, and in 1764 by Horrebow from Copenhagen. The orbit of the satellite was calculated, and later it was even given a name: Neith.

In fact the observers were seeing nothing more than telescopic 'ghosts'; the brilliant light of Venus brings out tiny telescopic imperfections. 'Neith' does not exist, and has never done so. If Venus had a satellite of any appreciable size, it would certainly have been found by now.

Anniversary

1711: Birth of Mikhail Lomonosov, the first Russian astronomer. His father was a fisherman! He had a somewhat chequered career, and was once imprisoned for insulting his colleagues; he was a poet and a chemist as well as an astronomer. He drew up the first accurate map of the Russian Empire, and championed the Copernican theory and Newtonian gravitation. It was he who, at the 1761 transit of Venus, concluded that the planet has a dense atmosphere. He died in 1765.

1991: Brilliant display of aurora seen over almost the whole of the British Isles.

November 9

The Ptolemæus Chain

Before taking final leave of Venus, let us pause to turn back briefly to the Moon; after all, this is a 'lunar day'. We have already discussed the Sinus Medii or Central Bay, site of two Surveyor landings, No. 4 (which failed) and No. 6 (which succeeded). Rather south of the Central Bay is a chain of

Anniversary

1967: Landing of Surveyor 6 on the Moon, at latitude 0°.5 N longitude 1°.4 W (in the Sinus Medii). It returned 29 000 images, and transmitted until 14 December.

Future Points of Interest

2003: Total eclipse of the Moon. Mid-eclipse 01.20 GMT. Totality lasts for only 11 minutes.

three huge walled plains which are of special interest: Ptolemæus, Alphonsus and Arzachel.

Ptolemæus, the largest and northernmost, is 92 miles in diameter. It has a darkish, comparatively smooth floor which makes it identifiable under any angle of illumination; there is no central mountain, but the floor contains one deep crater, the 5.5 mile Ammonius. The walls of Ptolemæus are fairly continuous, with peaks rising in places to 9000 feet.

Adjoining it to the south is Alphonsus, 80 miles across, with a lighter floor which contains a minor central mountain, a mass of fine detail and a system of rills. It was in Alphonsus that N. Kozyrev, in 1958, observed a red glow which indicated minor lunar activity. Here too are the remains of the Ranger 9 probe. Still further south is the 60-mile Arzachel, with high walls rising to 13 500 feet, a prominent central mountain and a deep crater. There are also some rills.

It is interesting to note the gradation in type between these three walled plains; it is difficult to doubt that they were formed by the same process even though their ages are obviously different. Just outside the chain is the 27-mile Alpetragius, with high, terraced walls and an enormous rounded central mountain with two symmetrical summit pits. This whole area is fascinating to study with even a small telescope.

November 10

Conditions on Venus

Photographs of Venus taken with ordinary telescopes show virtually nothing apart from the characteristic phase. The shadings which can sometimes be seen visually are almost impossible to record photographically without using filters, and even then they are very indefinite.

Images obtained from the surface by the Russian landers, and the radar results from the American probes, show that Venus is intensely hostile. A huge plain covers over 60% of the surface; there are two major uplands, Ishtar and Aphrodite; high mountains, reaching to at least 5 miles above the surrounding terrain; valleys, craters, and volcanoes which are almost certainly active. Windspeeds at the actual surface are low, though in that dense atmosphere – with a ground pressure 90 times that of the Earth's air at sea-level – even a sluggish wind will have tremendous force.

Why are Venus and the Earth so different? The reason must surely be the Earth's greater distance from the Sun. It is thought that in the early days of the Solar System, the Sun was less luminous than it is now, and in this case Venus and the Earth may have started to evolve along similar lines. But then the Sun became more powerful; the oceans of Venus boiled away, the carbonates were driven out of the rocks, and in a relatively short time, astronomically speaking, Venus changed from a potentially fertile world into a furnace-like environment remarkably like the conventional idea of hell. Any life which might have appeared there was ruthlessly destroyed. It is sobering to reflect that if the Earth has been a mere 25 00 000 miles closer to the Sun it would have suffered the same fate, and you would not now be reading this book.

November 11

Perseus and the Sea-Monster

It is time to return to the northern sky, so let us concentrate upon Perseus, which is well on view during November evenings; Australians will have to be content with seeing it very low over the horizon. It lies between Andromeda and Capella, and is unmistakable even though it contains no first-magnitude star.

Anniversary

1875: Birth of V.M. Slipher, director of the Lowell Observatory from 1917 to 1953. Vesto Melvin Slipher made vitally important observations of the spectra of the objects now known to be galaxies; for these

(Continued opposite)

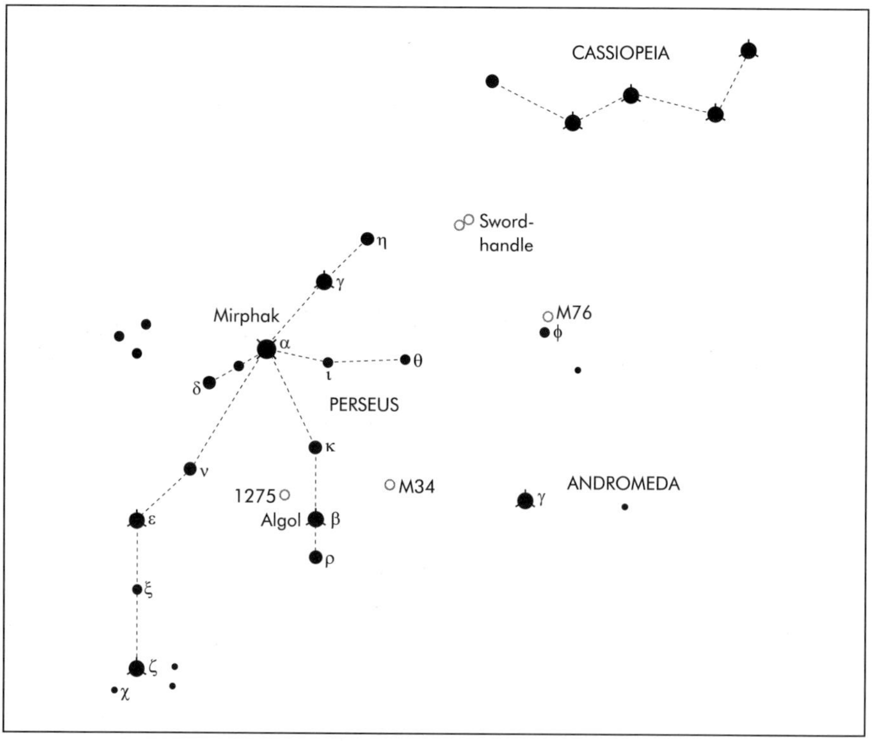

Perseus					
Greek letter	Name	Magnitude	Luminosity (Sun=1)	Distance (light-years)	Spectrum
α Alpha	Mirphak	1.8	6000	620	F5
β Beta	Algol	2.1 (max)	105	95	B8
γ Gamma	–	2.9	58	110	G8
δ Delta	–	3.0	650	326	B5
ϵ Epsilon	–	2.9	2600	680	B0
ζ Zeta	Atik	2.8	16 000	1100	B1
ρ Rho	Gorgonea Terti	3.2 (max)	120	196	M4

observations he used the 24-inch Lowell refractor. He also made many valuable observations of the planets, and it was he who engaged Clyde Tombaugh to come to Flagstaff and undertake what proved to be a successful search for a new planet (Pluto). Slipher died in 1969.

There is a famous legend about it. Perseus was a gallant hero who was returning from a dangerous mission; he was wearing winged shoes kindly loaned to him by the god Mercury, and had been engaged in decapitating a hideous Gorgon named Medusa, who had a woman's body and hair made of snakes – and whose glance could turn any living creature to stone. On his way Perseus spied the Princess Andromeda chained to a rock, where she had been left to make a meal for a particularly unpleasant monster which had been ravaging the kingdom. Perseus lost no time; he showed the monster Medusa's head, so that the beast was promptly petrified. In the best tradition, Perseus married Andromeda, and they lived happily ever after. This is one of the few mythological legends with a pleasant ending!

The leading stars of Perseus are listed above.

Perseus has an outline rather like the mirror image of the Greek letter lambda (λ). The Milky Way is very rich here, and it is worth sweeping the whole area with binoculars.

Future Points of Interest

1998: Mercury at eastern elongation. Occultation of Regulus by the Moon (22 h GMT).

November 12

The Demon Star

Anniversary

1980: Voyager 1 passed Saturn at just over 77 000 miles. Good images were obtained of the planet and also the satellites Titan, Dione, Rhea and Mimas.

One of the most famous stars in the sky is Algol or Beta Persei, known as the Demon and marking the head of the Gorgon Medusa. To astrologers it was the most evil star in the heavens, and certainly it behaves in a most unusual way, though apparently nobody noticed this before the Italian astronomer G. Montanari did so in 1669.

For most of the time Algol shines as a star of magnitude 2.1, but every 2.9 days it begins to fade; in four hours its magnitude has fallen to 3.4, and it remains dim for twenty minutes before starting to recover. This pattern repeats itself with absolute regularity. In 1782 John Goodricke, a young

English astronomer of Dutch extraction, realized why. Algol is not truly variable at all; it is a binary, with one component much brighter than the other. When the fainter star passes in front of the binary, Algol 'winks'. There is a much slighter 'wink' when the fainter star is hidden, but to detect this instruments are needed. Technically Algol is termed an eclipsing binary, though it should more logically be referred to as an occulting binary.

The primary star is over 100 times as luminous as the Sun; the secondary is larger but less powerful – the eclipses are not total. The real separation is of the order of 6 500 000 miles, and there are several dimmer stars in the system.

November 13

The Light-curve of Algol

Algol is a particularly good target for the naked-eye observer, because a full cycle can be followed over the course of a night, and the changes are obvious even in a few minutes. Times of minima are given in yearly handbooks and in periodicals such as *Sky and Telescope*, so that one always knows what to expect.

Fortunately there is no shortage of comparison stars. During the fades (or rises) we have Zeta Persei or Atik (magnitude 2.8) and Epsilon (2.9); when Algol is fainter, we have Kappa Persei or Misam (3.8), though Algol never falls to a magnitude as low as this. Near maximum, use Alpha Persei or Mirphak (1.8) and also Gamma Andromedæ (2.1). The star to avoid is Rho Persei, which lies in line with Kappa and Algol itself, because Rho is a red semi-regular variable with a range of magnitudes 3.2 to 4 and a very rough period of around 40 days. (It is, of course, worth following Rho also; Kappa makes an ideal comparison star.) It is interesting to draw up a light-curve; with practice, naked-eye estimates can be made to an accuracy of a tenth of a magnitude.

Anniversary

1971: Mariner 9 entered orbit round Mars. It returned 7329 images, giving the first views of the giant volcanoes. Contact was maintained until 27 October 1972.

Future Points of Interest

1998: Occultation of Mars by the Moon (18 h GMT).

November 14

The Sword-Handle in Perseus

Still with Perseus, our next target is the Sword-Handle, on no account to be confused with the Sword of Orion. Orion's Sword is marked by the great gaseous nebula M42. The

Perseus Sword-Handle is a double cluster, not listed in Messier's catalogue but given in the NGC as Nos 884 and 869. In the Caldwell catalogue they are listed together as C14; they are also known as Chi and h Persei.

The best way to find them is to use the W of Cassiopeia, taking a line from Gamma through Delta and prolonging it (23 January; I fear that this will be of no use to the southern-hemisphere observer, who will simply have to do his best). The clusters are visible with the naked eye, but are easily confused with the adjacent rich areas of the Milky Way. Binoculars or telescopes bring out the cluster superbly; they are in the same low-power field, and between them there is a reddish star. Each cluster is around 70 light-years in diameter, and they are of the order of 8000 light-years away.

Though they are undoubtedly associated, they are not identical twins; NGC 869 may be rather the less evolved of the two, and is also slightly closer, by a matter of a few hundred light-years. There is nothing else quite like this pair in the sky, and the two will repay study. They are, moreover, ideal subjects for the astrophotographer.

November 15

Anniversary

1738: Birth of William Herschel, possibly the greatest of all telescopic observers and discoverer of the planet Uranus.

Future Points of Interest

1999: Transit of Mercury (mid-transit 21 h 42 m GMT).

2000: Mercury at western elongation. Official start of the Leonid meteor shower – but do not expect too much tonight!

Objects in Perseus

We have not yet finished with Perseus, which is exceptionally rich in interesting objects. There are, for instance, some good double stars. Eta, some distance from Mirphak in the direction of Cassiopeia, is of magnitude 3.8; it has an 8.5-magnitude companion at a separation of 28 seconds of arc, so that this is an easy pair. Epsilon also has an 8th-magnitude companion, at 8.8 seconds of arc. There are dark rifts in the Milky Way, and there are two Messier objects, M34 and M76, plus a Caldwell object, C24 (NGC 1275).

M34 (RA 2 h 42 m, declination +42° 47′) is an open cluster discovered by Messier himself in 1764; it forms a triangle with Algol and Rho Persei, and lies just off a line joining Algol to Gamma Andromedæ. It is easy to find with binoculars, and can be glimpsed with the naked eye; it contains well over 60 stars. The distance is 1450 light-years.

M76 is the faintest of all the Messier objects. It was discovered by Méchain in 1780, and confirmed by Messier shortly afterwards. It is a planetary nebula, nicknamed the Little Dumbbell but much less symmetrical than the better-known M27, the Dumbbell in Vulpecula (9 September). It lies close to the fourth-magnitude star Phi Persei, almost midway between Gamma Andromedæ and Delta Cassiopeiæ in the W. The distance is 8200 light-years. It is decidedly elusive – but since Méchain and Messier found it with their small telescopes, no doubt modern observers can do the same. The position is RA 1 h 42 m, declination +51° 34′.

C24 (NGC 1275) is at RA 3 h 20 m, declination +41° 31′. It is an unusual galaxy, probably at least 300 million light-years away; it is a radio source, and radio astronomers know it as Perseus A. It is also an X-ray emitter. At an integrated magnitude of only 11.6 it is not easy to find, but it is worth seeking out.

November 16

Tracing the Equator

While we are waiting for the Leonids, it may be worth tracing the celestial equator once more, so for a northern-hemisphere observer we can begin in the west, where the Summer Triangle is disappearing into the twilight and Altair has already set, assuming that we are observing fairly late in the evening. Of course the equator passes through Pisces, south of the Square of Pegasus, and runs south of Alpha Piscium (29 September). It cuts through Cetus, between Menkar and Mira and almost touching the fourth-magnitude Delta Ceti, which is at declination +0° 20′ (5 October). It crosses a barren area, and then enters Orion, passing close to Mintaka in the Belt (–0° 18′) before cutting through the Milky Way in Monoceros, and then running some 5° south of Procyon before entering Hydra, south of the serpent's head and narrowly missing the reddish fourth-magnitude Iota Hydræ (–1° 9′). This is about as far as we can easily trace it tonight.

Anniversary

1996: Loss of the ambitious Russian probe Mars 6, which never achieved orbit, and merely fell in the sea shortly after blast-off. The Russians' ill-luck with the Red Planet continued.

Future Points of Interest

Watch for Leonid meteors tonight. The brief peak sometimes comes during the night of 16–17 November.

November 17

Leonid Night

November 17 marks the peak of the Leonid meteor shower. In most years it will be unspectacular, with a very low ZHR, but every 33 years there tend to be magnificent displays – as in 1799, 1833 and 1866. These are the years when the parent comet, Tempel – Tuttle, returns to perihelion – and the comet is due back once more in 1999.

It was the great meteor storm of 1833 that led to the identification of definite meteor showers. In America, H.A. Newton traced the Leonids back to the 'Year of the Stars', AD 902, and predicted another display of comparable splendour

Anniversary

1840: Birth of the fourth Earl of Rosse, the second of the astronomers of Birr Castle. Unlike his father, he made little use of the great reflector, but concentrated upon measuring the tiny quantity of heat received from the Moon. He died in 1908, and the reflector was not brought back into use until 1997.

Anniversary

1970: Lunokhod 1 landed on the Moon (taken there by the Soviet probe Luna 17). It travelled over a distance of 6.5 miles, guided by Russian controllers, and sent back over 200 panoramic pictures and 20 000 images. It operated for almost a year. The probe came down in the Mare Imbrium; it remains there, awaiting recovery.

in 1866; it duly took place. In the previous year E. Tempel and H. Tuttle had independently discovered the comet now named after them, and several astronomers, including G.V. Schiaparelli, realised that it must be this comet which was the Leonid parent.

Planetary perturbations meant that there were no major storms in 1899 or 1933, but in 1966 the Leonids were back; for a while – unfortunately during daylight in Britain – the ZHR reached 60 000. There seems every hope that there will be a great deal of Leonid activity in the years to either side of the dawning of the new millennium, and observers will be wise to be very much on the alert.

Future Points of Interest

Maximum of the Leonid meteor shower.

November 18

Canopus

So far as the Leonid meteors are concerned, northern observers have the advantage; but it cannot be denied that the far-southern stars are more brilliant than those of the far north. In particular, Britons are forever deprived of the glory of the great ship Argo, with its brilliant leader Canopus.

Argo – named for the ship which carried Jason and his companions in their quest of the Golden Fleece – was one of the original 48 consetellations, but was so immense that it was eventually dismantled into a keel (Carina), a poop (Puppis), sails (Vela) and a mast (Malus); the mast has disappeared, but the keel, the poop and the sails remain. Almost nothing of the old ship can be seen from most of Europe or the northern United States, though a little of Puppis protrudes briefly above the British horizon.

Canopus – once known as Alpha Argûs, now as Alpha Carinæ – is high up for southern observers during November evenings. Its magnitude is –0.7, so that it is brighter than any other star apart from Sirius, but unlike Sirius it owes its eminence to its great luminosity; according to the Cambridge catalogue (which I am following here) it is equal to 200 000 Suns, and it is 1200 light-years away. Other catalogues reduce these values somewhat, but in any case there is no doubt that Canopus is a true cosmic searchlight. It has an F-type spectrum, and so in theory ought to be slightly yellowish, but I have never seen a trace of colour in it; to me (and I suspect to most people) it looks pure white.

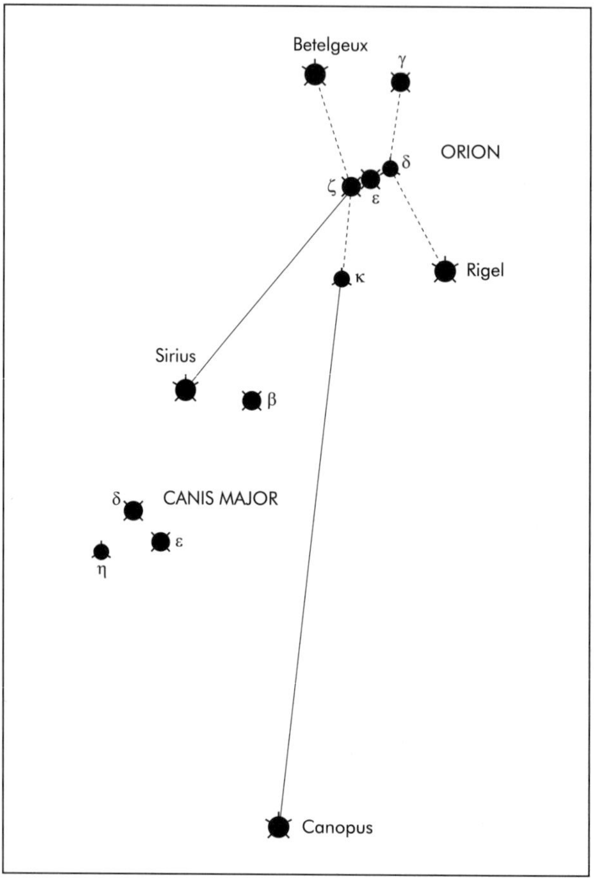

Stellar diameters are hard to measure. According to the best available estimate, Canopus is around 72 000 000 miles across, and it is of course very massive. It is squandering its energy at a furious rate, and cannot possibly go on shining for nearly as long as our mild Sun will be able to do.

The declination is –52° 42′. It can be seen from Alexandria, where Claudius Ptolemy lived and drew up his famous star catalogue, but never rises from Athens. This was one of the first proofs that the Earth is a globe, rather than being a plane.

November 19

The Andromedid Meteors

Quite apart from the Leonids, there is – or was – another interesting November meteor shower. Its maximum is due on

Anniversary

1969: Apollo 12 landed on the Moon, taking astronauts Charles Conrad and Alan Bean to the surface. The

(Continued opposite)

landing site was at latitude 3° 11′ S, longitude 23° 23′ W, in the Oceanus Procellarum. The astronauts were able to walk over to the old Surveyor 3 probe, which had landed in 1967, and bring parts of it back to Earth for analysis.

19 or 20 November, but it is now so sparse as to be virtually unnoticeable. The radiant lies in Andromeda.

The parent comet is dead. It was originally seen in 1772 by the French astronomer Montaigne; recovered in 1806 by J.L. Pons, and then in 1826 by an Austrian army officer, Wilhelm von Biela, who was an enthusiastic amateur astronomer. Biela worked out the orbit, and concluded that the period was between 6 and 7 years. Henceforth it has been referred to as Biela's Comet.

It was missed in 1839, because it was badly placed in the sky, but returned in 1845. Then, in January 1846, it split in two. The twins returned on schedule in 1852, but this was their farewell. They were missed in 1859, again badly placed, but when they failed to appear in 1866 it was obvious that they had disintegrated. In 1872 a major meteor shower was seen emanating from the place where the comet ought to have been; there was another fairly good display in 1886, but since then the activity has become feebler and feebler, so that it may in time cease altogether. Look for the 'Bieliids' tonight and tomorrow night by all means, but do not be disappointed if you see none.

Future Points of Interest

2000: Saturn at opposition.

Maximum of the Andromedid meteor shower.

November 20

Anniversary

1889: Birth of Edwin Powell Hubble.

Hubble and the Galaxies

Edwin Hubble, the American astronomer after whom the Space Telescope is named, was born in Marshfield, Missouri, on 20 November 1889, and educated in Chicago. After graduating from Chicago University he spent a brief period at Oxford University in England, and then returned to the United States, where he practised law for some months before taking up a research post at the Yerkes Observatory.

In 1917 he joined the US Army, and served until 1919, when he became a staff member at the Mount Wilson Observatory. Here he used the 100-inch Hooker reflector to detect short-period variables in the 'spiral nebulæ', and proved that these were in fact independent galaxies (24 October). He measured the red shifts in their spectra, and established that all the galaxies beyond our Local Group are receding from us, so that the entire universe is expanding; he classified the galaxies, and during his surveys of the sky also made many discoveries of gaseous nebulæ. It is fair to say that all subsequent research has been based upon Hubble's work – in which he was greatly helped by his assistant, Milton Humason, who began his observatory career as a mule driver and ended it as an observer second only to Hubble himself. Hubble died in California on 28 September 1953.

November 21

Auriga, the Charioteer

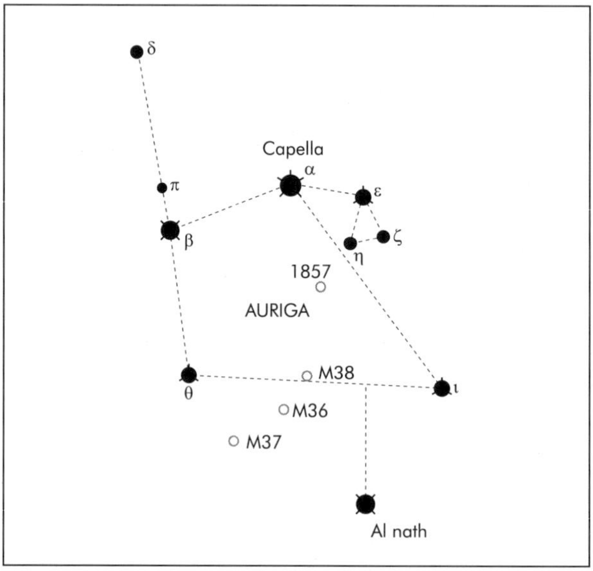

During summer evenings, Vega is almost overhead as seen from Britain. During winter evenings this position is occupied by Capella, in Auriga, the Charioteer. There can be little fear of confusion between the two, even though in brightness they are virtually equal; Vega is blue and Capella yellow, and moreover Capella has a characteristic little triangle of fainter stars (the 'Kids') close by.

Auriga itself is a prominent constellation, whose main stars make up a quadrilateral. Alnath, now known officially as Beta Tauri, used to be called Gamma Aurigæ. This was certainly more logical, since it clearly belongs to the Auriga pattern, while Taurus has no particular shape at all. Excluding Alnath, the leaders of Auriga are shown below.

Auriga

Greek letter	Name	Magnitude	Luminosity (Sun=1)	Distance (light-years)	Spectrum
α Alpha	Capella	0.1	90+70	43	G8
β Beta	Menkarlina	1.9	50	46	A2
δ Delta	–	3.7	60	163	K0
ϵ^2 Epsilon^2	Almaaz	3.0 (max)	200 000	4600	F0
ζ Zeta	Sadatoni	3.7 (max)	700	520	K4
η Eta	–	3.2	450	200	B3
θ Theta	–	2.6	75	82	A0
ι Iota	Hassaleh	2.7	700	267	K3

Auriga belongs to Orion's retinue. In mythology it represents Erechthonius, son of Vulcan (the blacksmith of the gods), who became King of Athens and, incidentally, invented the four-horse chariot.

There should be no problem in identifying Auriga. Quite apart from the brilliance of Capella, the quadralateral consisting of Capella, Beta, Theta and Iota is unmistakable.

November 22

Capella

Capella is the northernmost of the first-magnitude stars. At its declination of +46° it is circumpolar over Britain, and can be seen from almost every inhabited country; only from the southernmost tip of New Zealand, and the Falkland Islands, does it remain below the horizon.

Like the Sun, Capella has a G-type spectrum, but here the resemblance ends. The Sun is a yellow dwarf; Capella is a yellow giant – or, rather, a pair of yellow giants, because it is an excessively close binary. The primary has a diameter of 11 000 000 miles and a mass just over three times that of the Sun; the secondary is 6 000 000 miles across with a mass just under three times that of the Sun (remember that a smaller star is always denser than a larger one). The real separation is a mere 70 000 000 miles, and the orbital period is 104 days; the orbits are almost circular.

From a range of over 40 light-years it is very hard to see the two components separately, and telescopes of ordinary size show Capella as a single point of light. Its double nature was originally found, in 1899, by changes in its spectrum caused by the orbital motion. There is a third member of the system – a dim red dwarf which is itself double; it is over a tenth of a light-year from the main pair, and is none too easy to locate.

Capella and Vega are on opposite sides of the Pole Star, and about equally distant from it, so that when Vega is high up Capella is low down, and vice versa. When they are about equally high, as in April evenings, it is interesting to compare them. The colours are clearly different, but with the naked eye it is impossible to tell which is the brighter once extinction has been taken into account.

November 23

The Mysterious 'Kid'

Our target for tonight is a very strange star indeed. It lies close to Capella, and is officially known as Epsilon Aurigæ; its old proper name – Almaaz – is hardly ever used.

The little triangle of stars adjoining Capella is known as the Hædi, or Kid's. Epsilon Aurigæ lies at the apex; the others are Eta (3.2) and Zeta, which is much fainter.

Epsilon looks like an ordinary star – but it is not. It is exceptionally luminous and remote, and may be at least the equal of Canopus; in colour it is slightly 'off-white', though few people would call it yellow.

Every 27 years it begins to fade, and falls to magnitude 3.8, remaining at minimum for over a year before recovering. This last happened some time ago now; fading began on 22 July 1982, minimum extended between 11 January 1983 and 16 January 1984, and the whole sequence did not end until 25 June 1984. Clearly some unseen companion is passing in front of the star and eclipsing it, but the companion has never been seen; it shows no spectrum and is not detectable either in the ultra-violet, infra-red or radio range. It was once believed to be a huge, rarefied star not yet hot enough to shine, but it is now thought more likely to be a smaller, hot star surrounded by a shell of obscuring dust. The next fade is not due until 2009, with minimum between 2011 and 2012; we may learn more about it then, but in the meantime the curious Kid hides its secrets well.

Future Points of Interest

2003: Total eclipse of the Sun. Mid-eclipse falls at 23 h GMT. It will be best seen from the Antarctic.

November 24

Zeta Aurigæ

It is pure coincidence that Zeta Aurigæ or Sadatoni, the faintest of the Kids, is another unusual eclipsing binary. It is much less extreme than Epsilon, and we know much more about it. There is no connection between the two; Zeta is only about 500 light-years away, while Epsilon is nine times as remote.

With Zeta the interval between fadings is 972 days, and we can see the spectra of both components. The primary is a luminous K-type red giant, while the secondary is smaller, hotter and blue. When the blue star starts to pass behind its red companion, there is a period when its light comes to us after having passed through the outer layers of the red giant, and from this we can learn a great deal. Minimum lasts for 38 days, instead of over a year as with Epsilon. The magnitude range is from 3.7 to 4.1, so that the changes are slow and comparatively slight.

November 25

Clusters in Auriga

Auriga is a rich constellation, crossed by the Milky Way. It contains a number of open clusters within binocular range, and of these three are in Messier's list: M36, M37 and M38. They lie in a slightly bent line; M38 is around midway between two of the stars of the quadrilateral, Theta and Iota.

M37 is the brightest of the three (RA 5 h 52 m, declination +32° 33′). Its integrated magnitude is 5.6, and it lies not far off a line joining Theta Aurigæ to Beta Tauri. Messier himself discovered it in 1764; it is 3600 light-years away. Powerful binoculars are needed to resolve it into stars. M36 (RA 5 h 36 m, declination +34° 08′) is smaller and less bright; Lord Rosse called it 'a coarse cluster'. M38 (RA 5 h 29 m, declination +35° 50′) is almost as large as M37, and looser; Admiral Smyth described it as 'an oblique cross with a pair of large stars in each arm and a conspicuous single one in the centre'. Almost in the same line is another open cluster, NGC 1857, not in either the Messier or Caldwell catalogues; its position is RA 5 h 20 m, declination +39° 21′. It is much smaller than the three Messier objects but is easy enough to locate. Certainly the star-cluster enthusiast will find much to interest him in the Charioteer.

November 26

The Celestial Crane

With apologies to northern observers, it is time to 'go south' once more and discuss Grus, the Crane, much the most imposing of the four southern birds (the others are Pavo, the Peacock; Phœnix, the Phœnix; and Tucana, the Toucan). All are out of view from British latitudes, but at this time of the year they are very high up as seen from countries such as Australia. Frankly this is a somewhat confusing area, and needs sorting out, but at least Grus is distinctive. Its main stars are listed overleaf.

With a little imagination, the outline of Grus can be said to conjure up the picture of a bird in flight. Alnair and Al

Grus					
Greek letter	Name	Magnitude	Luminosity (Sun=1)	Distance (light-years)	Spectrum
α Alpha	Alnair	1.7	230	68	B5
β Beta	Al Dhanab	2.1	800	173	M3
γ Gamma	–	3.0	250	228	B8

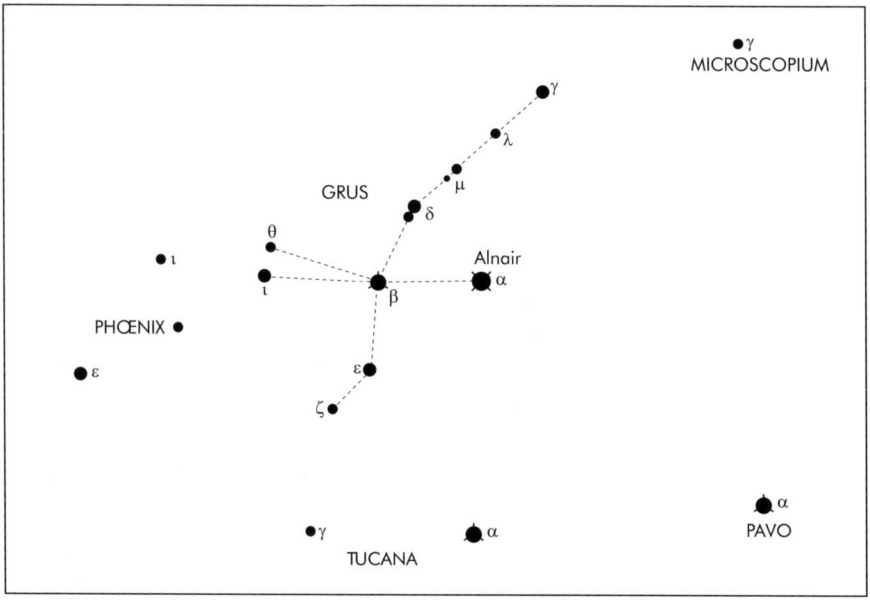

Dhanab provide good colour contrast; Alnair is bluish-white, Al Dhanab warm orange. Mu and Delta, between Gamma and Al Dhanab, look like wide doubles, but in each case there is no real association. Grus is rich in galaxies, but all of these are below tenth magnitude.

Adjoining Grus is the small constellation of Microscopium, the Microscope. It contains no star brighter than magnitude 4.6, and is entirely unremarkable.

November 27

Phœnix

The second bird, again out of view from Britain, is Phœnix, added to the sky by Bayer in 1603. It commemorates a mythical bird which periodically burns itself to death, rising once more from the ashes as good as new. It is less distinctive than Grus, and has only one reasonably bright star, the orange Alpha (Ankaa) which lies slightly off the mid-point of a line joining

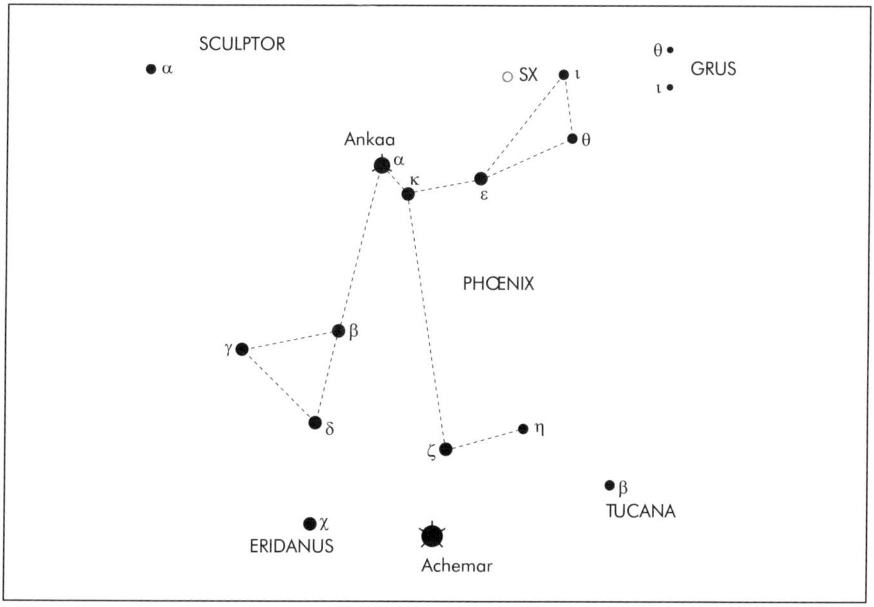

Fomalhaut in the Southern Fish to Achernar in Eridanus (2 November). The leaders of Phœnix are listed below.

Ankaa lies so close to Kappa (magnitude 3.9) that they have been taken for a wide double, but there is no true connection. Nearby is a triangle made up of Beta, Gamma and Delta (magnitude 3.9). Beta is a good double; the components are almost equal at magnitudes 4.0 and 4.2, and the separation is 1.4 seconds of arc.

Zeta Phœnicis, close to Achernar, is an eclipsing binary of the Algol type. Its magnitude range is from 3.6 to 4.1, so that it can be followed with the naked eye; good comparison stars are Delta (3.9), Eta (4.4) and Epsilon (3.9). The period is 1 day 16 hours; as with Algol itself (13 November) a full cycle can be followed in one night's observing. Also in this constellation is SX Phœnicis, a true, pulsating variable with the remarkably short period of 79 minutes. The range is from magnitude 6.8 to 7.5, and it is interesting to watch it perform; binoculars will suffice, though a wide-field telescope is more convenient. Obviously a good comparison chart is needed. The position is RA 23 h 47 m, declination –41° 35′, close to the 4.7-magnitude star Iota.

Phœnix

Greek letter	Name	Magnitude	Luminosity (Sun=1)	Distance (light-years)	Spectrum
α Alpha	Ankaa	2.4	60	78	K0
β Beta	–	3.3	58	130	G8
γ Gamma	–	3.4	5000	910	K5
ζ Zeta	–	3.6 (max)	105	220	B8

November 28

Peacock and Indian

The next bird is Pavo, the Peacock. The brightest star, Alpha Pavonis, is above the second magnitude; perhaps surprisingly it has never been given a proper name, though air navigators tend to refer to it simply as 'Peacock'. The best way to locate it is to start at Alpha Centauri, the brighter of the Pointers, pass a line through the red Alpha Trianguli Australe (18 April) and continue across a rather barren area until you come to Alpha Pavonis. In fact Alpha is rather divorced from the rest of the constellation, which consists mainly of a line of stars between magnitudes 3 and 4. The leaders are listed below.

Delta, in the 'line', is less than 19 light-years away, and in size, mass, temperature and luminosity it is very like our Sun; whether or not it has a system of planets we do not know. Kappa is a short-period variable with a magnitude range of 3.9 to 4.8 and a period of 9.1 days; it is what is called a Type II Cepheid or W Virginis star, much less luminous than a classical Cepheid of the same period. Good comparison stars for it are Zeta (4.0), Pi (4.4) and Nu (4.6) Pavo contains a bright

Future Points of Interest

1997: Mercury at eastern elongation.

2000: Jupiter at opposition, now 20° north of the equator.

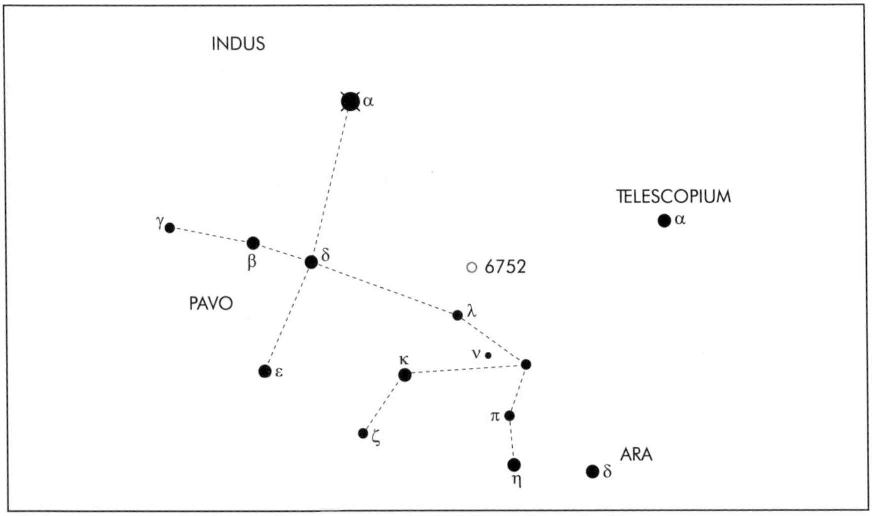

Pavo					
Greek letter	Name	Magnitude	Luminosity (Sun=1)	Distance (light-years)	Spectrum
α Alpha	–	1.9	700	230	B3
β Beta	–	3.4	28	100	A5
δ Delta	–	3.6	1	19	G5
η Eta	–	3.6	82	162	K1
κ Kappa	–	3.9 (max)	4	75	F5

globular cluster, C93 (NGC 6752) at RA 19 h 11 m, declination −59° 59′, which makes a triangle with Alpha and Beta.

Indus, the Indian, adjoins Pavo; it is very dim and formless, occupying much of the triangle formed by Alpha Pavonis, Alnair in Grus, and Alpha Tucanæ. The main object of interest is Epsilon Indi, which is very close (11.3 light-years). It is a red dwarf, only 14% as powerful as the Sun, and therefore the feeblest star visible with the naked eye. There are strong indications that it is attended by an unseen body which may well be a planet.

November 29

Tucana

We must complete our survey of the southern birds before moving on, and we are now left only with Tucana, the Toucan, which contains the Small Cloud of Magellan as well as two magnificent globular clusters. It adjoins Grus, Pavo and Achernar in Eridanus, and has one fairly bright star, Alpha – magnitude 2.9 – with a K-type spectrum so that it is decidedly orange. It is 114 light-years away, and could match 105 Suns. Beta is a very wide, easy double.

I have already discussed the Small Cloud (28 January). Almost silhouetted against it is 47 Tucanæ, the finest of all globulars with the exception of Omega Centauri; it has even been said to be the more imposing of the two, because it can be fitted into a simple telescopic field whereas Omega

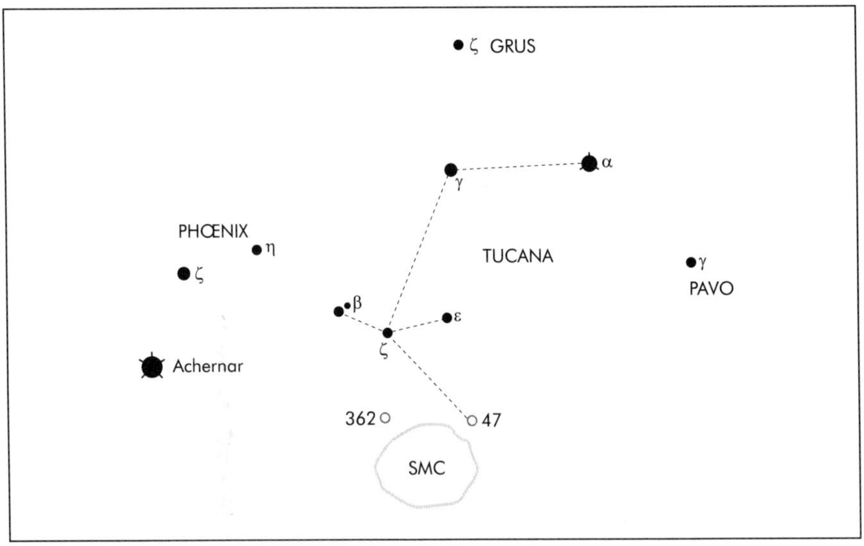

Centauri cannot. It is 15 000 light-years away, as against about 190 000 light-years for the Cloud; look at it with binoculars, and you will see that its surface brightness is much greater than that of the Cloud. Its position is RA 0 h 24 m, declination –72° 5′. Not far from it is C104, NGC 362: RA 1 h 3 m, declination –70° 51′, which is not markedly inferior to it, though it is below naked-eye visibility, and is less condensed than 47 Tucanæ and less easy to resolve into stars.

November 30

The Brightest Lunar Crater

When the Moon is full or nearly full, even a major crater can become obscure – simply because of the lack of shadows. However, there are a few features which always stand out. One of these is Aristarchus, on the Oceanus Procellarum or Ocean of Storms. It is only 23 miles in diameter, but it has brilliant walls and central peak, so that it can never be mistaken; it can be seen when illuminated only by earthshine, or when the Moon is totally eclipsed. No less a person than Sir William Herschel once mistook it for an erupting lunar volcano.

Aristarchus is also notable in another way. Almost every long-term lunar observer has seen occasional glows or local obscurations, indicating very mild activity; these are known as TLP or Transient Lunar Phenomena. More have been seen in and near Aristarchus than anywhere else on the Moon.

Close by Aristarchus is a crater of similar size, Herodotus. The two are not alike; Herodotus is not so bright, and has a much greyer floor. Extending from it is a long, winding valley, discovered by J.H. Schröter and therefore named after him. Altogether, this whole area is one of the most intriguing on the entire lunar surface.

Future Points of Interest

2001: A 'blue moon'. This is the second full moon of the month (see 26 September).

2002: Penumbral eclipse of the Moon (mid-eclipse 01.47 GMT). 86% of the Moon enters the penumbral zone, so that the eclipse should be fairly noticeable.

December

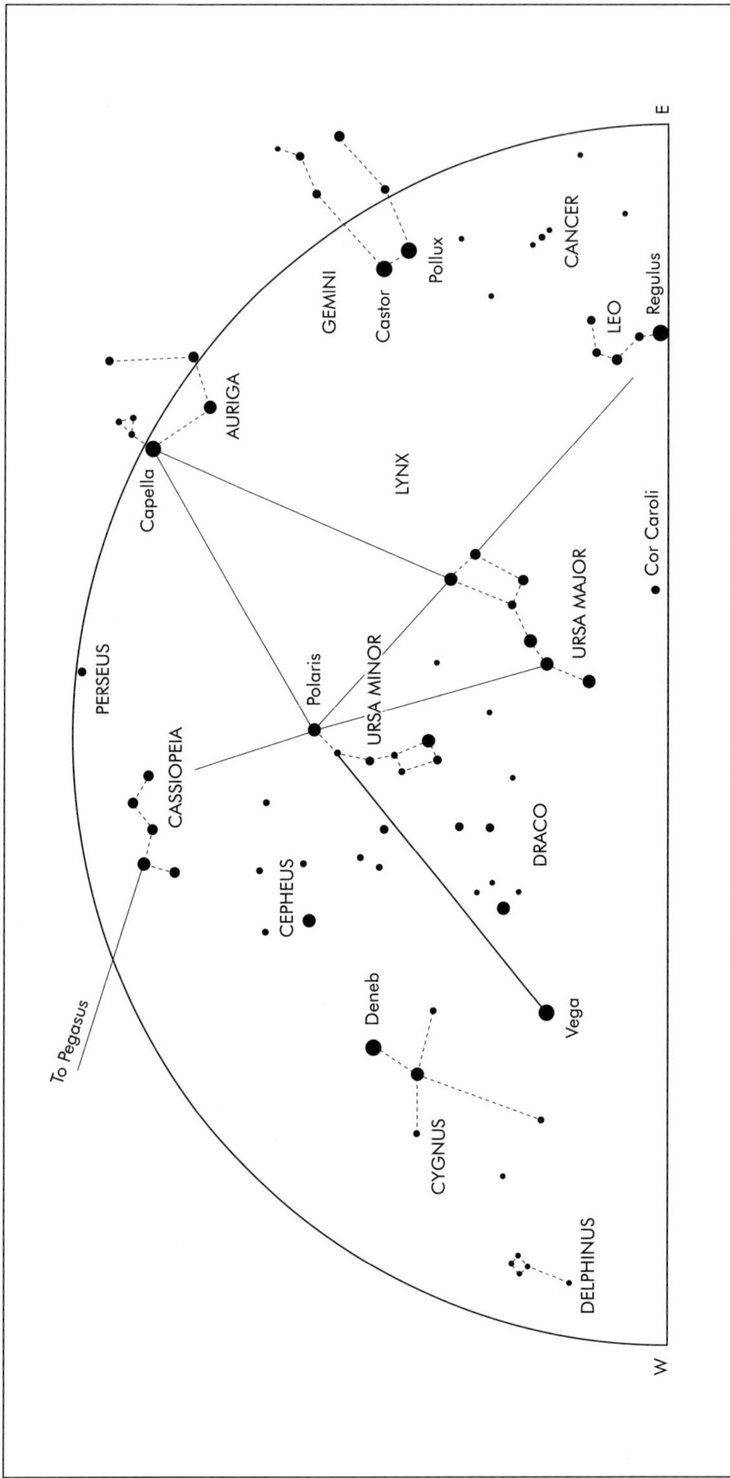

The Northern Hemisphere in December
Northern aspect (50° N 15 December 10pm)

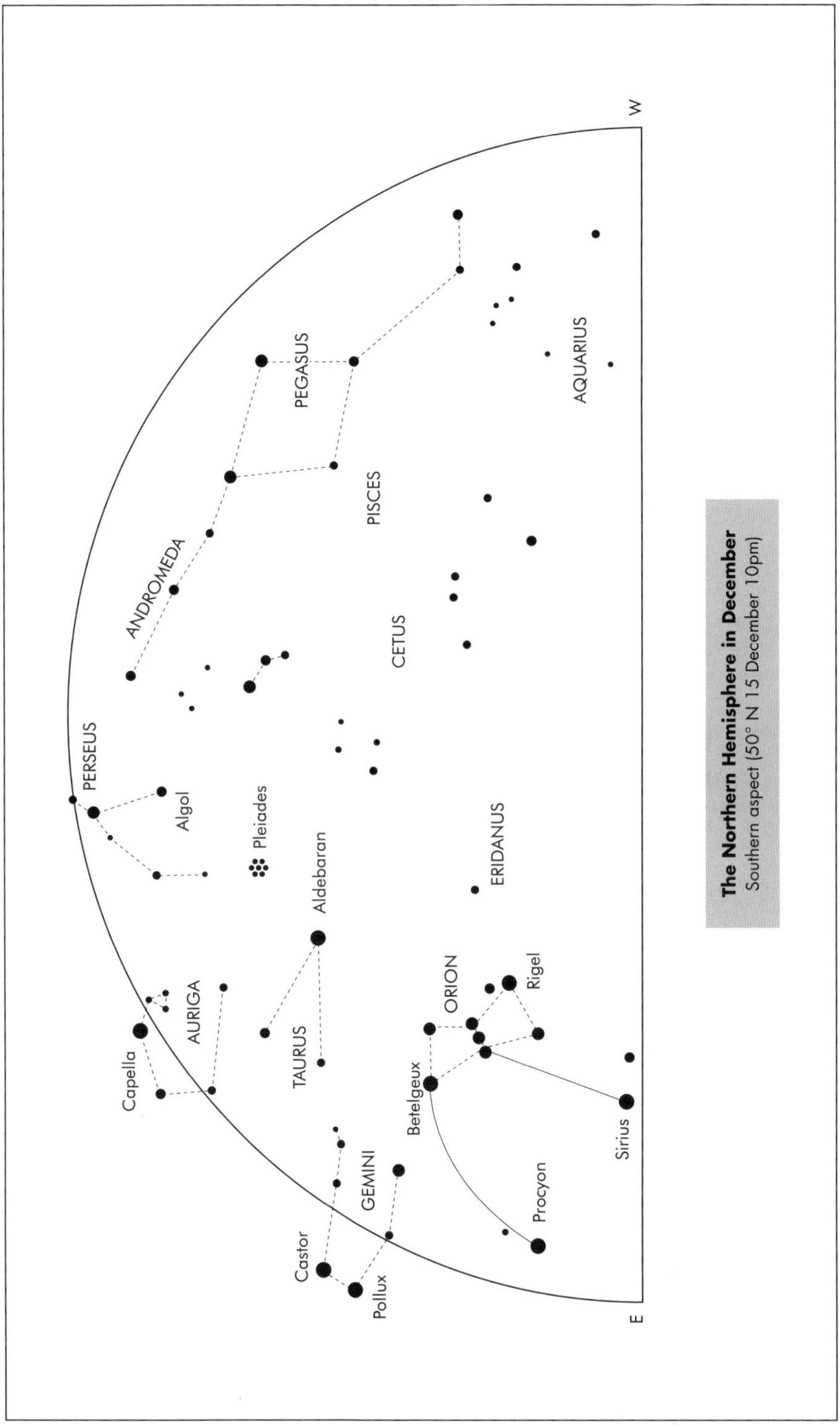

The Northern Hemisphere in December
Southern aspect (50° N 15 December 10pm)

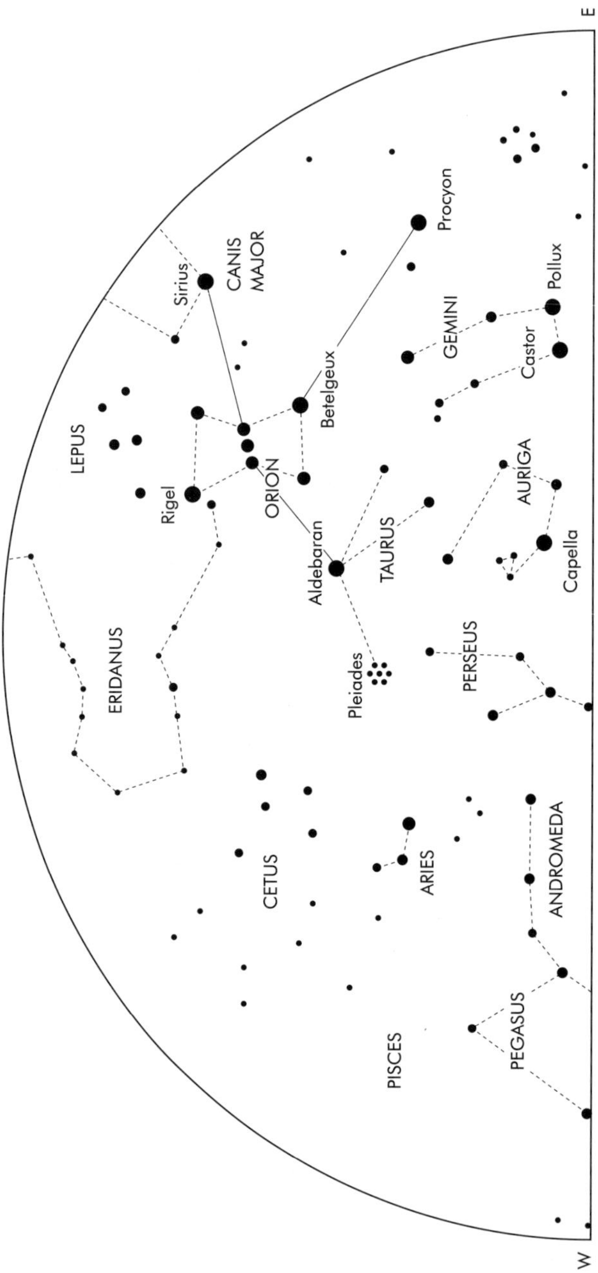

The Southern Hemisphere in December
Northern aspect (35° S 15 December 10pm)

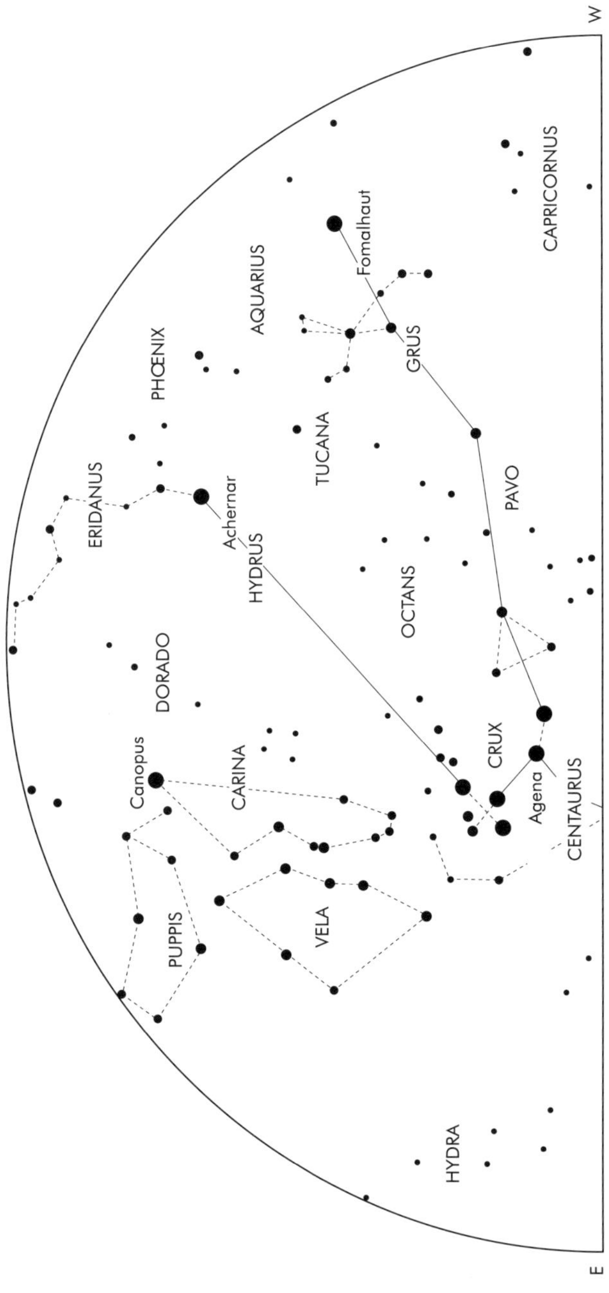

The Southern Hemisphere in December
Southern aspect (35° S 15 December 10pm)

December 1

The December Sky

Northern observers have both their main 'guides' on view. Orion attains a respectable altitude by mid-evening, and Ursa Major is well above the north-east horizon, even though the Great Bear does appear to be standing on its tail. Capella in Auriga is very close to the overhead point, which means that Vega is skirting the northern horizon and we have certainly lost the Summer Triangle, though in fact Deneb, like Vega, never actually sets over the British Isles. Cassiopeia is high in the north-west.

Pegasus remains visible in the west, though before very long it will start to merge into the evening twilight. Sirius has made its entry, and around midnight Leo appears over the eastern horizon. The Milky Way is at its best, running right across the sky from Cygnus through Cassiopeia, Perseus, Auriga and Gemini down to the horizon.

Southern observers lack Ursa Major, but Orion is excellently placed together with the Hunter's retinue; Achernar is close to the overhead point, and Canopus is also high and Pegasus well above the horizon. Capella is very low in the north, and the Southern Cross also is badly placed during December evenings. The Milky Way is superb, and so too are the Clouds of Magellan; it is sad that there are no comparable Clouds in the northern hemisphere of the sky.

December 2

The Giant Planet

Jupiter is the giant of the Solar System, and is more massive than all the other planets combined. It is on view for several months in every year, so this may be a good time to discuss it in some detail. During the period covered here, oppositions are timed for August 1997, September 1998, October 1999, November 2000, January 2002 and February 2003. Note that 2001 is left out – merely because Jupiter's synodic period of 399 days is over a month longer than an Earth year.

Around the turn of the century Jupiter is in the northern hemisphere of the sky. At the 1997 opposition the declination is −17°; the planet then moves north, reaching declination +23° by 2002. The opposition magnitude stays much the same, at between −2 and −2.5. Of all the planets, only Venus – and on rare occasions, Mars – can become more brilliant than this. It is small wonder that the ancients named the planet in honour of the ruler of Olympus.

Anniversary

1974: Pioneer 11 flew past Jupiter at 28 840 miles, sending back splendid data. It then went on to encounter Saturn in September 1979. Contact with it was lost in 1996, and it is now on its way out of the Solar System.

Anniversary

1972: Pioneer 10 passed Jupiter at 81 660 miles. It is now on its way out of the Solar System.

Future Points of Interest

1999: Mercury at western elongation.

2001: Saturn at opposition.

December 3

Jupiter as a World

Jupiter is truly immense. Its globe could swallow up more than 1300 globes the volume of the Earth, but it is 'only' 318 times as massive as the Earth, because its constitution is very different. There is a silicate core at a high temperature, perhaps as much as 50 000 °C, surrounded by layers of liquid hydrogen, overlaid by the cloudy layers which we can see. Hot though the core is, Jupiter is not a 'failed star'. For stellar-type nuclear reactions to begin, we need a temperature of 10 000 000 °C.

Telescopically, the first thing to notice about Jupiter is that its yellowish disk is obviously flattened. This is because of the quick axial rotation. The orbital period is 11.9 Earth years, but the Jovian 'day' is less than 10 hours long, so that the equator bulges out. The equatorial diameter is 89 424 miles; the polar diameter 83 100 miles, so that the flattening is very marked. The Earth also is an 'oblate spheroid', but with us the difference between the polar and equatorial diameters is only 27 miles, as against over 6000 miles for Jupiter. The axial inclination is only 3°, so that Jupiter is almost 'upright' in its orbit.

The upper atmosphere is made up chiefly of hydrogen; there are hydrogen compounds such as ammonia and methane, together with a substantial amount of helium. The upper atmosphere temperatures are very low – of the order of −150 °C; but remember, Jupiter's mean distance from the Sun is 483 000 000 miles, well over five times that of the Earth.

Future Points of Interest

2002: Total eclipse of the Sun; mid-eclipse 08 h GMT. Totality lasts for 2 m 4 s. The eclipse will be seen from South Africa, the Indian Ocean and Australia.

December 4

The Belts and Spots of Jupiter

Even a small telescope will show surface details on Jupiter. The cloud belts are very prominent, and are always present. Generally there are two main belts, one to either side of the equator (the North Equatorial Belt or NEB, and the South Equatorial Belt or SEB), with others at higher latitudes. Spots, wisps and festoons are also to be seen, and the details can be amazingly complex. Moreover Jupiter is a world in constant turmoil, with violent winds, so that the surface is always changing. One cannot hope to draw up anything like a permanent map.

Because of the quick spin, the surface features appear to be carried across the disk from one side to the other; the shifts are obvious after only a few minutes' observation. Jupiter does not rotate in the way that a solid body would do. There is a strong equatorial current, and the rotation period here – in System I, the region between the two equatorial belts – is 9 h 50 m 30 s; for the rest of the planet (System II) the period is 9 h 55 m 41 s. To complicate matters still further, various discrete features such as spots have rotation periods of their own, so that they drift about in longitude.

Amateurs do excellent work in measuring the various rotation periods. The method is to time the moment when the planet's rotation brings the feature to the central meridian, which is easy to define because of the polar flattening. The rotation period can then be calculated from tables. A modest telescope – say of 6 inches aperture – is quite adequate for this sort of programme.

December 5

Jupiter's Great Red Spot

The most famous feature on Jupiter is the Great Red Spot, which has been under observation for most of the time since the seventeenth century. It often disappears for a while, but it always comes back. It was once believed to be the top of a volcano poking through the clouds; then it was thought to be a solid or semi-solid body floating in the outer gas, but the spacecraft have proved that it is a whirling storm – a phenomenon of Jovian 'weather'. Presumably it has persisted for so long because of its exceptional size, but we cannot be sure that it is permanent. At its peak, around 1878, it measured 30 000 miles by 7000 miles, so that its surface area was greater than that of the Earth; the present dimensions are more like 14 000 miles by 7000 miles. The cause of the colour is not certainly known, but may be due to phosphorus. For some years around 1878 it was described as brick-red.

Even when the Spot is not on view, we can detect the so-called Red Spot Hollow in the southern edge of the South Equatorial Belt. The rotation period of the Spot itself differs from its surroundings, so that it drifts in longitude, but the latitude remains virtually constant at about 22° S.

December 6

Jupiter in the Space Age

Six space-craft have now passed by Jupiter. Pioneer 10 made its pass in 1973; Pioneer 11 in 1974; Voyagers 1 and 2 in

1979; Ulysses (essentially a probe for studying the solar poles) in 1992, and Galileo in 1995. The Pioneers and Voyagers provided a tremendous amount of information; a thin, dark ring was detected, and it was confirmed that Jupiter is surrounded by zones of deadly radiation. There is a very strong magnetic field. But in July 1994 there was a different sort of encounter; a comet was seen to crash into the planet!

The comet, Shoemaker–Levy 9 (named after its co-discoverers) was discovered photographically in 1993, and found to be in orbit round Jupiter. It was then realised that it would actually impact. It had already been broken up, and between 16 and 22 July 1994 the fragments rained down, producing great disturbances in the Jovian atmosphere; the vast dark blotches could be seen with a very small telescope, and persisted for months. As well as being spectacular, the impacts provided new data about the nature of Jupiter's clouds as well as of the doomed comet itself.

The Galileo probe – named after the great Italian astronomer of the seventeenth century – was launched in 1989. It consisted of an orbiter and an entry probe, which were separated during the final approach. On 7 December 1995 the entry probe crashed into the Jovian clouds, and transmitted until it was destroyed by the increased pressure. The orbiter entered a closed path round Jupiter, and began a systematic survey of the family of satellites.

Anniversary

1905: Birth of G.P. Kuiper.

December 7

Space Pioneer

We have been discussing the various spacecraft to Jupiter. It is therefore fitting to pause for a moment to pay tribute to one of the great pioneers of this sort of research: Gerard Peter Kuiper, who was born in Holland on 7 December 1905. He became an assistant astronomer at Leiden University, but in 1933 emigrated to the United States, where he remained for the rest of his life. In 1960, following spells at the Yerkes and Mount Wilson Observatories, he went to the University of Arizona to became head of the Lunar and Planetary Laboratory, retaining this post right to the time of his death in 1973.

Kuiper was deeply involved in all the early space programmes, and made many valuable contributions. He also pioneered the setting-up of telescopes at high altitude, and it is due to him that the great observatory on Mauna Kea in Hawaii was established. It was sad that he died prematurely before seeing the full effects of his work. The first crater to be identified on Mercury is named after him.

December 8

Jupiter's Family

Jupiter has an extensive retinue of satellites. Four of these are large; they were observed by Galileo Galilei in 1610, with his primitive telescope, and although he was not actually the first to see them they are always known as the Galileans. Their individual names are Io, Europa, Ganymede and Callisto. The remaining twelve attendants are very small, and quite beyond the range of ordinary telescopes.

Details of the Galileans are given below.

All are planet-sized; Ganymede is actually larger than Mercury, though less massive. But for the glare of Jupiter, they would be easy naked-eye objects.

All the Galileans move more or less in the plane of the Jovian equator, and with even a small telescope it is fascinating to follow their movements from night to night – even from hour to hour. They may pass behind Jupiter, and be occulted; they may pass in transit across the planet's disk, together with their shadows; they may move into Jupiter's own cone of shadow, and be eclipsed. All these phenomena are listed in publications such as the *Handbook* of the British Astronomical Association.

The Galileans appear as small disks, but it was not until the space-craft era that we obtained reliable maps of their surface features.

December 9

The Galileans

Before the Space Age, it was tacitly assumed that the Galileans must be very similar to each other, probably icy and cratered. In fact they have been found to be a very varied group.

Ganymede and Callisto are indeed icy and cratered; there is evidence of past tectonic activity on Ganymede, but not much on Callisto. Europa is quite different. There are almost no craters, and the surface is smooth and icy, with a network

Future Points of Interest

2003: Mercury at eastern elongation.

First maximum of the Puppid meteor shower. This is a weak, ill-defined southern shower, with a ZHR which seldom exceeds 10.

The Galilean Satellites

Name	Mean distance from Jupiter (miles)	Orbital period d	h	m	Diameter (miles)	Mean opposition magnitude
Io	262 000	1	18	28	2264	5.0
Europa	417 000	3	13	14	1945	5.3
Ganymede	666 000	7	3	43	3274	4.6
Callisto	1 170 000	16	16	32	2981	5.6

of shallow cracks looking rather like those in an eggshell. Io is violently volcanic, with a red, sulphury surface and sulphur volcanoes erupting all the time, hurling material high above the ground. Though the general surface temperatures are very low, the volcanic vents are fiercely hot. Apparently this activity is due to the fact that Io's interior is constantly 'churned' by the changing gravitational pulls exerted by Jupiter and the other Galileans.

Io has long been known to affect Jupiter's radio emission, and it is now known to be connected with the planet by a flux tube carrying up to 5 000 000 ampères. Since Io also moves in the thick of Jupiter's radiation zones, it must qualify as the most lethal world in the entire Solar System.

December 10

Taurus

Now let us return to the stars.

Taurus, the second constellation of the Zodiac, is one of the more conspicuous groups even though it has no really distinctive shape, and is certainly nothing like a bull. Between

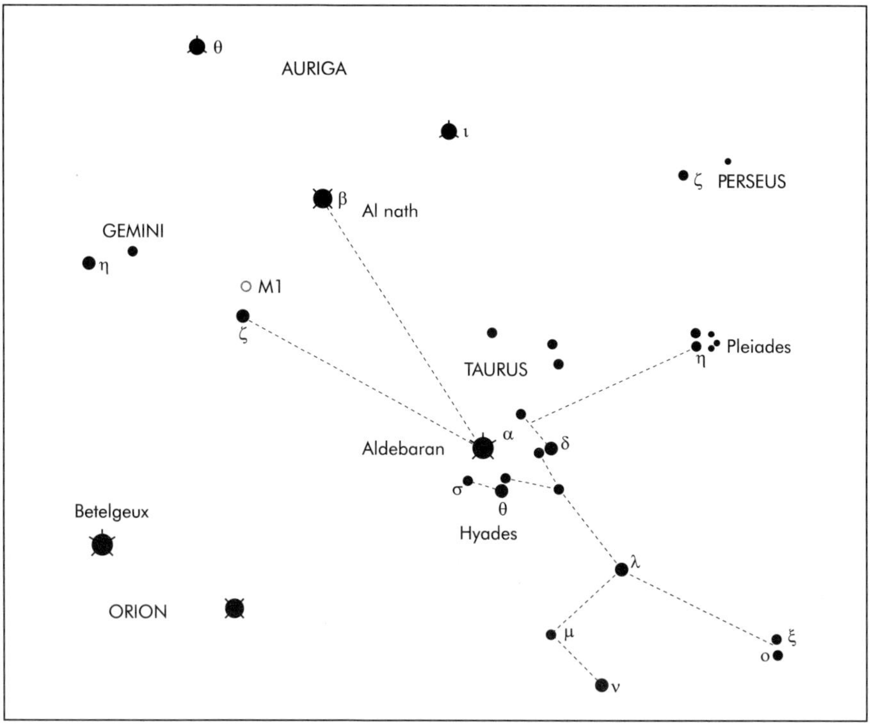

Taurus					
Greek letter	Name	Magnitude	Luminosity (Sun=1)	Distance (light-years)	Spectrum
α Alpha	Aldebaran	0.8	90	68	K5
β Beta	Al Nath	1.6	400	130	B7
ζ Zeta	Alheka	3.0	1300	490	B2
η Eta	Alcyone	2.9	400	238	B7
λ Lambda	–	3.3 (max)	450	330	B3

4000 and 2000 BC it contained the Vernal Equinox, so that if records had been drawn up at that period we might well refer to the First Point of Taurus rather than the First Point of Aries.

In mythology, Taurus represents the white bull into which Jupiter changed himself when he wanted to carry off Europa, daughter of the King of Crete, for the usual discreditable reasons. The constellation is easy to find, because its first-magnitude leader, the orange Aldebaran, lies in line with Orion's Belt. The main stars of Taurus are listed above.

It is interesting to compare Aldebaran with Betelgeux in Orion. Aldebaran is almost constant, but of course Betelgeux varies. Betelgeux is almost always the brighter of the two; but when comparing them, remember to allow for extinction.

Al Nath used to be included in Auriga, as Gamma Aurigæ (21 November). Taurus is noted mainly for three objects: two magnificent open clusters, the Hyades and the Pleiades, and the supernova remnant always known as the Crab Nebula (16 December).

December 11

The Hyades

Aldebaran is now very prominent in the evening sky, from either the northern or the southern hemisphere. Superficially it looks very much like Betelgeux, though in fact it is not nearly so remote or powerful.

Extending in a sort of V-formation from Aldebaran are the stars of the Hyades, one of the finest of all open clusters. Mythologically the Hyades were the daughters of Atlas and Æthra, and were half-sisters of the Pleiades (12 December). Aldebaran is not a cluster member. It simply happens to lie about half-way between the Hyades and ourselves, and this is a pity, as its bright light tends to overpower the cluster.

Of the true Hyades stars, Epsilon Tauri (magnitude 3.5) and Gamma (3.6) are of type K, and light orange in hue. Delta (3.8) makes up a wide pair with the fainter star 64

Anniversary

1863: Birth of Annie Jump Cannon, one of America's outstanding woman astronomers. She worked at Harvard College Observatory, mainly on the classification of stellar spectra; the present system is due largely to her. She also discovered five novæ and over 300 variable stars. She died in 1941.

1972: Landing on the Moon of the Apollo 17 module, carrying Eugene Cernan and Harrison Schmitt (Schmitt was the first professional geologist to go to the

(Continued opposite)

Moon.) The landing was at latitude 20° 12′ N, longitude 30° 45′ E, in the Taurus-Littrow area. Thanks to Schmitt's expert knowledge, this was the most scientifically valuable of all the Apollo missions. Up to now, it has also been the last manned expedition.

Tauri (4.8). Theta Tauri is a naked eye double, made up of a white star of magnitude 3.4 and a K-type orange companion of magnitude 3.8; with binoculars, the colour contrast between the two Thetas is striking. In fact, binoculars probably give the best view of the Hyades, since the cluster is much too large to be fitted into the more restricted field of a telescope.

December 12

The Seven Sisters

Undoubtedly the Pleiades hold pride of place among open clusters. They have been known since very ancient times, and almost every country has legends about them. Homer, in the *Odyssey* writes that Ulysses 'sat at the helm and never slept, keeping his eyes upon the Pleiads', and there are three references in the Bible – for example, 'Canst thou bind the sweet influences of the Pleiades, or loose the bands of Orion?' (Job 38: 31).

The main mythological story follows a fairly conventional pattern. The Pleiades were seven beautiful maidens, who were walking in the woods when they were spied by the hunter Orion, who gave chase; his intentions were clearly anything but honourable, so Jupiter intervened, changed the girls into stars and swung them up into the sky – no doubt to Orion's intense disgust!

The cluster's nickname of the Seven Sisters indicates that seven stars should be visible with the naked eye, and this is true for people with average eyesight; but keen-sighted people can see more, and the record, apparently held by the nineteenth-century German astronomer Eduard Heis, is said to be nineteen. If you can manage a dozen, you are doing very well indeed. As with the Hyades, the Pleiades are perhaps best seen with binoculars.

Future Points of Interest

Maximum of the Geminid meteor shower.

December 13

The Geminids

This is the night of the peak of the annual Geminid meteor shower. The radiant lies close to Castor, the senior though fainter Twin,

Generally speaking, meteors are best seen after midnight, when the darkened hemisphere of the Earth is on the 'leading' side and the meteors dash into the atmosphere at greater speeds, but with the Geminids the peak activity is usually around 22 hours GMT on 13 December so that it is useful to maintain watch all night if you want any really useful results.

The Geminids can usually be relied upon to give a good display. The maximum ZHR is around 75, but can be as much as 90, though it must be agreed that all in all the Geminids are not as spectacular as the August Perseids. They are of medium speed, and there are many fireballs. The shower was exceptionally rich in 1990 and again in 1991.

There is no known parent comet, but the orbit is much the same as that of the tiny asteroid 3200 Phæthon. It is only about three miles in diameter, but at perihelion moves to within about 13 000 000 miles of the Sun – closer than the orbit of Mercury – though at aphelion it recedes to almost 120 000 000 miles, beyond the orbit of the Earth. Its period is 1.43 years. If Phæthon really is the Geminid parent, it must be a comet which has lost all its volatiles, and this is certainly a possibility, though it is by no means a certainty.

December 14

The Stars of the Pleiades

The brightest member of the Pleiades cluster is Alcyone or Eta Tauri, magnitude 2.9. For some reason or other Johann von Mädler, the German astronomer who, with his colleague Beer, drew up the first really good map of the Moon put forward the theory that Alcyone was the central star of the Galaxy. Few people believed him.

Next in order come Electra, Atlas, Merope, Maia, Taygete, Celæno, Pleione and Asterope. This makes a total of nine named stars. However, Atlas and Pleione are very close together, and Pleione is somewhat variable, while Asterope is on the limit of naked-eye visibility for people with normal sight. All these stars are hot and bluish-white, and most of them are quick spinners; indeed Pleione rotates about a hundred times as rapidly as the Sun, and must be egg-shaped. The cluster is young, with an estimated age of only a few million years.

All in all, there are at least 500 stars in the cluster, and there is extensive nebulosity, very difficult to see telescopically but surprisingly easy to photograph.

One can never tire of looking at the Pleiades; this is incomparably the finest of all open clusters. It was included in Messier's catalogue, as M45, and it is 1432/5 in the NGC.

Anniversary

1962: Mariner 2 passed Venus at 20 648 miles, and sent back the first reliable data from close range. Contact was maintained until 4 January 1963.

Future Points of Interest

2001: Annular eclipse of the Sun (mid-eclipse 21 h GMT). Annularity lasts for 3 m 54 s. The eclipse will be seen from Central America and the Pacific area.

Anniversary

1970: First pictures received direct from the surface of Venus, from the Russian space-craft Venera 7.

December 15

Lambda Tauri

The brightest Algol-type eclipsing binary is Algol itself (12 November). Next comes Lambda Tauri, which can also be followed with the naked eye and is well on view during evenings in December. It is not too far from the equator (declination +12° 5 and so is at a convenient altitude from either hemispheres of the Earth.

To find it, use the stars of the Hyades as a sort of arrowhead; they will show the way to Lambda Tauri, which, perhaps rather surprisingly, has never been given a proper name of any sort. The maximum magnitude is 3.3. Every 3.95 days it fades down to 3.8, and after a very brief minimum recovers. The range is only half a magnitude, less than with Algol, and here too the eclipses are partial, with only about 40% of the primary being hidden by the secondary component. The shallower minimum, when the secondary is eclipsed by the primary, amounts to less than a tenth of a magnitude.

Good comparison stars are available; it is easy to use Gamma Tauri (3.6), Xi (3.7), Nu (3.9) and Mu (4.3).

The distance of Lambda Tauri is 330 light-years. It is more luminous than Algol, and the combined mass of the two components may be as much as eight times that of the Sun. The real separation is only about 8 500 000 miles.

Anniversary

1857: Birth of Edward Emerson Barnard, an American astronomer who worked at the Lick and the Yerkes Observatories. He was an outstanding observer; he discovered several comets as well as Amalthea, the fifth satellite of Jupiter; he identified the nearby star now named after him, and he specialised in studies of dark nebulæ. He died in 1923.

December 16

The Crab Nebula

Before moving on from Taurus we must concentrate upon M1, the Crab Nebula, which has probably been studied more intensively than any other object in the sky. In fact it is not a true nebula at all. It is the remnant of a brilliant supernova seen in the year 1054, and reported mainly by Chinese and Korean astronomers; it became as bright as Venus, and was visible with the naked eye in broad daylight. It stayed on view for over a year, but then faded below naked-eye visibility. Its remnant was recovered by an English amateur, John Bevis, in 1731, and confirmed by Messier in 1758, who commented that 'it contains no star; it is a whitish light, elongated like the flame of a taper'. The nickname was given to it in 1846 by Lord Rosse, using his 72-inch reflector at Birr Castle in Ireland (17 June).

Most books say that it is invisible in binoculars. This is not true; strong binoculars show it as a dim patch close to Zeta Tauri. However, a telescope is needed to show it clearly, and to make out the almost incredible complexity it is essential to use either photography or electronic devices. Visually, it is frankly a disappointment.

We have identified the actual remnant of the supernova; it is a pulsar, deep inside the nebulosity – a neutron star, spinning round 30 times every second; this is indeed the Crab's 'power-house'. Also, the Crab sends out radiation in every part of the electromagnetic spectrum, from long radio waves through to ultra-short gamma-rays. It is expanding, like all supernova remnants, and photographs taken over the years show marked changes in its structure.

It is well worth trying to photograph the Crab; there are many astronomers who regard it as the most important object in the sky from a theoretical point of view. The distance from us is 6000 light-years, so that the actual outburst took place well before there were any astronomers on Earth able to record it.

December 17

The Ringed Planet

Saturn, outermost of the planets known in ancient times, is on average 886 000 000 miles from the Sun; it is a slow mover, and during the period covered in this book it comes to opposition late in every year. The 1997 and 1998 oppositions fall in August; 1999 and 2000 in November, and 2001–2004 in December, so that this is a good time to turn our attention toward the Ringed World. By the turn of the century it is well north of the celestial equator; its declination at the time of the 2002 opposition is +22°, in the region of Gemini, so that it is well placed in the evenings throughout northern winter. From countries such as Australia, of course, it is much lower down.

In some ways – though by no means all – Saturn resembles Jupiter. It is considerably smaller, with an equatorial diameter of 74 914 miles, and it is a quick spinner – the mean rotation period is 10 h 14 m – so that it is flattened at the poles; though it has 744 times the volume of the Earth it is only 95 times as massive, and the mean density of its globe is less than that of water. Its internal make-up is not unlike that of Jupiter, but the inner core is smaller and less hot. The orbital period is 29.5 years.

As with Jupiter, there are belts and spots, but Saturn is a much less active world, and there is nothing remotely comparable with the Great Red Spot on Jupiter.

With the naked eye Saturn looks like a bright star; at its best it may reach magnitude −0.3, brighter than any star apart from Sirius and Canopus. Binoculars may show that there is something unusual about its shape, but to appreciate the full beauty of Saturn you do need a telescope.

Future Points of Interest

2002: Saturn at opposition.

December 18

The Rings of Saturn

What singles out Saturn from anything else in the sky is its system of rings. It is true that Jupiter, Uranus and Neptune also have ring systems, but these are dark and obscure, and they pale before Saturn's.

The rings are very extensive; from one side to the other they measure almost 170 000 miles. Yet they are less than a mile thick, and this means that when they are placed edgewise-on to us they almost disappear; even large telescopes will only reveal them as wafer-thin lines of light. This last happened in 1995, and the next edgewise presentation will not occur until 2009, so that throughout the period covered here the rings will be well displayed. The planet's south pole is uncovered, with the rings hiding part of the northern hemisphere.

The rings may look solid, but no solid or liquid ring could exist in such a region; the gravitational pull of Saturn would promptly tear it to pieces (even if it could form in the first place). The rings are made up of myriads of small, icy particles, spinning round Saturn in the manner of dwarf moons. They may represent the débris of an old icy moon which wandered too close and paid the supreme penalty, but it is more likely that they were produced from material which never condensed into a satellite.

There are three main rings. The outer two, A and B, are separated by a division named in honour of G. D. Cassini, the Italian astronomer who discovered it in 1675; in Ring A there is a smaller division, Encke's, which is not a difficult object when the system is well displayed. Closer-in there is Ring C, the Crêpe or Dusky Ring, discovered by Bond in 1850; a four-inch telescope will show it, but it is semitransparent and can be elusive. There are several more faint rings outside the main system, but ordinary telescopes will not show them; in the main we rely upon space-research methods.

A three-inch telescope will give a good view of the rings, and nobody seeing them for the first time cannot fail to be enthralled. There is nothing to rival the glory of Saturn.

December 19

Storms on Saturn

The rings of Saturn are so magnificent that they tend to divert the observer's attention away from the globe. This can be unfortunate, because even though Saturn is a much quieter world than Jupiter it can sometimes surprise us with violent outbreaks.

One of these occurred in August 1933, when a brilliant white spot near the equator was discovered by an English amateur astronomer, W.T. Hay (you may remember him better as Will Hay, the stage and screen comedian). It became very prominent, and clearly marked an uprush of material from below the main cloud layer; eventually it lengthened into a white streak, and then disappeared. A similar white spot was discovered in September 1990 by an American amateur, S. Wilber, and behaved in much the same way. It was well imaged by the Hubble Space Telescope, and for some time was truly spectacular. We never know when a new outbreak may occur, so that it is always worth while to keep a close watch.

Saturn has a magnetic field, 1000 times stronger than that of the Earth though still twenty times weaker than that of Jupiter; neither are the radiation zones so dangerous. Predictably, Saturn is a source of radio emissions.

December 20

Space-craft to Saturn

Three space-craft have so far passed by Saturn. Pioneer 11 was the first, in September 1979; it flew past at a range of 13 000 miles, and sent back useful data, but the encounter was something of an afterthought by NASA; it was found that after the rendezvous with Jupiter, Pioneer still had enough fuel reserve to take it back across the Solar System to a meeting with Saturn. Voyager 1 followed in November 1980, and Voyager 2 in August 1981, at 63 000 miles. The images returned were both spectacular and surprising. The rings were much more complex than had been expected; there were thousands of narrow ringlets and minor divisions, and there were thin rings even inside the Cassini Division. In the brightest ring, B, there were curious radial 'spokes' of darker material, and this ought to have been anticipated, because they can be seen with powerful Earth-based telescopes. The fainter rings outside the main system were also imaged; one of them, Ring F, proved to be curiously 'clumpy' and twisted.

Nowadays the Hubble Space Telescope is able to obtain good, detailed images of Saturn and its system, but it is fair to say that there are still some features of the ring system which are decidedly puzzling. The next scheduled probe, Cassini–Huygens, is not due to arrive at Saturn until the year 2004.

Anniversary

1968: Launch of Apollo 8, the first vehicle to take men (Borman, Lovell and Anders) round the Moon.

December 21

The Satellites of Saturn

The satellite system of Saturn differs from that of Jupiter. There is one large moon – Titan – and several of medium size, together with a number of dwarfs discovered by the Voyagers. The satellites known before the Voyager missions are listed below.

Titan can be seen with any telescope, and even strong binoculars will show it. A three-inch telescope will show Iapetus and Rhea and, with some difficulty, Dione and Tethys; the others require larger apertures. Phœbe has retrograde motion, and is almost certainly a captured asteroid. Eclipse, transit and occultation phenomena occur, as with the satellites of Jupiter, but are much more difficult to observe, even in the case of Titan.

The Voyager images show that apart from Titan, the satellites are icy and cratered, but they are not alike; for example Mimas has one huge crater, Enceladus has smooth, almost crater-free areas, and Hyperion is shaped rather like a hamburger; its orbital period is not the same as its revolution period. Iapetus has one hemisphere which is bright and one which is dark; it is always brightest when west of Saturn, and the more reflective area is turned toward us. Apparently darkish material has welled up from inside the globe, but its precise nature is not known. Perhaps the main interest in observing Saturn's satellites is seeing how many you can count.

Satellites of Saturn

Name	Mean distance from Saturn (miles)	Orbital period d	h	m	Longest Diameter (miles)	magnitude
Mimas	115 000	0	22	37	261	12.9
Enceladus	148 000	1	8	53	318	11.8
Tethys	183 000	1	21	18	650	10.3
Dione	235 000	2	17	41	696	10.4
Rhea	328 000	4	12	25	950	9.7
Titan	760 000	15	22	41	3201	8.3
Hyperion	920 000	21	6	38	224	14.2
Iapetus	2 200 000	79	7	56	892	10 (max)
Phœbe	8 050 000	550	10	50	143	16.5

December 22

The Cassini Mission

Before saying adieu to Saturn, we must certainly say something more about Titan, which is one of the most intriguing members of the Sun's family.

Patches have subsequently been recorded by the Hubble Space Telescope, but we still know little about the surface of Titan. It is very cold, at a temperature of −168 °C; there may be oceans – not of water, but of methane or ethane.

If all goes well, we will learn more in 2004. The Huygens probe, carried to Saturn by the Cassini spacecraft, is scheduled to come down on Titan and send back information direct from the surface. The chances of any life there are minimal, but it is very likely that Titan will provide us with plenty of shocks.

December 23

The Ursids – and Tuttle's Comet

Tonight is the maximum of the Ursid meteor shower. The Ursids have their radiant in Ursa Minor, not far from Kocab (7 January) – much the most northerly radiant of all official showers – and they are not usually rich; the normal ZHR is no more than 5, though there can be occasional really good displays, as in 1945 and 1986.

We know the parent of the stream; it is Tuttle's Comet, originally discovered in 1790 by Pierre Méchain. An orbit was computed, with a period of 13 years, but the comet was not seen again until 1858, when it was recovered by Horace Tuttle from the Harvard College Observatory. Since then it has been seen at every return except that of 1953, when it was very badly placed; the current period is 13.5 years. It is never bright enough to be seen with the naked eye, and seldom develops much in the way of a tail.

It is interesting that Horace Tuttle was concerned with the discoveries of three comets associated with meteor streams, the others being Swift–Tuttle (the Perseids) and Temple–Tuttle (the Leonids). Tuttle himself was an extraordinary character. He was a hero of the American War of Independence and a skilled astronomer, but he was also at one stage dismissed from the US Navy for embezzling funds. Astronomy certainly does not lack colourful personalities.

Anniversary

1966: Landing of the Russian probe Luna 13 on the Moon, at latitude 18°.9 N, longitude 63° W (in the Oceanus Procellarum). Images sent back; chemical studies undertaken. Contact was lost on 27 December.

Future Points of Interest

Maximum of the Ursid meteor shower.

December 24

The Barwell Meteorite

Just before 4.15 p.m. on 24 December 1965, people living in parts of Leicestershire saw a rare phenomenon. A meteorite

Anniversary

1761: Birth of Jean Louis Pons, whose first Observatory post, at Marseilles, was that of caretaker. He was self-taught, and concentrated upon hunting for comets; in all he found 36, and ended his career as Director of the Museum Observatory in Florence. He died in 1831.

1965: Fall of the Barwell Meteorite, in Leicestershire.

flashed across the sky and shone brilliantly, together with sound effects, before it broke up and scattered fragments around the village of Barwell.

One large piece landed in the drive of a house, and made a large hole. Another hit the bonnet of a car, and the driver, thinking that some boy had thrown a stone, threw the fragment aside. Yet another piece entered a house via an open window, and was later found nestling coyly in a vase of artificial flowers. Nobody was hurt – in fact, there is no authenticated case of anyone being seriously injured by a tumbling meteorite – but if a fragment had struck a human being the results would certainly have been dire.

The Barwell Meteorite was stony, like so many others, and certainly came from the asteroid belt; as we have seen (5 May) there is no connection between meteorites and shooting-star meteors. The original weight of the Barwell object seems to have been around 46 kilogrammes. Since 1965 there have been only two known British meteorites. On 25 April 1969 an object shot across the sky and scattered pieces of material near Bovedy, in Northern Ireland, though the main mass presumably fell in the sea; and on 5 May 1991 a tiny meteorite fell at Glatton, in Cambridgeshire.

Anniversary

1642: Birth of Isaac Newton, arguably the greatest of all mathematicians.

Future Points of Interest

2000: Partial solar eclipse (mid-eclipse 18 h GMT). 72% of the Sun is covered, but the eclipse will be well seen only from the Arctic.

December 25

The Star of Bethlehem

Christmas is with us once more, and every year the same question is asked: What was the Star of Bethlehem? Is there any scientific explanation for it?

Unfortunately we have so little to guide us. The Star is mentioned only once in the Bible, in the Gospel according to St Matthew, and nowhere else; contemporary astronomers give us no leads, and moreover we are by no means certain of our dates. The one thing we do know for certain is that Christ was not born on 25 December AD 1. Our AD dates are reckoned according to the calculations of a Roman monk, Dionysius Exiguus, who died in 556. He computed the date of Christ's birth to be 754 years later than the founding of Rome, and the system has become so firmly established that it will never be altered now, even though it is certainly wrong; Christ was born earlier than Exiguus imagined. Moreover, 25 December was not celebrated as Christmas Day until the fourth century – by which time the real date had been forgotten, so that our Christmas is wrong too.

So let us briefly go through the various explanations which have been offered.

(1) Venus. This can be ruled out at once. The movements of the planets were well known, and if the wise men had

been deceived by Venus they would not have been very wise. On similar grounds we can also reject Jupiter, Mars, or any other star or planet.

(2) A planetary conjunction – the close proximity in the sky of two planets, such as Jupiter and Saturn. This is the familiar 'red herring', but it is equally unsatisfactory. Such a conjunction would not have been spectacular; it would have lasted for some time, and everyone would have seen it.
(3) A comet. Again this would have been visible for some time; it would not have moved in the way described by St Matthew, and it would surely have been mentioned elsewhere. Halley's Comet came back at least several years too early to be accepted as a candidate.
(4) A nova or supernova. This has been regarded as a possibility, but again it would not have moved, and would have been reported elsewhere.

Various other theories are even less plausible (even an occultation of Jupiter by the Moon has been proposed!) but if we want to find a solution which fits at least some of the facts as described in the Bible, it seems that we must consider two meteors, seen at different times but moving in the same direction. At least this would explain why they were seen only by the wise men, and why they moved quickly across the sky. Otherwise it is best to admit that we do not know – and probably never will.

December 26

The False Cross

We must pay one more visit to the 'deep south' before the end of the year. During December evenings Canopus is almost overhead from countries such as Australia; but there is much else of interest in the now-separated parts of the old ship Argo). Carina (the Keel), Vela (the Sails) and Puppis (the Poop) are all rich; there is also Pyxis (the Compass) which has replaced the old Mast. The leading stars are listed opposite.

Note the False Cross, which is made up of Epsilon and Iota Carinæ and Kappa and Delta Velorum. It is often mistaken for the Southern Cross, but it is larger, more symmetrical and not so bright. As with the true Cross, however, three of its stars are white and the fourth – in this case Epsilon Carinæ – orange-red.

The Milky Way flows through this area, and there are many clusters and nebulæ; it is worth sweeping with binoculars. Two small constellations Volans (the Flying Fish) and Pictor (the Painter) intrude into Carina.

Future Points of Interest

2002: Mercury at eastern elongation.

Second maximum of the ill-defined Puppid meteor shower.

Carina, Vela and Puppis

Greek letter	Name	Magnitude	Luminosity (Sun=1)	Distance (light-years)	Spectrum
Carina					
α Alpha	Canopus	–0.7	200 000	1200	F0
β Beta	Miaplacidus	1.7	130	85	A0
ϵ Epsilon	Avior	1.9	600	200	K0
θ Theta	–	2.8	3800	750	B0
ι Iota	Tureis	2.2	7500	800	F0
υ Upsilon	–	3.0	520	320	A7
Vela					
γ Gamma	Regor	1.8	3800	520	WC7
δ Delta	Koo She	2.0	50	68	A0
κ Kappa	Markeb	2.5	1320	390	B2
λ Lambda	Al Suhail al Wazn	2.2	5000	490	K5
μ Mu	–	2.7	58	98	G5
Puppis					
ζ Zeta	Suhail Hadar	2.2	60 000	2400	O5.8
π Pi	–	2.7	110	130	K5
ρ Rho	Turais	2.8	525	300	F6
τ Tau	–	2.9	60	82	K3

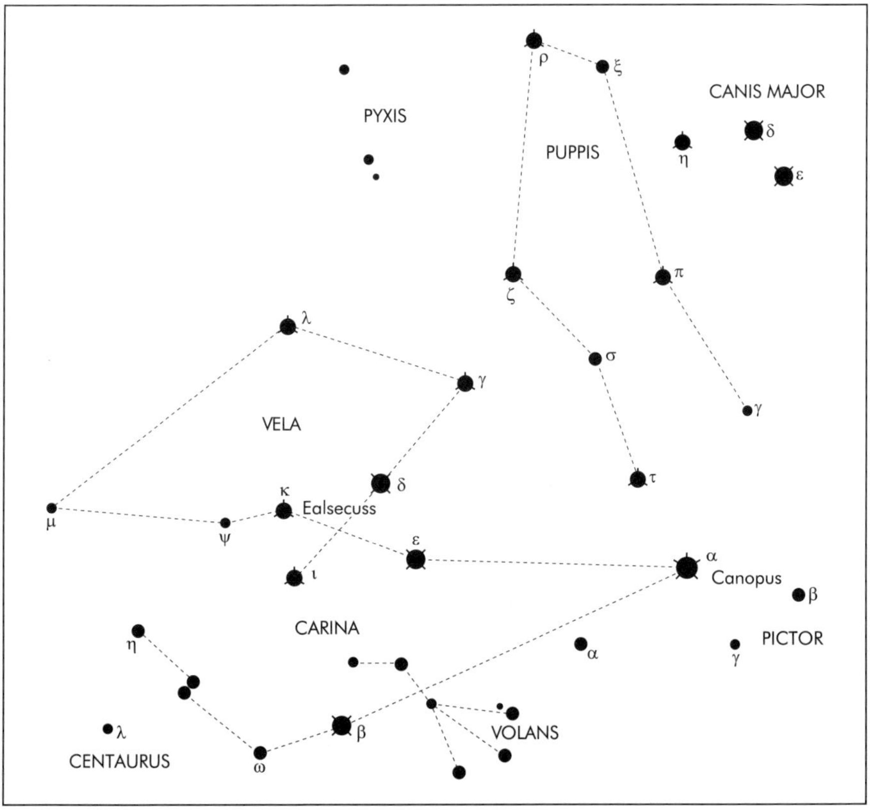

December 27

Eta Carinæ

If you are observing from a latitude which is sufficiently far south, look tonight at the region of one of the most peculiar of all stars, Eta Carinæ in the ship's Keel. Its declination is -60°, so that I am afraid you will have to be south of latitude 40° N. This rules out almost the whole of Europe, though Eta Carinæ does just rise briefly above the horizon of Athens.

It lies in a very rich area, near the False Cross (December 26). Its present magnitude is below 6, but it is the most erratic of all variable stars, and for a while in the 1830s and 1840s it outshone even Canopus. Telescopically it looks like an 'orange blob', quite unlike a normal star, and it is associated with nebulosity. The distance is about 6400 light-years, and at its peak it may have been 6 000 000 times as luminous as the Sun, making it the most powerful star known. It is still as powerful today, but it has become shrouded by dust, so that its light is absorbed by the dust and then reradiated as infra-red; Eta Carinæ is the brightest infra-red source in the sky outside the Solar System. It is super-massive and highly unstable; in the future it will undoubtedly explode as a supernova – and if so it will indeed be spectacular, though we may have to wait for many thousands of years yet. It is always worth watching, because there is always the chance that it will again flare up. Magnificent pictures of it and its nebula have been taken with the Hubble Space Telescope.

Anniversary

1571: Birth of Johannes Kepler, the German astronomer who was the first to prove that the planets move round the Sun in elliptical orbits. He made improvements to the refracting telescope, but his main contributions were theoretical, and all subsequent work has really been based on his three Laws of Planetary Motion. He died in 1630.

December 28

Beta Pictoris: a Planetary Centre?

Our very last foray into the far south takes us back to the area of Canopus. Close by is the little constellation of Pictor, the Painter, which has no bright star but which does contain one remarkably interesting object.

Beta Pictoris is an A-type star; according to the Cambridge catalogue it is 78 light-years away and 68 times as luminous as the Sun. Its declination is –51°, so that it does not rise from anywhere in Europe. In 1983 the Infra-Red Astronomical Satellite IRAS found it to have a large 'infra-red excess', in the same way as Vega (27 July), and subsequently this material was detected optically by R. Terrile and B. Smith at the Las Campanas Observatory in Chile.

Of course we cannot be sure; but Beta Pictoris does seem to be an exceptionally promising candidate as a planetary

Anniversary

1882: Birth of Sir Arthur Eddington, one of the greatest British mathematical astronomers. He spent much of his career at Cambridge University. He was a pioneer of theories of the evolution and constitution of stars, and an outstanding relativist. He was also an excellent writer of popular works, and also one of the first major BBC astronomers. He died in 1945.

centre. For this reason alone it is worth identifying, though it does look so very ordinary. It is rather sobering to reflect that at this very moment, some astronomer in the Beta Pictoris system may be turning his telescope toward the dim yellow star known to us as the Sun.

December 29

Keeping Watch on the Sun

Throughout this book I have a feeling that the Sun has been rather neglected, though we have discussed topics such as the solar cycle (14 February) and Spörer's Law (23 October). However, there is one interesting project which can be undertaken with almost any telescope. It may be of no real scientific value, but it gives great personal satisfaction. This involves compiling your own graph of solar activity.

The amount of activity is given by what is termed the Wolf or Zürich number, originally worked out by E. Wolf of Zürich in 1852. It depends upon the numbers of spots and spot-groups visible, and the formula is a very simple one: $R = k(10g + f)$, where R is the Zürich number, g is the number of spot-groups seen, f is the total number of individual spots seen, and k is a constant depending on the equipment and site of the observer. (k is not usually far from unity, so for our present purpose it can safely be ignored.) The Zürich number may range from zero for a clear disk up to over 200 near spot-maximum.

The last minimum fell in 1996, when there were many spotless days. The next maximum is due around the turn of the century, so that throughout the period covered in this book the solar observer is due for an interesting time. If you make daily spot-counts, you will soon see how your graph is developing.

Remember, though, never to look direct through any telescope, even with the addition of a dark filter; if you want to observe sunspots, use the method of projection. I am not ashamed for repeating this warning almost *ad nauseam*. So far as the Sun is concerned, one mistake is one mistake too many.

December 30

A Last Look Round

As we near the end of our tour, it seems fitting to take one last look round the evening sky – this time from the northern hemisphere, where we have both our two main 'markers', Orion and the Great Bear. Orion is dominant; check on the brightness of Betelgeux – we have discussed it earlier (15 January) but one can never be sure how it will behave, and it may now be either markedly dimmer or markedly brighter than it was at the start of the year. Sirius glitters from the low south; pure white though it may be, it does seem to flash different colours. Capella is almost at the zenith; check on Algol in Perseus to see whether it is undergoing one of its minima. Look back at Mizar in the Bear, and the glow of Præsepe in the Crab; and above all, do not fail to look back at the Pleiades, unlike anything else in the sky.

There is plenty to see. And another year lies close ahead.

Future Points of Interest

2001: Penumbral eclipse of the Moon. Mid-eclipse, 10 h 30 m GMT. 89% of the Moon's disk will enter the penumbra, so that the eclipse ought to be fairly easy to detect with the naked eye, if of course the Moon is above the horizon at your observing site.

December 31

The End of the Century

The Old Year is nearly over; tomorrow morning we welcome the New Year. This is always a special occasion, but there will be added interest in the near future, because we will welcome not only the New Year but also the New Century – indeed, the New Millennium.

There is some popular confusion here, because it is often claimed that the first day of the twenty-first century will be 1 January 2000. It will not; it will be 1 January 2001.

To see why this must be so, look back in history. Our calendar is decidedly chaotic, and has been revised several times, but there is one indisputable fact: there was no year 0. The year 1 BC was followed immediately by AD 1. Therefore, the new century started on the first day of AD 1; and if you think about this, it is easy to see why the first day of the coming century must be 1 January 2001.

Not that it really matters; no doubt both dates will be celebrated in the appropriate fashion. It is impossible to look ahead and predict what the new century has in store for us; we can only hope that it will be more peaceful than the century to which we are preparing to say farewell.

My best wishes to you all.

Future Points of Interest

2003: Saturn at opposition.

Appendix A

The 88 Constellations

Latin name	*English name*	*First-magnitude star(s)*
Andromeda	Andromeda	
Antlia	The Air-Pump	
Apus	The Bee	
Aquarius	The Water-Bearer	
Aquila	The Eagle	Altair
Ara	The Altar	
Aries	The Ram	
Auriga	The Charioteer	Capella
Boötes	The Herdsman	Arcturus
Cælum	The Sculptor's Tools	
Camelopardalis	The Giraffe	
Cancer	The Crab	
Canes Venatici	The Hunting Dogs	
Canis Major	The Great Dog	Sirius
Canis Minor	The Little Dog	Procyon
Capricornus	The Sea-Goat	
Carina	The Keel	Canopus
Cassiopeia	Cassiopeia	
Centaurus	The Centaur	Alpha Centauri, Agena
Cepheus	Cepheus	
Cetus	The Whale	
Chamæleon	The Chameleon	
Circinus	The Compasses	
Columba	The Dove	
Coma Berenices	Berenice's Hair	
Corona Australis	The Southern Crown	
Corona Borealis	The Northern Crown	
Corvus	The Crow	
Crater	The Cup	
Crux Australis	The Southern Cross	Acrux, Beta Crucis
Cygnus	The Swan	Deneb
Delphinus	The Dolphin	
Dorado	The Swordfish	
Draco	The Dragon	
Equuleus	The Little Horse	
Eridanus	The River	Achernar

Latin name	*English name*	*First-magnitude star(s)*
Fornax	The Furnace	
Gemini	The Twins	Pollux
Grus	The Crane	
Hercules	Hercules	
Horologium	The Clock	
Hydra	The Watersnake	
Hydrus	The Little Snake	
Indus	The Indian	
Laceta	The Lizard	
Leo	The Lion	Regulus
Leo Minor	The Little Lion	
Lepus	The Hare	
Libra	The Balance	
Lupus	The Wolf	
Lynx	The Lynx	
Lyra	The Lyre	Vega
Mensa	The Table	
Microscopium	The Microscope	
Monoceros	The Unicorn	
Musca Australis	The Southern Fly	
Norma	The Rule	
Octans	The Octant	
Ophiuchus	The Serpent-Bearer	
Orion	The Hunter	Rigel, Betelgeux
Pavo	The Peacock	
Pegasus	The Flying Horse	
Perseus	Perseus	
Phœnix	The Phœnix	
Pictor	The Painter	
Pisces	The Fishes	
Piscis Australis	The Southern Fish	Fomalhaut
Puppis	The Poop	
Pyxis	The Compass	
Reticulum	The Net	
Sagitta	The Arrow	
Sagittarius	The Archer	
Scorpius	The Scorpion	
Sculptor	The Sculptor	
Scutum	The Shield	
Serpens	The Serpent	
Sextans	The Sextant	
Taurus	The Bull	Aldebaran
Telescopium	The Telescope	
Triangulum	The Triangle	
Triangulum Australe	The Southern Triangle	
Tucana	The Toucan	
Ursa Major	The Great Bear	
Ursa Minor	The Little Bear	
Vela	The Sails	
Virgo	The Virgin	Spica
Volans	The Flying Fish	
Vulpecula	The Fox	

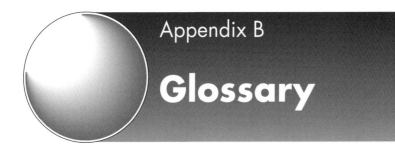

Appendix B
Glossary

Absolute magnitude. The **apparent magnitude** that a star would have if it could be observed from a standard distance of 10 **parsecs** (32.6 light-years).

Achromatic object-glass. An **object-glass** which has been corrected so as to eliminate **chromatic aberration** or false colour as much as possible.

Albedo. The reflecting power of a planet or other non-luminous body. The Moon is a poor reflector; its albedo is a mere 7% on average.

Altazimuth mounting for a telescope. A mounting on which the telescope may swing freely in any direction.

Altitude. The angular distance of a celestial body above the horizon.

Ångström unit. One hundred-millionth part of a centimetre.

Aphelion. The furthest distance of a planet or other body from the Sun in its orbit.

Apogee. The furthest point of the Moon from the Earth in its orbit.

Apparent magnitude. The apparent brightness of a celestial body. The lower the magnitude, the brighter the object: thus the sun is approximately −27, the Pole Star +2, and the faintest stars detectable by modern techniques around +30.

Asteroids. One of the names for the minor planet swarm.

Astrolabe. An ancient instrument used to measure the altitudes of celestial bodies.

Astronomical unit. The mean distance between the Earth and the Sun. It is equal to 149 598 500 km.

Aurora. Auroræ are 'polar lights'; Aurora Borealis (northern) and Aurora Australis (southern). They occur in the Earth's upper atmosphere, and are caused by charged particles emitted by the Sun.

Bailly's beads. Brilliant points seen along the edge of the Moon just before and just after a total solar eclipse. They are caused by the sunlight shining through valleys at the Moon's limb.

Barycentre. The centre of gravity of the Earth-Moon system. Because the Earth is 81 times as massive as the Moon, the barycentre lies well inside the Earth's globe.

Binary star. A stellar system made up of two stars, genuinely associated, and moving round their common centre of gravity. The revolution periods range from millions of years for very widely-separated visual pairs down to less than half an hour for pairs in which the components are almost in contact with each other. With very close pairs, the components cannot be seen separately, but may be detected by spectroscopic methods.

Black hole. A region round a very small, very massive collapsed star from which not even light can escape.

BL Lacertæ objects. Variable objects which are powerful emitters of infra-red radiation, and are very luminous and remote. They are of the same nature as quasars.

Cassegrain reflector. A reflecting telescope in which the secondary mirror is convex; the light is passed back through a hole in the main mirror. Its main advantage is that it is more compact than the Newtonian reflector.

Celestial sphere. An imaginary sphere surrounding the Earth, whose centre is the same as that of the Earth's globe.

Cepheid. A short-period **variable star**, very regular in behaviour; the name comes from the prototype star, Delta Cephei. Cepheids are astronomically important because there is a definite law linking their variation periods with their real luminosities, so that their distances may be obtained by sheer observation.

Chromosphere. That part of the Sun's atmosphere which lies above the bright surface or photosphere.

Circumpolar star. A star which never sets. For instance, Ursa Major (the Great Bear) is circumpolar as seen from England; Crux Australis (the Southern Cross) is circumpolar as seen from New Zealand.

Conjunction. (1) A planet is said to be in conjunction with a star, or with another planet, when the two bodies are apparently close together in the sky. (2) For the inferior planets, Mercury and Venus, inferior conjunction occurs when the planet is approximately between the Earth and the Sun; superior conjunction, when the planet is on the far side of the Sun and the three bodies are again lined up. Planets beyond the Earth's orbit can never come to inferior conjunction, for obvious reasons.

Corona. The outermost part of the sun's atmosphere, made up of very tenuous gas. It is visible with the naked eye only during a total solar eclipse.

Cosmic rays. High-velocity particles reaching the Earth from outer space. The heavier cosmic-ray particles are broken up when they enter the upper atmosphere.

Cosmology. The study of the universe considered as a whole.

Counterglow. The English name for the sky-glow more generally called by its German name of the **Gegenschein**.

Culmination. The maximum altitude of a celestial body above the horizon.

Declination. The angular distance of a celestial body north or south of the celestial equator. It corresponds to latitude on the Earth.

Dichotomy. The exact half-phase of the Moon or an **inferior planet**.

Doppler effect. The apparent change in wavelength of the light from a luminous body which is in motion relative to the observer. With an approaching object, the wavelength is apparently shortened, and the spectral lines are shifted to the blue end of the spectral band; with a receding body there is a red shift, since the wavelength is apparently lengthened.

Double star. A star made up of two components – either genuinely associated (binary systems) or merely lined up by chance (optical pairs).

Driving clock. A mechanism for driving a telescope round at a rate which compensates for the axial rotation of the Earth, so that the object under observation remains fixed in the field of view.

Dwarf novæ. A term sometimes applied to the U Geminorum (or SS Cygni) variable stars.

Earthshine. The faint luminosity on the night side of the Moon, frequently seen when the Moon is in its crescent phase. It is due to light reflected on to the Moon from the Earth.

Eclipse, lunar. The passage of the Moon through the shadow cast by the Earth. Lunar eclipses may be either total or partial. At some eclipses, totality may last for approximately $1\frac{3}{4}$ hours, though most are shorter.

Eclipse, solar. The blotting-out of the Sun by the Moon, so that the Moon is then directly between the Earth and the Sun. Total eclipses can last for over 7 minutes under exceptionally favourable circumstances. In a partial eclipse, the Sun is incompletely covered. In an annular eclipse, exact alignment occurs when the Moon is in the far part of its orbit, and so appears smaller than the Sun; a ring of sunlight is left showing round the dark body of the Moon. Strictly speaking, a solar 'eclipse' is the **occultation** of the Sun by the Moon.

Eclipsing variable (or Eclipsing Binary). A **binary star** in which one component is regularly **occulted** by the other, so that the total light which we receive from the system is reduced. The prototype eclipsing variable is Algol (Beta Persei).

Ecliptic. The apparent yearly path of the Sun among the stars. It is more accurately defined as the projection of the Earth's orbit on to the celestial sphere.

Elongation. The angular distance of a planet from the Sun, or of a satellite from its primary planet.

Equator, celestial. The projection of the Earth's equator on to the **celestial sphere**.

Equatorial mounting for a telescope. A mounting in which the telescope is set up on an axis which is parallel with the axis of the Earth. This means that one movement only (east to west) will suffice to keep an object in the field of view.

Equinox. The equinoxes are the two points at which the **ecliptic** cuts the **celestial equator**. The Vernal Equinox or First Point of Aries now lies in the constellation of Pisces;

the Sun crosses it about 21 March each year. The autumnal equinox is known as the First Point of Libra; the Sun reaches it about 22 September yearly.

Escape velocity. The minimum velocity which an object must have in order to escape from the surface of a planet, or other celestial body, without being given any extra impetus.

Extinction. The apparent reduction in brightness of a star or planet when low down in the sky, so that more of its light is absorbed by the Earth's atmosphere. With a star 1° above the horizon, extinction amounts to 3 magnitudes.

Eyepiece (or Ocular). The lens, or combination of lenses, at the eye-end of a telescope.

Faculæ. Bright, temporary patches on the surface of the Sun.

Fireball. A very brilliant **meteor**.

Flares, solar. Brilliant eruptions in the outer part of the Sun's atmosphere. Normally they can be detected only by spectroscopic means (or the equivalent), though a few have been seen in integrated light. They are made up of hydrogen, and emit charged particles which may later reach the Earth, producing magnetic storms and displays of auroræ. Flares are generally, though not always, associated with sunspot groups.

Flare stars. Faint Red Dwarf stars which show sudden, short-lived increases in brilliancy, due possibly to intense flares above their surfaces.

Galaxies. Systems made up of stars, nebulæ, and interstellar matter. Many, though by no means all, are spiral in form.

Galaxy, the. The system of which our Sun is a member. It contains approximately 100 000 million stars, and is a rather loose spiral.

Gamma-rays. Radiation of extremely short wavelength.

Gegenschein. A faint sky-glow, opposite to the Sun and very difficult of observe. It is due to thinly-spread interplanetary material.

Gibbous phase. The phase of the Moon or planet when between half and full.

Great circle. A circle on the surface of a sphere whose plane passes through the centre of that sphere.

Green Flash. Sudden, brief green light seen as the last segment of the Sun disappears below the horizon. It is purely an effect of the Earth's atmosphere. Venus has also been known to show a Green Flash.

Gregorian reflector. A telescope in which the secondary mirror is concave, and placed beyond the focus of the main mirror. The image obtained is erect. Few Gregorian telescopes are in use nowadays.

H.I and H.II regions. Clouds of hydrogen in the Galaxy. In H.I regions the hydrogen is neutral; in H.II regions the hydrogen is ionized, and the presence of hot stars will make the cloud shine as a nebula.

Heliacal rising. The rising of a star or planet at the same time as the Sun, though the term is generally used to denote the time when the object is first detectable in the dawn sky.

Herschelian reflector. An obsolete type of telescope in which the main mirror is tilted, thus removing the need for a secondary mirror.

Hertzsprung–Russell diagram (usually known as the H–R diagram). A diagram in which stars are plotted according to spectral types and their **absolute magnitudes.**

Horizon. The great circle on the celestial sphere which is everywhere 90 degrees from the observer's zenith.

Hubble's constant. The rate of increase in the recession of a galaxy with increased distance from the Earth.

Inferior planets. Mercury and Venus, whose distances from the Sun are less than that of the Earth.

Infra-red radiation. Radiation with wavelength longer than that of visible light (approximately 7500 Ångströms).

Interferometer, stellar. An instrument for measuring star diameters. The principle is based upon light-interference.

Kepler's laws of planetary motion. These were laid down by Johannes Kepler, from 1609 to 1618. They are: (1) The planets move in elliptical orbits, with the Sun occupying one focus. (2) The radius vector, or imaginary line joining the centre of the planet to the centre of the Sun, sweeps out equal areas in equal times. (3) With a planet, the square of the sidereal period is proportional to the cube of the mean distance from the Sun.

Libration. The apparent 'tilting' of the Moon as seen from Earth. There are three librations: latitudinal, longitudinal and diurnal. The overall effect is that at various times an observer on Earth can see a total of 59% of the total surface of the Moon, though, naturally, no more than 50% at any one moment.

Light-year. The distance travelled by light in one year: 9.4607 million million kilometres or 5.8 million million miles.

Local group. A group of more than two dozen galaxies, one member of which is our own **Galaxy**. The largest member of the Local Group is the Andromeda Spiral, M.31.

Lunation. The interval between successive new moons: 29 d 12 h 44 m. (Also known as the Synodic Month.)

Magnetosphere. The region of the magnetic field of a planet or other body. In the Solar System, only the Earth, Jupiter, Mercury, Saturn, Uranus and Neptune are known to have detectable magnetospheres.

Main Sequence. A band along an **H-R Diagram**, including most normal stars except for the giants.

Maksutov telescope. An astronomical telescope involving both mirrors and lenses.

Mass. The quantity of matter that a body contains. It is not the same as 'weight'.

Mean sun. An imaginary sun travelling eastward along the celestial equator, at a speed equal to the average rate of the real Sun along the **ecliptic**.

Meridian, celestial. The great circle on the **celestial sphere** which passes through the **zenith** and both celestial poles.

Meteor. A small particle, friable in nature and usually smaller than a sand grain, moving round the Sun, and visible only when it enters the upper atmosphere and is destroyed by friction. Meteors may be regarded as cometary débris.

Meteorite. A larger object, which may fall to the ground without being destroyed in the upper atmosphere. A meteorite is fundamentally different from a **meteor.** Meteorites are not associated with comets, but may be closely related to asteroids.

Micrometeorite. A very small particle of interplanetary material, too small to cause a luminous effect when it enters the Earth's upper atmosphere.

Micrometer. A measuring device, used together with a telescope to measure very small angular distances – such as the separations between the components of double stars.

Nebula. A cloud of gas and dust in space. Galaxies were once known as 'spiral nebulæ' or 'extragalactic nebulæ'.

Neutron. A fundamental particle with no electric charge, but a mass practically equal to that of a **proton.**

Neutron star. The remnant of a very massive star which has exploded as a **supernova.** Neutron stars send out rapidly-varying radio emissions, and are therefore called 'pulsars'. Only two (the Crab and Vela pulsars) have as yet been identified with optical objects.

Newtonian reflector. A reflecting telescope in which the light is collected by a main mirror, reflected on to a smaller flat mirror set at an angle of 45°, and thence to the side of the tube.

Nodes. The points at which the orbit of the Moon, a planet or a comet cuts the plane of the **ecliptic**; south to north (Ascending Node) or north to south (Descending Node).

Nova. A star which suddenly flares up to many times its normal brilliancy, remaining bright for a relatively short time before fading back to obscurity.

Object-glass (or Objective). The main lens of a refracting telescope.

Obliquity of the ecliptic. The angle between the **ecliptic** and the celestial equator. 23 d 26 m 45 s.

Occultation. The covering-up of one celestial body by another.

Opposition. The position of a planet when exactly opposite to the Sun in the sky; the Sun, the Earth and the planet are then approximately lined up.

Orbit. The path of a celestial object.

Orrery. A model showing the Sun and the planets, capable of being moved mechanically so that the planets move round the Sun at their correct relative speeds.

Parallax, trigonometrical. The apparent shift of an object when observed from two different directions.

Parsec. The distance at which a star would have a parallax of one second of arc: 3.26 **light-years**, 206 265 **astronomical units**, or 30.857 million million kilometres.

Penumbra. (1) The area of partial shadow to either side of the main cone of shadow cast by the Earth. (2) The lighter part of a sunspot.

Perigee. The position of the moon in its orbit when closest to the Earth.

Perihelion. The position in orbit of a planet or other body when closest to the Sun.

Perturbations. The disturbances in the orbit of a celestial body produced by the gravitational effects of other bodies.

Phases. The apparent changes in shape of the Moon and the inferior planets from new to full. Mars may show a **gibbous phase**, but with the other planets there are no appreciable phases as seen from Earth.

Photosphere. The bright surface of the Sun.

Planetary nebula. A small, dense, hot star surrounded by a shell of gas. The name is ill-chosen, since planetary nebulæ are neither planets nor nebulæ!

Poles, celestial. The north and south points of the celestial sphere.

Populations, stellar. Two main types of star regions: I (in which the brightest stars are hot and white), and II (in which the brightest stars are old Red Giants).

Position angle. The apparent direction of one object with reference to another, measured from the north point of the main object through east, south and west.

Precession. The apparent slow movement of the celestial **poles**. This also means a shift of the celestial equator, and hence of the equinoxes; the Vernal **Equinox** moves by 50 sec of arc yearly, and has moved out of Aries into Pisces. Precession is due to the pull of the Moon and Sun on the Earth's equatorial bulge.

Prime meridian. The meridian on the Earth's surface which passes through the Airy Transit Circle at Greenwich Observatory. It is taken as longitude 0°.

Prominences. Masses of glowing gas rising from the surface of the Sun. They are made up chiefly of hydrogen.

Proper motion, stellar. The individual movement of a star on the celestial sphere.

Proton. A fundamental particle with a positive electric charge. The nucleus of the hydrogen atom is made up of a single proton.

Quadrant. An ancient astronomical instrument used for measuring the apparent positions of celestial bodies.

Quadrature. The position of the Moon or a planet when at right-angles to the Sun as seen from the Earth.

Quasar. A very remote, super-luminous object. Quasars are now known to be the cores of very active galaxies, though the source of their energy is still a matter for debate.

Radial Velocity. The movement of a celestial body toward or away from the observer; positive if receding, negative if approaching.

Radiant. The point in the sky from which the meteors of any particular shower seem to radiate.

Retardation. The difference in the time of moonrise between one night and the next.

Retrograde motion. Orbital or rotational movement in the sense opposite to that of the Earth's motion.

Right ascension. The angular distance of a celestial body from the vernal equinox, measured eastward. It is usually given in hours, minutes and seconds of time, so that the right ascension is the time-difference between the **culmination** of the Vernal **Equinox** and the culmination of the body.

Saros. The period after which the Earth, Moon and Sun return to almost the same relative positions: 18 years 11.3 days. The Saros may be used in eclipse prediction, since it is usual for an eclipse to be followed by a similar eclipse exactly one Saros later.

Schmidt camera (or Schmidt telescope)**.** An instrument which collects its light by means of a spherical mirror; a correcting plate is placed at the top of the tube. It is a purely photographic instrument.

Scintillation. Twinkling of a star; it is due to the Earth's atmosphere. Planets may also show scintillation when low in the sky.

Sextant. An instrument used for measuring the altitude of a celestial object.

Sidereal period. The revolution period of a planet round the Sun, or of a satellite round its primary planet.

Sidereal time. The local time reckoned according to the apparent rotation of the **celestial sphere.** When the vernal **equinox** crosses the observer's **meridian**, the sidereal time is 0 hours.

Solar wind. A flow of atomic particles streaming out constantly from the Sun in all directions.

Solstices. The times when the Sun is at its maximum **declination** of approximately $23\frac{1}{2}$ degrees; around 22 June (summer solstice, with the Sun in the northern hemisphere of the sky) and 22 December (winter solstice, Sun in the southern hemisphere).

Specific gravity. The density of any substance, taking that of water as 1. For instance, the Earth's specific gravity is 5.5, so that the Earth 'weighs' 5.5 times as much as an equal volume of water would do.

Spectroheliograph. An instrument used for photographing the Sun in the light of one particlar wavelength only. The visual equivalent of the spectroheliograph is the spectrohelioscope.

Spectroscopic binary. A binary system whose components are too close together to be seen individually, but which can be studied by means of spectroscopic analysis.

Speculum. The main mirror of a reflecting telescope.

Superior planets. All the planets lying beyond the orbit of the Earth in the Solar System (that is to say, all the principal planets apart from Mercury and Venus).

Supernova. A colossal stellar outburst, involving (1) the total destruction of the white dwarf member of a binary system, or (2) the collapse of a very massive star.

Synodic period. The interval between successive **oppositions** of a **superior planet.**

Tektites. Small, glassy objects found in a few localized parts of the Earth. Nobody is yet certain whether or not they come from the sky!

Terminator. The boundary between the day- and night-hemispheres of the Moon or a planet.

Transit. (1) The passage of a celestial body across the observer's meridian. (2) The projection of Mercury or Venus against the face of the Sun.

Transit instrument. A telescope mounted so that it can move only in **declination**; it is kept pointing to the meridian, and is used for timing the passages of stars across the meridian. Transit instruments were once the basis of all practical timekeeping. The Airy transit instrument at Greenwich is accepted as the zero for all longitudes on the Earth.

Twilight. The state of illumination when the sun is below the horizon by less than 18 degrees.

Umbra. (1) The main cone of shadow cast by the Earth. (2) The darkest part of a sunspot.

Van Allen zones. Zones of charged particles around the Earth. There are two main zones; the outer (made up chiefly of **electrons**) and the inner (made up chiefly of protons).

Variable stars. Stars which change in brilliancy over short periods. They are of various types.

White dwarf. A very small, very dense star which has used up its nuclear energy, and is in a very late stage of its evolution.

Year. (1) Sidereal: the period taken for the Earth to complete one journey round the Sun (365.26 days). (2) Tropical: the interval between successive passages of the Sun across the Vernal Equinox (365.24 days). (3) Anomalistic: the interval between successive perihelion passages of the Earth (365.26 days; slightly less than 5 minutes longer than the sidereal year, because the position of the perihelion point moves along the Earth's orbit by about 11 seconds of arc every year). (4) Calendar: the mean length of the year according to the Gregorian calendar (365.24 days, or 365 d 5 h 49 m 12 s).

Zenith. The observer's overhead point (altitude 90°).

Zodiac. A belt stretching round the sky, 8° to either side of the **ecliptic**, in which the Sun, Moon and principal planets are to be found at any time. (Pluto is the only planet which can leave the Zodiac, though many asteroids do so.)

Zodiacal Light. A cone of light rising from the horizon and stretching along the **ecliptic**; visible only when the Sun is a little way below the horizon. It is due to thinly spread interplanetary material near the main plane of the Solar System.

Appendix C

The Greek Alphabet

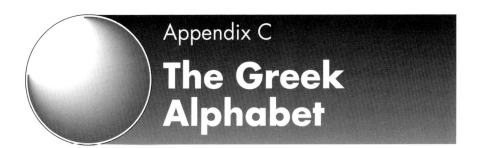

The twenty-four letters of the classical Greek alphabet are given below:

Name	Lower case	Upper case
alpha	α	A
beta	β	B
gamma	γ	Γ
delta	δ	Δ
epsilon	ε	E
zeta	ζ	Z
eta	η	H
theta	θ	Θ
iota	ι	I
kappa	κ	K
lambda	λ	Λ
mu	μ	M
nu	ν	N
xi	ξ	Ξ
omicron	o	O
pi	π	Π
rho	ρ	P
sigma	σ	Σ
tau	τ	T
upsilon	υ	Υ
phi	ϕ	Φ
chi	χ	X
psi	ψ	Ψ
omega	ω	Ω

Index